BIOTECHNOLOGY AND PLANT PROTECTION

Viral Pathogenesis

&

Disease Resistance

BIOTECHNOLOGY AND PLANT PROTECTION

Viral Pathogenesis & Disease Resistance

College Park, MD 19 – 21 October 1992

Donald D Bills
(Agricultural Research Service, USA)

Shain-dow Kung
(The Hong Kong University of Science & Technology)

UMS/USDA/Monsanto

World Scientific
Singapore • New Jersey • London • Hong Kong

Published by

World Scientific Publishing Co Pte Ltd
5 Toh Tuck Link, Singapore 596224
USA office: 27 Warren Street, Suite 401-402, Hackensack, NJ 07601
UK office: 57 Shelton Street, Covent Garden, London WC2H 9HE

British Library Cataloguing-in-Publication Data
A catalogue record for this book is available from the British Library.

BIOTECHNOLOGY AND PLANT PROTECTION
Viral Pathogenesis and Disease Resistance

ISBN-13 978-981-02-2274-1
ISBN-10 981-02-2274-2

BIOTECHNOLOGY AND PLANT PROTECTION

Viral Pathogenesis
&
Disease Resistance

PROCEEDINGS OF THE
FIFTH INTERNATIONAL SYMPOSIUM

Jointly Sponsored by the United States Department of Agriculture,
the University of Maryland System,
and the Monsanto Company

Editorial Board

Co-Editors-in-Chief: DONALD D. BILLS, USDA, ARS, Beltsville, MD
SHAIN-DOW KUNG, Hong Kong University of Science and Technology

Editors: C. JACYN BAKER, USDA, ARS, Beltsville, MD
ROBERT OWENS, USDA, ARS, Beltsville, MD
STEVE HUTCHESON, University of Maryland, MBI-CAB
DAVID STRANEY, University of Maryland, MBI-CAB
DAVID FISCHOFF, Monsanto Company
DILIP SHAH, Monsanto Company

PREFACE

We have made many changes in the way we grow plants since the start of agriculture about 10,000 years ago. But the most radical changes, which increased grain production by nearly three-fold, occurred within the last 50 years. Plant breeders introduced new varieties of wheat, corn, and rice, while the amount of irrigated land tripled, and the use of fertilizer increased nine-fold. Chemists developed substances to control pests, and this further accentuated productivity.

Along with these changes, the population of the earth more than doubled in the last 50 years, increasing from less than 3 billion to about 6 billion--enough new mouths to eat the increased food supply. The population of the earth is growing at a rate of about 90 million per year, a number slightly greater than the population of the United States west of the Mississippi river. The need for even more food production is evident.

Increased numbers of people and intensive industrial and agricultural activities created problems that are causing concern as we near the end of the 20th Century. Environmentalists insist that practices that pollute must be restricted or stopped, and we can expect that legislators and regulators will not allow the continued use of many of the pesticides that allowed food production to increase so dramatically.

An important aspect of successful agriculture is the control of plant diseases that reduce productivity, quality, and profitability. Application of exogenous chemicals and development of endogenous resistance are two general approaches to controlling plant diseases. As the former falls under attack and increased regulation, the latter becomes more achievable through biotechnology. In this volume the authors explore the application of biotechnology to understanding viral pathogenesis and the nature of plant resistance to viral disease.

More important, they show how this information foreshadows the development of plants with increased native resistance to disease.

In organizing this symposium, our intent was to bring together the leaders in the field for an exchange of the best current knowledge. We believe our intentions were fulfilled, and most of the credit must go to the excellent speakers who shared their diverse knowledge and experience with us. The remainder of the credit must be divided between the members of the Program Committee, who had the fine judgment to assemble a panel of such well-qualified speakers, and the members of the Arrangements Committee, who provided a favorable environment for our symposium.

We are indebted to the Agricultural Research Service of the USDA, the University of Maryland, and the Monsanto Company, who sponsored the symposium and provided financial support and other facilities and services that made the Fifth International Symposium on Biotechnology and Plant Protection a success. The symposium provides a continuing example of a successful and mutually beneficial cooperation between a public university, a governmental agency, and a private industry.

Donald D. Bills: *Agricultural Research Service, USDA*
Shain-dow Kung: *The Hong Kong University of Science and Technology*

ACKNOWLEDGMENT

We gratefully acknowledge the efforts of the members of the Program Committee and the Arrangements Committee. Special appreciation is due to Dr. Robert Owens, who chaired the Program Committee; and Mrs. Helen Phillips, who chaired the Arrangements Committee. We particularly thank Mrs. Dana Lamberti and Mrs. Claudia Volz for their conscientious effort in preparing and proofreading the camera-ready copy for this volume.

Donald D. Bills
Shain-dow Kung

PROGRAM COMMITTEE

C. Jacyn Baker, *USDA-ARS*
Robert Owens, *USDA-ARS*
Steve Hutcheson, *UMS/MBI-CAB*
David Straney, *UMS/MBI-CAB*
David Fischoff, *Monsanto Company*
Dilip Shah, *Monsanto Company*

ARRANGEMENTS COMMITTEE

Helen Phillips, *UMS/MBI-CAB*
Patricia Moore, *UMS/MBI-CAB*
Greg Silsbee, *UMS/MBI-CAB*
Dana Lamberti, *USDA-ARS*
Claudia Volz, *USDA-ARS*

CONTENTS

Disassembly Mechanisms of Plant RNA Viruses: Fine-Tuning and Possible Sites for Transgenic Protection

T. M. A. Wilson
Department of Virology
Scottish Crop Research Institute
Invergowrie, Dundee DD2 5DA
United Kingdom

Although widely accepted, the cotranslational disassembly hypothesis for positive-strand RNA viruses of plants remains poorly studied *in vivo*. The trigger(s) for and site(s) of nucleocapsid uncoating are still unknown. Recent data from *in vitro* and *in vivo* work on animal alphaviruses confirm that 60S ribosomal subunits and 28S rRNA bind the nucleocapsid protein and can disassemble incoming virions released from endosomes. Plant ribosomal proteins (about 20) also bind plant viral coat proteins (CP's) which may help direct virions to the cellular translation machinery or provide the uncoating trigger prior to active cotranslational disassembly. Some plant viruses possess characteristics which suggest an interaction with extra- or intra-cellular receptors which may also help to destabilize the structure. About 200 5'-nucleotides must be uncoated prior to cotranslational disassembly of TMV *in vitro*. The presumed role of Ca^{2+} ion chelation, shell expansion, and specific RNA extrusion in initiating the cotranslational disassembly of isometric virions has recently been questioned after site-directed mutagenesis of turnip crinkle virus. Some, but not all, metatope and cryptotope monoclonal antibodies against TMV CP inhibit cotranslational virus disassembly *in vitro*. Transgenic plants expressing such MAb's are currently being screened and tested for resistance to TMV infection. Further subtleties in the early events of infection have been revealed using TMV or "designer pseudovirions" and CP-transgenic tobacco mesophyll protoplasts or epidermal cells; nevertheless, this remains an area of research with more questions than answers.

Introduction

Having evolved an extremely stable, protective protein coat which prevents damage to their single-stranded RNA genome during the accumulation of progeny particles in infected cells, and during transfer between susceptible hosts, viruses are confronted with a structural paradox. How to uncoat and release their genetic material at the correct time and place to initiate an infection?

For many animal RNA viruses, data exist for uptake by receptor-mediated endocytosis and a consequent conformational change in the virion to favor disassembly within the cell. Plant viruses, on the other hand, rely on gaining entry into host cells through physical wounds caused by a variety of fungal or animal vectors. Thus, there seems little opportunity for the structure or stability of the virion to be modified as an immediate consequence of the cell entry process. Subtle modification of the capsid within the vector, between acquisition and transmission, is a possibility as will be discussed below. Indeed, if virus is placed on the surface of a host leaf without causing any mechanical damage, no infection occurs. Viruses cannot penetrate the waxy cuticular layers or the thick cellulose cell walls. Similarly, mesophyll protoplasts simply incubated with virus particles alone do not become infected. Chemicals (e.g., PEG or PLO) or physical stresses (e.g., electroporation) which cause transient damage to the plasmalemma are required to achieve even the most modest infection frequency among those protoplasts which survive such treatment.

Models have been proposed in which cell wall components and/or the phospholipid bilayer of the plasmalemma bring about extracellular virus disassembly and release the genome into the cytoplasm, where translation and the replication cycle can begin. Teleologically and experimentally, these models have received little support.

Eight years ago, it was shown that slightly destabilized particles of tobacco mosaic virus (TMV) could act efficiently as templates in cell-free translation systems, encoding (at least) their most 5'-proximal polypeptide (Wilson, 1984a; 1984b; 1986). The process was described as cotranslational disassembly; and intermediate complexes, called "striposomes," could be isolated following translational arrest with cycloheximide. In 1986, we showed that striposome-like complexes also existed *in vivo* in TMV-inoculated tobacco epidermal cells (Shaw *et al.*, 1986). More extensive studies on their physico-chemical properties have confirmed the predicted composition of these "*in vivo* striposomes" (Wu and Shaw, unpublished results; Wilson, 1988). *In vitro* data showing cotranslational disassembly have been obtained with virions of different morphology from several taxonomic groups (reviewed in Mundry *et al.*, 1991). However, no other plant virus has been studied for evidence of striposome formation *in vivo*. While it makes pragmatic and elegant sense for a virus to use the energy of virus-specific protein synthesis to destabilize, or sequentially disassemble its nucleocapsid, and thereby prolong the protective effect of the coat protein (CP) on the genome, two pivotal questions remain: How does the process start?; and How do polycistronic genomic RNA segments uncoat completely, if they must also conform to the classic "Kozak Rule" for eukaryotic mRNA translation (Kozak, 1989)? At best, we have only circumstantial evidence (described below) to address the first of these questions.

Intrinsic Structural Features of Viruses as Triggers for Nucleocapsid Disassembly

Traditionally, plant virus particles are considered to arise by a process of self-assembly in which all the necessary information to generate the final structure is contained within the CP and nucleic acid molecules themselves. Since both components are asymmetrical and contain highly conserved sequence information, it has been shown for several viruses that a discrete assembly pathway exists; the best studied examples being tobacco mosaic virus (TMV) (Bloomer and Butler, 1986) and turnip crinkle virus (TCV) (Wei *et al.*, 1990), representing helical rod-shaped and icosahedral spherical plant viruses, respectively. Although there is clearly a polarity or asymmetry to virus assembly in both cases, to ensure formation of a closed final structure, conformational changes in either component (CP or ssRNA) during assembly remove any strict requirement to adhere to the principle of micro-reversibility during disassembly. For example, in common strain TMV, assembly initiates at a short (75 nt) RNA sequence having a G residue at every third base position but which, overall, must also fold into a stem-loop structure of the correct length and stability to insert into, but be melted by, the first oligomeric 20S aggregate of TMV CP (Turner *et al.*, 1989). In a rod-shaped virus, all RNA secondary structure is lost, and only the influence of primary base sequence can modulate the stability of protein-RNA interactions. In isometric nucleocapsids RNA:RNA as well as protein:protein and protein:RNA interactions are possible.

Rod-Shaped Virus Particles: It has been shown that, in a sub-population of common strain (U1 or *vulgare*) TMV particles, at least the genomic 5'-leader sequence is easily uncoated by a wide variety of chemical treatments (buffers between pH 7.5-9.5; urea; dimethyl sulphoxide; or ionic detergents such as SDS; reviewed in Mundry *et al.*, 1991). Recently, we found that extensive dilution and sedimentation in pH 7.5 buffer removed CP for 9-10 nm at the concave, 5'-end ofTMV, exposing approximately 190-220 nts (Mundry *et al.*, 1991). The sub-population of rods in which this happened increased with increasing pH. A variety of techniques confirmed this conclusion; perhaps the most visual being the production of disome initiation complexes, formed in the presence of sparsomycin, an inhibitor of peptidyl transferase (Figure 1). It is known that the complete U1 RNA leader sequence (68 nt) and the first AUG codon of TMV RNA are required to obtain such complexes. Therefore, these must have been uncoated either during pretreatment or by 40S subunit scanning. Other results excluded the latter possibilty (Mundry *et al.*, 1991). Similar uncoating results were obtained with the U2 strain of TMV (Mundry *et al.*, 1991) (synonym: tomato mild green mosaic virus), although the latter was found to be more stable to disassembly in alkali. For reasons discussed below, this observation may be significant, as all four G residues in the U2 RNA leader are 5'-terminal or 5'-proximal and spaced in a triplet frame (Solis and Garcia-Arenal, 1990). Thus, it seems that, in TMV at least, there is an intrinsically unstable region of RNA-protein interactions, or a "trigger," to initiate particle disassembly. Conveniently, this exposes the highly efficient 5'-leader sequence and first AUG codon for the non-structural 126 kDa replicase protein - an RNase T1-resistant 70 nt sequence classically known as "omega" in common strain TMV RNA.

From X-ray crystallography (Namba *et al.*, 1989), it is known that a guanine in the first (5') nucleotide position of the trinucleotide binding site of each CP subunit confers added stability through "additional" H-bonds with the side chains of Arg122 and Asp115; hence, the propensity of genomic TMV RNA to have a G at every third base position. Therefore, the lack of G in most genomic Tobamovirus RNA leaders sequenced to date could account for the observed structural instability, at least to mild chemical treatments *in vitro*. The presence of four G residues, in triplet phase, in the more stably packaged TMV U2 strain RNA leader tends to confirm this hypothesis, provided the overall phase of RNA encapsidation from the internal U2 origin-of-assembly sequence (OAS; Figure 2) places each G in the 5'-most binding site on the first, second, fifth, and seventh CP subunits, respectively, from the 5'-end of the virion. In all cases, the m^7Gppp (cap) structure clearly cannot be accommodated by TMV CP and would be exposed to bind translation initiation factor eIF-4E. Although the origin-of-assembly sequence (OAS) on TMV U2 RNA is easily identified by homology with U1, and U2 CP has Arg122 and Asp115 conserved, there is a discrepancy in the sequence published for the U2 5'-leader (ΔA_{32}) which could affect this simple assembly phasing/virus instability model (Solis and Garcia-Arenal, 1990; Kukla *et al.*, 1979).

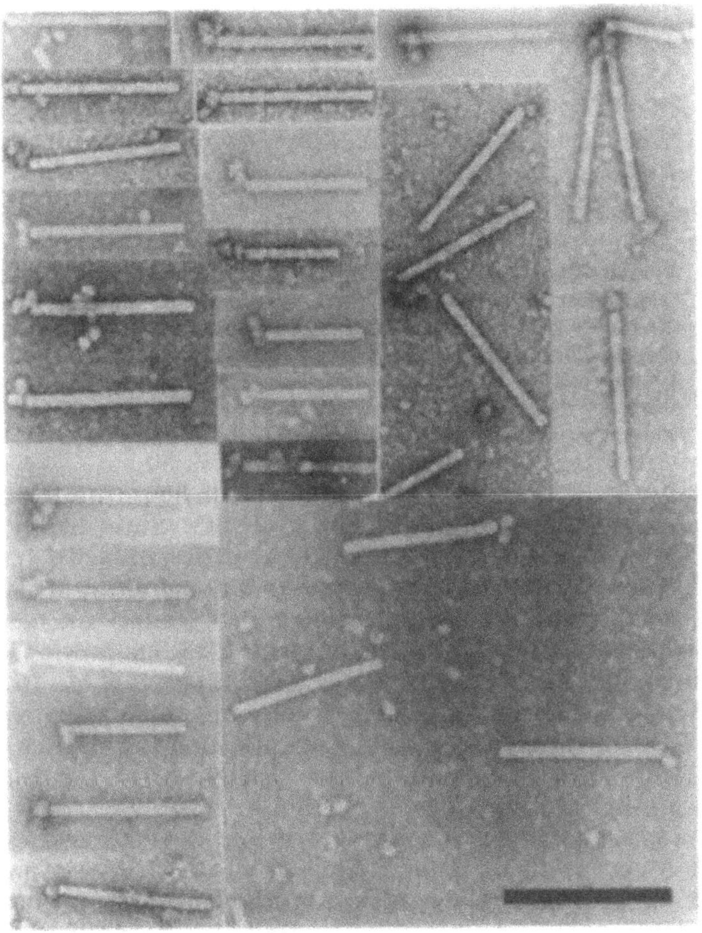

Figure 1. Disome Initiation Complexes Between 80S Ribosomes and TMV Particles as Evidence for Complete Uncoating of the 5'-Leader Prior to Cotranslational Disassembly. Selected initiation complexes from sparsomycin-arrested rabbit reticulocyte and wheat germ cell-free translation reactions programmed with pH 7.5-treated TMV U1 particles. Scale bar = 300 nm. Reproduced with permission from K. W. Mundry *et al.* (1991). Copyright 1991, Society for General Microbiology.

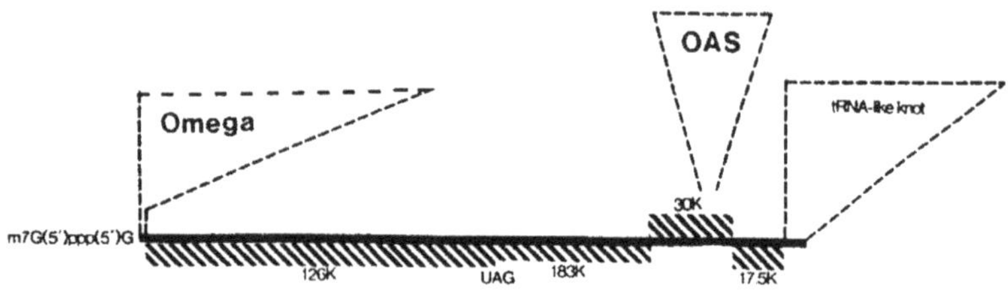

Figure 2. Genetic Map for U1 Strain TMV RNA (6395 nts). The location of the origin-of-assembly sequence (OAS), the unfolded 5'-leader (~"omega") and the highly folded 3'-(tRNA-like) regions are shown. Coding regions for the 126 kDa and UAG-readthrough 183 kDa replicase components, the 30 kDa cell-to-cell movement protein (MP), and the 17.5 kDa CP are shown as hatched areas above or below the bold line (ssRNA) to indicate which two of the three possible reading frames they use.

Another inconsistency which exists in the literature regards the polarity of disassembly of the C_c strain of TMV (Sunn-hemp mosaic virus). We have shown that dilution and centrifugation of C_c particles at pH 7.2 is sufficient to initiate cotranslational virus disassembly (Wilson and Watkins, 1985), implying 5'-leader uncoating as discussed above. However, treatment with SDS has previously been shown to expose the 3'-terminus of C_cRNA (Beachy *et al.*, 1976). To date, we have no information on the G-content and 5'-leader sequence of TMV C_cRNA, which may also have implications for data described below using transcapsidated hybrid virions in U1 CP-transgenic tobacco protoplasts (Osbourn *et al.*, 1989).

Among potexviruses, it is presumed that the OAS resides at the extreme 5'-terminus of papaya mosaic virus RNA and has been mapped to an 8-fold repeat of the motif GCAAA (Lok and Abou-Haidar, 1986). This must present a serious barrier, rather than a trigger, for initiating cotranslational virus disassembly. Conversely, the 5'-end of potato virus X RNA has a 83 nt untranslated leader with no G's in the first 41 bases (Smirnyagina *et al.*, 1991). This sequence can also act as a translational enhancer for foreign mRNA expression *in vitro* and so could be imagined to initiate the cotranslational disassembly of PVX, just as "omega" does for TMV.

Elegant work by Verduin and colleagues (Roenhorst *et al.*, 1989) has shown TMV striposome complexes *in vitro* in which up to twenty-four 80S ribosomes were bound to one virion. By correlating the relative lengths of the residually packaged portions of the TMV genome with the number of attached ribosomes (Figures 3 and 4) it was estimated that each 80S complex occupied 140-150 nts of TMV RNA. The first of 24 ribosomes had, therefore, exposed somewhere between 3360-3700 nts [the first amber stop codon is located after the 126 kDa ORF (Figure 2) at genome co-ordinate 3420].

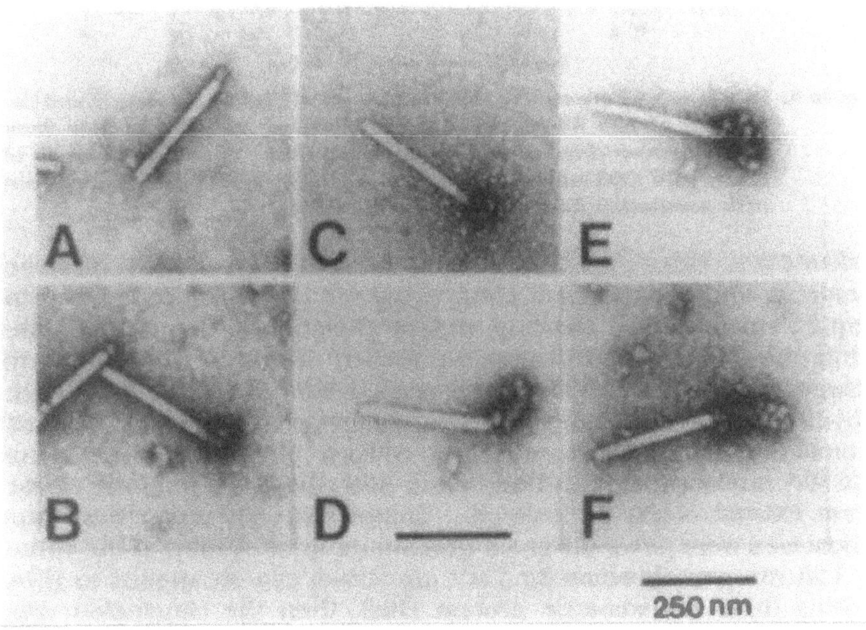

Figure 3. Selected Striposome Complexes Formed Between pH8-Treated TMV and Wheat Germ 80S Ribosomes *in vitro*. Panels A-F show increasing numbers of ribosomes associated with progressively shorter TMV rods. Scale bar = 250 nm. Reproduced with permission from B.J.M Verduin, 1992.

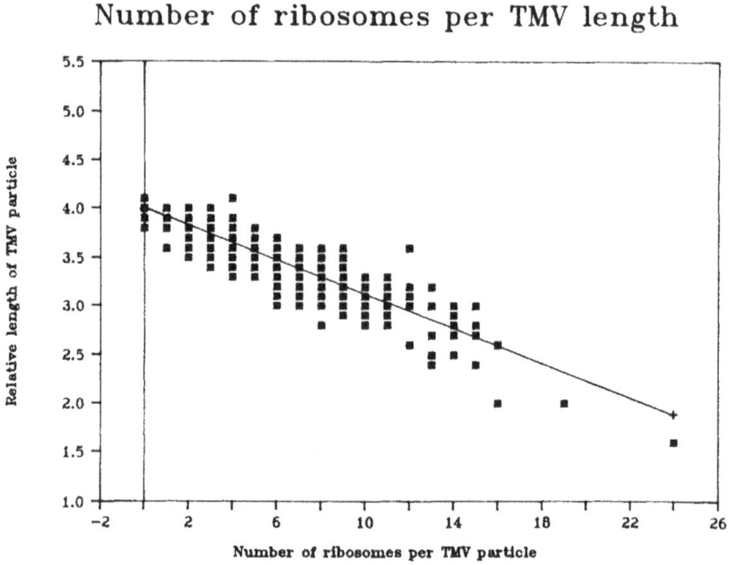

Figure 4. Relationship Between the Residual Length of TMV Nucleocapsid and the Number of 80S Ribosomes Bound. Electron microscopic data from glutaraldehyde-fixed complexes is summarized. The absolute length of intact TMV (300 nm) is plotted here as relative length 4.0. Reproduced with permission from B.J.M. Verduin, 1992.

Icosahedral Virus Particles: Cotranslational disassembly of some isometric and of bacilliform plant viruses *in vitro* has been shown to require nucleocapsid swelling and/or chelation of bound Ca^{2+}ions, while the 80S ribosomes remain attached during at least the early stages (Brisco *et al.*, 1985; Brisco *et al.*, 1986a; Brisco *et al.*, 1986b; Shields *et al.*, 1989). Again, Verduin's group showed that arrested translation complexes containing cowpea chlorotic mottle virus (CCMV) nucleocapsids and up to four 80S ribosomes from the wheat germ extract could be isolated. Complexes with more than four ribosomes were never observed (Roenhorst *et al.*, 1989). If the value of 150 nts per ribosome for TMV (as above) can be applied to RNA exiting from an isometric protein shell, then the conclusion was drawn that when more than 600 nts were released, the CCMV capsid became unstable and collapsed. This value is interesting since one molecule of CCMV RNA4 (800 nts) must be encapsidated with one copy of CCMV RNA3 to achieve particle stability (a "head-full") (Verduin, 1992).

In cell-free translation reactions programmed with CCMV particles, the ^{35}S-methionine-labelled CP synthesized associated with wheat germ monosomes; whereas the product of RNA3, an ssRNA-binding protein, bound to polysomes (Verduin, 1992). Immunoblotting ("far western blotting") of denatured, fractionated ribosomal proteins revealed that CCMV particles and CP bound to approximately 20 ribosomal proteins (10-15 from the 60S subunit) (Verduin, 1992; Roenhorst, 1989). Similar results were obtained with the CP's of TMV, brome mosaic virus (BMV), southern bean mosaic virus (SBMV), and blackeye cowpea mosaic virus (BCMV) (Verduin, 1992; Roenhorst, 1989). These observations have raised speculation that one or more receptor(s) for plant viruses may exist on plant cell ribosomes and that these play a role, not only in directing virus association with the translation machinery, but also in actively destabilizing the nucleocapsid, as shown below in more detail for two alphaviruses in the animal Togavirus group. However, further speculation that saturation of these ribosomal receptors, in viral CP-transgenic plants, results in the virus-resistant phenotype (or in classic cross-protection) is not supported because of the apparent lack of specificity in CP-ribosome binding compared with the restricted range of viruses inhibited by a particular CP transgene. Furthermore, CP-mediated protection often occurs in transformed plants despite there being vanishingly low or undetectable amounts of CP present. Stoichiometric saturation of all available ribosome receptors (20 r-proteins per 80S) would require a very large amount of free viral CP subunits as "decoy" ligands.

For animal alphaviruses, receptor-mediated endocytosis is considered to be the favored route for cell entry and, after acidification of the endosome, about half of the cell-associated Sindbis or Semliki Forest Virus (SFV) particles deliver their nucleocapsids into the cytoplasm; the rest are degraded in lysosomes (reviewed in Singh and Helenius, 1992). This delivery process takes about 30 minutes. For SFV in BHK-21 cells, uncoating was found to be 100 percent efficient, and >80 percent of the CP was removed from the RNA only 1-2 minutes after nucleocapsid release into the cytoplasm. Sedimentation studies showed the alphavirus CP to be bound to 60S ribosomal subunits, with 3-6 CP subunits bound per ribosome throughout the cytoplasm. *In vitro* assembled nucleocapsids were also disassembled in a reticulocyte lysate (Singh and Helenius, 1992).

Rather than simply sequestering pre-released CP, the 60S ribosomal subunits were shown to be the cause of nucleocapsid disassembly by mixing purified SFV and ribosomes without a source of energy and without active RNA translation. The nucleocapsids decreased in size and became irregular in shape (Figure 5). SFV RNA remains associated with the vacuolar membrane during translation, probably bound by a few residual CP subunits (Figure 6). Further, *in vitro* studies showed that 28S rRNA alone could also uncoat SFV nucleocapsids. The Sindbis RNA encapsidation sequence shares 54

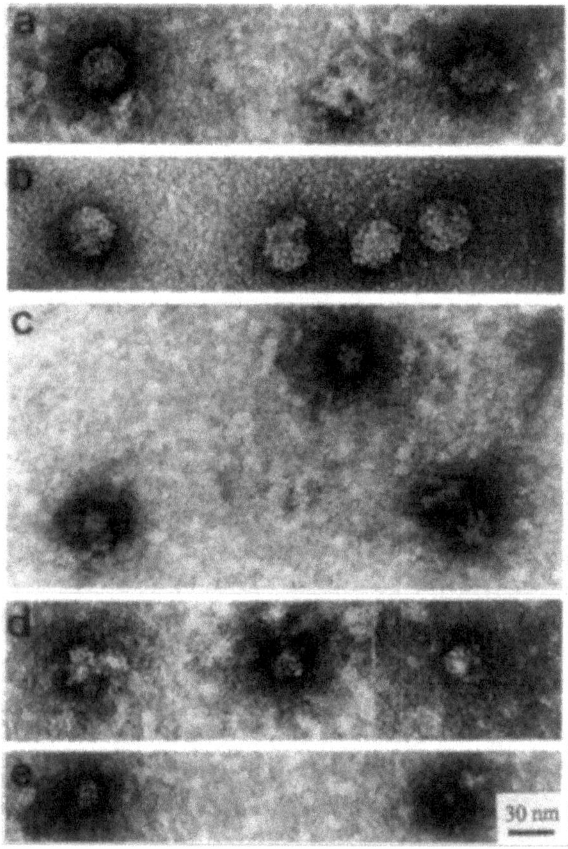

Figure 5. Electron Micrographs Showing the Structural Effects of Purified Ribosomes Added to Purified Semliki Forest Virus Particles *in vitro*. Intact and partially uncoated SFV nucleocapsids, negatively stained with 2 percent uranyl acetate. (a) and (b), intact SFV; (c)-(e), remnants of nucleocapsids as uncoating progresses in mixtures with purified ribosomes. Reproduced with permission from I. Singh and A. Helenius, 1992. Copyright 1992, Academic Press Ltd.

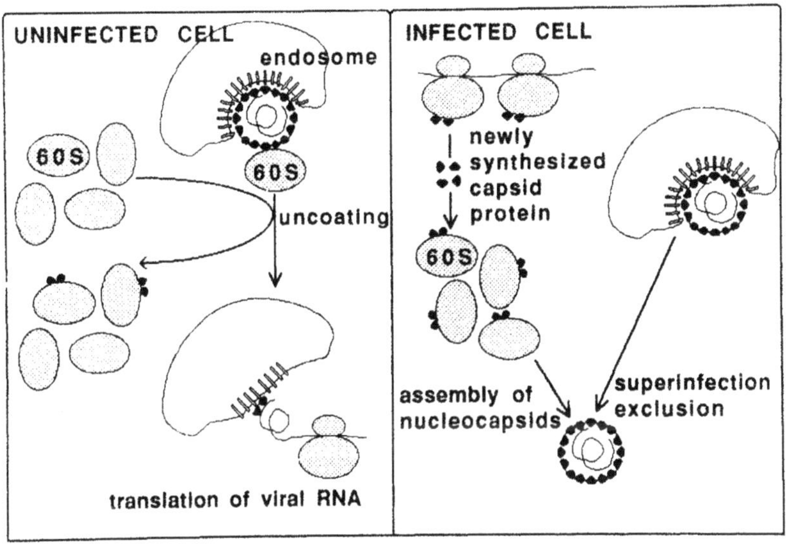

Figure 6. Schematic Diagram of the Control of Alphavirus Disassembly and Reassembly by Saturable Levels of Nucleocapsid Protein Receptors on Cellular 60S Ribosomal Subunits. See text for details. Reproduced with permission from I. Singh and A. Helenius, 1992. Copyright 1992, Academic Press Ltd.

percent identity with 400 nts on mammalian 28S rRNA. Vitrification and electron microscopy of mixtures of purified ribosomes and nucleocapsids has revealed pock-marked structures where each capsomere has lost some component (Singh and Helenius, 1992). Unlike most isometric virus capsids, alphavirus CP resembles chymotrypsin in X-ray structure (Choi *et al.*, 1991); and the positively charged amino-portion binds RNA on the inside of the shell; but some of this region is also exposed on the surface and may bind 28S rRNA. Clearly these viruses have exploited their requirement and affinity for the cellular translation machinery to achieve disassembly and release

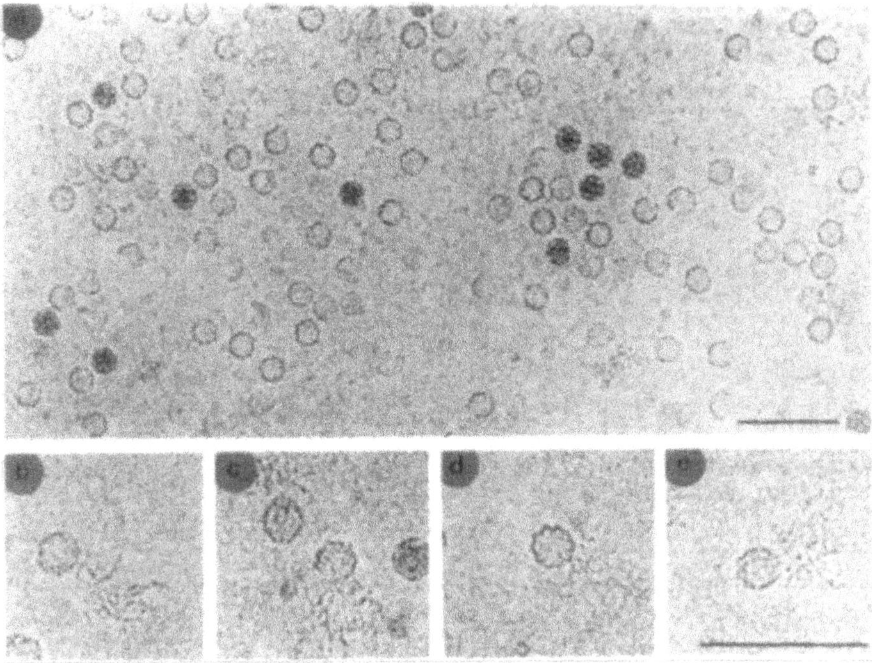

Figure 7. Cryo-electron Micrographs of Vitrified TYMV Particles Thawed to Room Temperature for 30 Seconds and Refrozen Rapidly. Many empty shells with a few intact (dense centered) particles can be seen in panel a. Panels b-e, selected particles showing a hole in the protein shell through which the filamentous RNA genome is escaping. Scale bars = 100 nm. Reproduced with permission from J. Witz, 1992. Copyright 1992, Society for General Microbiology.

of their RNA genome, although, in this case, active translation does not seem essential - in contrast to the sequential disassembly of rod shaped virions.

In general, we have observed a good correlation between the ability of isometric or bacilliform viruses to undergo cotranslational disassembly *in vitro* and their ability to swell by relaxation of protein-

protein interactions in their capsids (Roenhorst *et al.*, 1989; Brisco *et al.*, 1985; Brisco *et al.*, 1986a; Brisco *et al.*, 1986b; Shields *et al.*, 1989). Viruses which were permanently swollen (AlMV) or required only mildly alkaline pH (>7.5) to swell (CCMV and BMV) functioned well. Those stabilized by divalent cations (SBMV) required treatment with chelating agents. Others which did not swell under these conditions (TYMV and CPMV) were poor templates for cotranslational disassembly *in vitro*. Extremes of temperature or high pH are required to release 5-10 CP subunits from TYMV (one capsomere?), permitting escape of the RNA and formation of empty protein shells. A similar structural event occurs *in vivo* within 45 seconds of inoculating host leaves with purified TYMV particles (Matthews and Witz, 1985). Recently, cryo-electron microscopy has revealed TYMV particles in which the genomic RNA is escaping through a hole at the periphery of some capsids when vitrified virions were thawed to room temperature for just 30 seconds (Figure 7) (Adrian *et al.*, 1992). This result suggests that physico-chemical events or conformational changes in either the CP or RNA component alone are sufficient to "open the molecular trapdoor" and that the TYMV genome escapes even without the energy and polarity conferred by simultaneous translation. These authors (Adrian *et al.*, 1992) also discuss the role of a major co-operative transition in the distribution of TYMV RNA within the viral core and a small shell expansion (3-5 percent), caused by freezing, as intermediate steps presaging genome release. If the weak point in the protein shell is not random, but arises because of a specific interaction with the RNA then, despite the high symmetry, there is a functional polarity, comparable to the small isometric bacteriophage which contain one copy of the maturation (A) protein.

Another mechanism proposed to destabilize and trigger nucleocapsid disassembly is phosphorylation, as occurs with poliovirus, which also leaves empty capsids *in vitro* and *in vivo* (Lackmann *et al.*, 1987).

It has been suggested that reduction of disulphide bridges between envelope glycoproteins E1 and E2 in alphavirus particles occurs during direct penetration of the plasma membrane and is required to disassemble the icosahedral lattices, not only of the membrane glycoproteins themselves, but also of the underlying nucleocapsid protein, leading eventually to release of the RNA (Brown and Edwards, 1992). Clearly this uncoating mechanism contrasts with that described above following receptor-mediated endocytosis.

Structurally, the most amenable isometric plant RNA virus to study is turnip crinkle virus (TCV) which can be disassembled and reassembled *in vitro*. Six high affinity binding sites for CP (F_a-F_f) have been mapped on the 4.0 kb ssRNA genome (Wei *et al.*, 1990) and, during virus assembly, three prefabricated CP dimers (six CP subunits) bind to these sites in the crystallographic C/C configuration, with three of the six arm-domains intertwined in a β-annulus. This so-called "rp-complex" is thought to initiate virus self-assembly (Wei *et al.*, 1990). The six CP-binding sites are clustered between 0.7 kb of TCV RNA, which may help to condense the 0.7-1.0 kb from the 5'-end of the genome, and in the 3'-most genome into a small volume and make encapsidation more efficient. Unlike TMV, and the structurally closely-related tomato bushy stunt virus (TBSV), TCV capsids bind only one Ca^{2+}ion per CP subunit. This can be lost simply by dilution in buffer at pH 7.5, without chelation. In the expanded virus state, both TBSV and TCV CP subunits lose some of their contacts and openings appear (8 nm long and 2 nm wide) at each of the 60 faces of the icosahedral capsid. The N-terminal regions of 60 CP subunits, which are normally internal, (RNA binding domain and arm-domain) then protrude through these holes, making the capsid sensitive to proteases, which can now cleave off the N-terminal 70-100 amino acids. Most of the extruded R-domains may have no TCV RNA associated with them, but any/all of the six high affinity CP-binding sites on the RNA could be pulled out from the lumen of the capsid and exposed to the cytoplasm where they could initiate translation. Swollen TCV particles have been shown to act as translation templates *in vitro* for both the 5'-proximal 27 kDa polypeptide and the 3'-proximal CP, whereas naked TCV RNA produces only the expected 27 kDa product (Heaton, Laasko, and Morris, unpublished work; Heaton and Morris, 1992).

For many years it has been assumed that the highly conserved carboxyl-carboxylate pairs of amino acids in plant viral CP's, which are involved in Ca^{2+} binding, also control capsid swelling, assembly, and stability. These residues were believed to act as triggers for nucleocapsid disassembly in the low Ca^{2+} ion conditions which exist inside plant cells (Durham *et al.*, 1977; Durham, 1978). Therefore, the pH-dependent expansion and release of TCV RNA was thought to depend on two juxtaposed aspartic acid residues in the CP. Recently, full-length infectious transcripts of cloned TCV in which both these Asp residues were mutagenized to Asn were found to disassemble, replicate, and reassemble progeny mutant TCV particles normally in protoplasts. Only at the whole plant level was a defect in cell-to-cell and systemic spread recorded (Laasko and Heaton, unpublished work). Thus, the simple model proposed for virus swelling and concomitant RNA release *in vivo* requires further analysis. It will be interesting to see how these mutated virions behave in cell-free translation systems at pH 7.5.

Cellular "Receptors" or Sites Involved in Virus Destabilization *In Vivo*

Divalent metal ions are extremely inhibitory to the self-assembly of TMV or TMV-like structures *in vitro*. Paradoxically, but significantly, high levels of free Ca^{2+}ions also inhibit the specific infectivity of TMV particles, but not TMV RNA (Figure 8). Presumably this can be attributed to their effect in maintaining virus stability in the 1-2 percent of epidermal cells initially infected (wounded/repaired). Although the TCV result described above raises some questions, the structure and stability properties of TMV and TCV are very different.

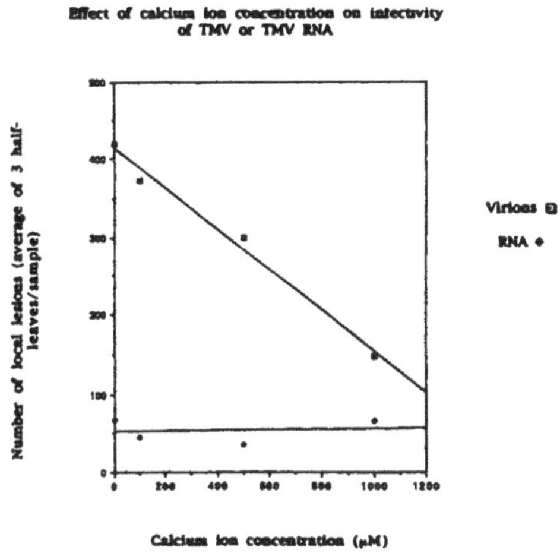

Figure 8. Effect of Calcium Ion Concentration on the Infectivity of TMV Particles or TMV RNA. TMV particles (open boxes) or TMV RNA (closed diamonds) in buffer containing different concentrations of $CaCl_2$ were applied to three separate half leaves per sample (100μl of 0.5 μg TMV/ml applied per half leaf). Results are average lesion counts per leaf. Reproduced with permission from J.K. Osbourn, 1991.

The question arises, does virus destabilization and/or disassembly *in vivo* result simply from the intrinsic structural properties of the nucleocapsid -a "molecular trapdoor;" a "coiled spring" of RNA; a conformational weak point opened up by capsid swelling; translation initiation factor-binding to an exposed 5'-cap structure, a surface VPg moiety, or a short exposed RNA leader sequence - and the prevailing chemical conditions within the cell (pH, cation concentration); or is there a specific site or receptor in, or on, the cell to which virions must bind in order to initiate an infection?

For plant viruses, there is little evidence for any intra- or extracellular receptor molecule, least of all a host- or cell-type-specific molecule. We have shown (Turner *et al.*, 1987) that native TMV particles microinjected into isolated oocytes of the toad *Xenopus laevis* will uncoat and transiently express their 126 kDa replicase protein, which must also happen during the early events of plant cell infection. Likewise, non-replicative, TMV-like pseudovirus particles uncoat and express their reporter mRNA "genomes" efficiently in a wide variety of plant and animal cells or cell-free extracts. Both these observations point to an intrinsic ability of the virus to be destabilized and (possibly) follow the cotranslational disassembly route in any metabolically active cell. Nevertheless, the remarkable stability of viruses such as TYMV and TBSV to extremes of pH, temperature, salt concentration, chelators, and even to the digestive enzymes of the human alimentary tract (Tomlinson *et al.*, 1982) contrast with the speed and efficiency with which they release their RNA genome following inoculation of a leaf (Matthews and Witz, 1985) or on thawing (Adrian *et al.*, 1992).

It has been suggested (Argos *et al.*, 1980), that TBSV CP resembles the lectin concanavalin A, and as such it may be significant that intact TBSV particles have recently been shown to agglutinate red blood cells (Heaton and Morris, 1992). Could this imply a role for a poly-, oligo-, or mono-saccharide-binding step to signal a conformational change in the TBSV capsid, leading to virus disassembly; and where does this occur? Monoclonal antibodies to discriminate between CP epitopes in compact or swollen TBSV particles, or anti-idiotypic antibodies may be able to identify this site immunocytochemically in newly inoculated epidermal cells or in electroporated protoplasts, at short times after infection.

Following several early reports on the fusigenic properties of plant viruses or their isolated CP's, a systematic study was undertaken with TMV, BMV, CCMV and SBMV, or their RNA-free CP's and model membranes (Datema *et al.*, 1987). Non-specific electrostatic interactions caused precipitates of lipid-protein complexes to form. Indirect hydrophobic interactions also produced lipid-only precipitates. CP dimers of all plant viruses induced vesicle fusion.

However, there was no indication of any specific event(s) likely to lead to virus uncoating. Plant membrane materials have also been tested for their ability to bind to and uncoat TMV (Kiho *et al.*, 1979a; Pustowoit *et al.*, 1984). The results were inconclusive concerning a possible physiological role for membranes; but together, PLO and phospholipids certainly stripped virions (Kiho *et al.*, 1979b). Recently, we have found that standard protocols for plasmid preparation from cleared lysates of *Escherichia coli* involving Triton X-100 have a similar effect on *in vivo* assembled pseudovirus particles or exogenous TMV, as monitored by sucrose density gradient analysis (Hwang and Wilson, unpublished observations).

Extensive studies with CCMV and cowpea protoplasts concluded that the most likely route of entry was through membrane lesions caused by osmotic shock of the protoplasts (Roenhorst *et al.*, 1988). Electron microscopy also revealed that mechanical inoculation of leaf surfaces caused lesions and beads of cytoplasm to be extruded and that, upon retraction and cell repair over the next 10-60 minutes, virions were passively drawn into the cell (Plaskitt *et al.*, 1988; Laidlaw, 1987). Hence the well-known correlation between turgor pressure and the efficiency of leaf inoculation (Yarwood, 1957).

Recent studies (Turnbull-Ross *et al.*, 1992) on the picornavirus-like plant virus, parsnip yellow fleck virus have revealed an RGD amino acid motif in the 31 kDa capsid protein (Figure 9) which is a homolog of VP1 in several picornaviruses, including foot-and-mouth disease aphthovirus (FMDV), where the same motif occurs. In several strains of coxsackievirus A9, a C-terminal extension of VP1 carries the RGD sequence; and in fibronectin and other extracellular proteins which interact with integrins, the same RGD motif occurs. In all cases, the RGD has been implicated in receptor-mediated attachment of the virus or matrix protein to the cell surface. Is this circumstantial evidence for a plant virus-cell membrane receptor interaction? PYFV is transmitted by aphids but requires the helper virus anthriscus yellows, so it is possible that the RGD has some role in transmission rather than virus localization or disassembly in wounded phloem cells. Work is currently underway to mutagenize this sequence.

Although there is an early association between plant viruses or viral dsRNA and chloroplast membranes, virus uncoating and replication occur in leaf epidermal cells without chloroplasts, so it seems unlikely that the chloroplast membrane is an essential site for virus disassembly. In newly inoculated epidermal cells, we have observed striposome complexes and associated gold-labelling of the expressed antigen dispersed throughout the cytoplasm, which argues against a localized receptor for early uncoating (Plaskitt *et al.*, 1988).

The existence of virus docking sites, or sub-cellular receptors continues to fascinate plant virologists, not least for the opportunity to isolate such components and dissect their gene(s), if appropriate, as shown recently for an animal virus receptor (Racaniello, 1992). Receptor mutagenesis could then provide a novel form of crop protection. There are numerous descriptive reports of virions attaching preferentially to particular cell membranes or to organelles; e.g., beet western yellows luteovirus attaches to the ER of phloem sieve cells (D'Arcy and De Zoeten, 1979), but the significance and molecular biology remain obscure.

Inhibition of Virus Disassembly in Transgenic Plants

Early Events in Viral Coat Protein Transgenic Plants: Since the first demonstration of CP-mediated resistance to virus inoculation in 1986 by Powell Abel and colleagues (Powell Abel *et al.*, 1986), the phenomenon has been reported using more than 30 different plant virus-CP systems. Apart from tomato spotted wilt virus (a plant Bunyavirus with a predominantly negative-sense, ssRNA genome and an ambisense component), all substantiated cases of CP-mediated resistance have involved viruses with positive-sense ssRNA genomes (reviewed by Beachy *et al.*, 1990; Harms, 1992; Harrison, 1992; Wilson, 1993).

Early reports with TMV suggested a direct correlation between the level of CP expressed transgenically and the degree of resistance observed. More recently, and particularly with potyviruses and luteoviruses, this correlation has not been sustained. Furthermore, some gene constructions which, by design or accident, failed to make any functional or detectable CP *in vivo* were shown still to confer good protection against the appropriate challenge virus. In these cases, it is likely that some form of RNA:RNA interference is occurring later during the virus replication cycle, rather than an immediate inhibition of virus disassembly. This phenomenon will be discussed in more detail elsewhere in this volume.

In 1957, Bawden and Pirie (1957) showed that excess free CP added to a TMV inoculum reduced its infectivity - a significant result which laid the foundations for some forms of viral cross-protection and provided the rationale behind transgenic CP-mediated protection. We showed that when high concentrations of TMV U1 CP (0.1-1.0 mg/ml) were added to cell-free translation reactions programmed with pH 8-pretreated TMV particles there was a substantial reduction (50-98 percent, between experminents) in the extent of cotranslational disassembly of pH 8-treated TMV particles (Tables 1 & 2 below; Wilson and Watkins, 1986). Exogenous TMV CP had much less effect on the translation of naked TMV RNA or on unrelated viral RNA's; e.g., turnip rosette virus RNA (Wilson and Watkins, 1986). We attributed the former to RNA recoating or inhibited uncoating rather than simply to nuclease degradation at the 5'-end of the destabilized sub-population of particles (Mundry *et al.*, 1991). Other proteins (e.g., BSA, cytochrome c) and other plant viral CP's (e.g., CCMV CP) had either a non-specific inhibitory effect on both TMV templates, which could be due to RNase contamination, or else caused no significant reduction in either template (Tables 1 & 2). In some experiments, BSA raised the translational activity of TMV RNA or particles.

It also appeared that compatible CP:CP interactions were required for maximum inhibition. As shown in Table 2, TMV U1 strain subunits were less effective at inhibiting the cotranslational disassembly of TMV C_C particles, but worked well against C_C RNA encapsidated in U1 CP. Unfortunately, we were never successful in producing large amounts of assembly-competent C_cCP to make the reciprocal hybrid virions (i.e., U1 RNA packaged in C_cCP) with which to extend these studies (Watkins, Osbourn, and Wilson; unpublished work).

Evidence that RNA recoating or CP exchange can occur *in vivo*, as suggested by Bawden and Pirie's result (1957), was obtained when the lower epidermis of tobacco leaves were inoculated with 30μg of native or pH9-pretreated TMV in the presence of 200μg U-^{14}C-labelled CP (approximately 10^6 cpm), a molar ratio of approximately 1:14,000. After 2 hours, before virus replication had added further unlabelled CP to the system, the epidermal cells were peeled off, frozen, and powdered in liquid nitrogen and the TMV recovered by ultracentrifugation. The amount of ^{14}C-CP which had become associated with virions was measured (as cpm) and compared with a control in which labelled CP was added to the powdered, frozen cells immediately before extraction and centrifugation. The results are summarized in Table 3.

In TMV CP-transgenic plants, up to 0.1 percent of the total soluble protein is CP (Powell Abel *et al.*, 1986). Since an average mature plant cell contains about 2 ng of protein and about 20 percent of this is "soluble" (Clowes and Juniper, 1968), it can be estimated

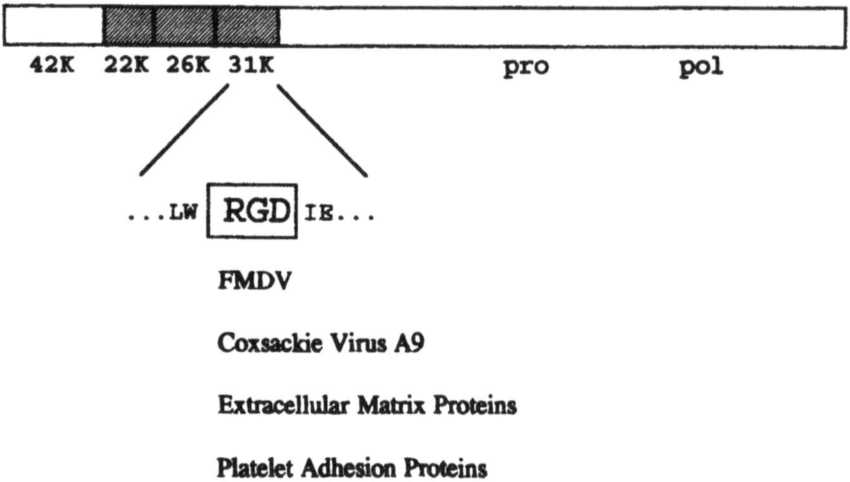

Figure 9. Map of the Polyprotein Encoded by Parsnip Yellow Fleck Virus RNA, a Picorna-Like Virus. PYFV has at least three structural proteins (22, 26, and 31 kDa) with an RGD motif in the largest in the context shown. Diagram prepared by and reproduced with permission from B. Reavy, 1992.

that a maximum of 1.2 x 10^7 CP subunits are available to recoat any incoming (parental) TMV particles.

Numerous lines of evidence suggest that inhibition of incoming virus disassembly operates in many cases of CP-mediated protection. Most obvious are those examples where inoculation with naked viral RNA, in amounts equivalent to the challenge virus used, as judged by biological infectivity assay, overcame the protection. Recently, in support of the "inhibition of uncoating model," Wu and colleagues (Wu *et al.*, 1990) showed that the quantitative efficiency of striposome formation in TMV CP-transformed tobacco mesophyll protoplasts was substantially inhibited compared with early events in PEG-inoculated vector-only transformed control protoplasts. Following from, and substantiating these data, western blots of CP-transformed mesophyll protoplast extracts failed to show any 126 kDa or 30 kDa TMV-encoded polypeptides from 0-48 hours post-electroporation in the presence of 30 μg/ml TMV U1 particles (Osbourn, 1990). Likewise,

Table 1. Inhibition of Translation of U1 Strain TMV or TMV RNA Templates by Added Proteins.

	Inhibition (Percent) of Total Protein Synthesis[a]	
Addition	U1 RNA[b]	U1 Particles[c]
0.1[d]BSA	0[e]	14
1.0 BSA	17	0[e]
0.1 Cyto c	34	31
0.1 CCMV CP	9	31
0.1 TMVP[f]	2	60
1.0 TMVP[f]	25	98

a Measured as percent reduction in average TCA-insoluble cpm of L-^{35}S-methionine in paired 3μl samples from duplicate 25μl standard rabbit reticulocyte incubations (where total protein (globin) concentration ~ 25 mg/ml) compared with the highest BSA-control value for each TMV template tested.
b Final concentration = 100μg/ml.
c Final concentration = 2.0 mg/ml, pretreated at pH8.0.
d Final concentrations in mg/ml (0.1 or 1.0).
e 1.0 mg/ml BSA was less inhibitory in incubations programmed with pH8-treated virions than with RNA. Reductions (percent) from the maximum value obtained were calculated for each template type.
f TMV U1 strain CP.

Table 2. Inhibition of Translation of Virus or Viral RNA Templates by Added Proteins.

	Inhibition (Percent) of Total Protein Synthesis[a]				
Addition	U1 RNA[b]	U1 Particles[c]	C_c RNA[b]	C_c Particles[c]	Hybrid Virus[c,d]
Buffer	0	0	0	0	0
1.0[e]BSA	0	13	8	0	7
0.1 TMVP[f]	5	21	11	16	11
1.0 TMVP[f]	30	50	34	41	62

a Measured as percent reduction in average TCA-insoluble cpm of L-^{35}S-methionine in paired 3μl samples from duplicate 25μl standard rabbit reticulocyte incubations (where total protein (globin) concentration ~25 mg/ml) compared with the buffer-only control for each template tested.
b Final concentration = 100μl/ml.
c Final concentration = 2.0 mg/ml, pretreated at pH8 (U1 and hybrid virions) or at pH7.2 (C_c virions).
d Hybrid virions = TMV C_c long RNA encapsidated in TMV U1 CP. Required pH8-, not pH7.2-, pretreatment to render them translatable.
e Final concentrations in mg/ml (0.1 or 1.0).
f TMV U1 strain CP.

Table 3. Exchange of Free ^{14}C-CP Subunits with TMV Particles Co-Inoculated onto Tobacco Lower Epidermis.

	Total cpm ^{14}C-CP Recovered in Virions after 2 hours		
	Expt. 1.	Expt. 2.	^{14}C-CP Subs./Virion[a]
Control[b]	1620	1432	0.02
TMV + ^{14}C-CP	5632	6008	0.08
pH9-TMV + ^{14}C-CP	6980	6520	0.09

a Average assuming all applied virions were equally susceptible to CP exchange.
b Background control, ^{14}C-CP added to peeled epidermal tissue in liquid nitrogen.

northern blots failed to detect any positive-sense TMV RNA in CP-transgenic mesophyll protoplasts electroporated with 30 µg/ml TMV but did show a faint signal after 24 hours when electroporated with 1 µg/ml TMV RNA (Osbourn, 1990).

Using non-replicative, pseudovirus particles containing mRNA for the reporter enzyme *E. coli* ß-glucuronidase (GUS), we confirmed, independently, that primary uncoating of incoming virus (or virus-like) particles can be inhibited by up to 98 percent in electroporated mesophyll protoplasts which express TMV CP (Osbourn *et al.*, 1989).

Control or CP-transgenic Xanthi tobacco leaves (line 3404, Powell Abel *et al.*, 1986) were mechanically inoculated on the lower epidermis with 1 µg/ml TMV. The epidermal cells, or the remaining mesophyll tissue were removed at 0, 18, or 36 hours post-inoculation and extracted. A faint signal, corresponding to shorter-than-genome length, positive-sense TMV RNA, was seen in northern blots of total RNA from the CP-transgenic epidermal tissue only at 18 hours post-inoculation. Strong TMV RNA signals were clear in all the non-transformed control leaf samples (Osbourn, 1990).

Immunosorbent electron microscopy (ISEM) was then used to trap and quantify TMV particles in sap extracted from the epidermal or mesophyll tissue of each class of tobacco at the time of inoculation and at 36 hours p.i. The results are given in Table 4. They indicate that, at least in the 3404 line of CP-transgenic tobacco, a 3-4-fold increase in TMV particles occurs in epidermal cells between 0-36 hours (the only time point studied). It was noted that almost all the rods seen in the CP-transgenic epidermal extracts were between one-third to one-fifth the unit length of TMV (300 nm). The simplest explanation for this, in conjunction with the northern blot result above, is an accumulation of the dicistronic, subgenomic (sg) mRNA encoding the 30 kDa polypeptide and having the TMV OAS. In turn, this requires that some of the initial virus inoculum was uncoated, made an undetectably low amount of replicase, and produced full-length minus-strand TMV RNA template, from which the sg mRNA was transcribed.

Progeny plants of transgenic line 3404 contained variable and sometimes low amounts of transgenically expressed CP. One could speculate that the CP level was at a threshold where some early steps of TMV infection occurred at low levels but that the full replication cycle was blocked. Using different tissue-specific promoters (*rbcS*, CaMV 35S, or *pal2*) to express the TMV CP gene *in planta*, Clark and colleagues (Clark *et al.*, 1990; Register and Nelson, 1992) have also shown a direct correlation between the level of CP expression in epidermal cells and the degree of resistance to TMV. Since TMV spreads from cell-to-cell *via* plasmodesmata in an unencapsidated form, and since inhibition of particle disassembly is a major site of action of CP-mediated protection in reducing the effective inoculum concentration, then the extent of interference which occurs in the initially invaded epidermal cells can assume a disproportionate significance in the overall phenotype.

Clearly compatible CP:CP interactions provided the best interference with particle disassembly *in vitro* and *in vivo*. However, we were also interested in 5'-leader sequence:CP interactions and their role in signalling particle uncoating. For this, we constructed a large number of T7 promoter transcription plasmids in which different artificial or viral leaders of various lengths were joined to the GUS coding region through an *Nco*I site. After the GUS gene, a 440bp fragment of TMV cDNA spanning the OAS was cloned such that transcripts made from suitably linearized plasmid DNA had a 3'-proximal OAS for *in vitro* encapsidation in TMV U1 CP (Sleat *et al.*, 1986).

In vitro translation experiments, analogous to those described and summarized in Tables 1 and 2, were set up using 20μg of each of these pseudovirus particles per reaction, in the presence or absence of an additional 20μg TMV CP. The results are summarized in Table 5 and suggest that the closer the 5'-leader sequence resembles that of U1 TMV RNA the more effective the CP-mediated inhibition of cotranslational disassembly *in vitro*. A broader range of leader-GUS transcripts or pseudovirus particles were then PEG-inoculated into control or CP-transgenic tobacco mesophyll protoplasts. The resulting GUS activities are given in Table 6. Again, the data suggest that pseudovirus particles containing 5'-leaders derived from the TMV sequence are preferentially sensitive to inhibition by the transgenically expressed CP. In this case, the ratio of expression of naked GUS mRNA in the two protoplast types serves as an internal standard for physiological or between-sample variation. Clearly these preliminary results and broader studies on the stability properties and expression of chimaeric and hybrid virions merit further attention.

Table 4. Evidence for Limited TMV Replication in Epidermal Cells of CP-Transgenic Tobacco Line 3404 by ISEM[a]

Time (hr)	Control Xanthi		CP-tg Xanthi (line 3404)	
	Epidermis[b]	Mesophyll	Epidermis	Mesophyll
0	20(5)[c]	1(1)	150(50)	340(80)
36	>2000[d]	>1000[d]	650(150)[e]	450(150)[e]

a Trapping of particles *ex* sap was on grids coated with rabbit polyclonal antiserum to TMV CP for 30 minutes at room temperature.
b Lower epidermis peeled off mesophyll immediately or 36 hours after inoculation with 1 μg/ml TMV U1, and both tissue types extracted immediately.
c Values represent the average total number of particles counted on 5 randomly chosen EM grid squares (blind experiment). Numbers in brackets are (+/- 1 S.D).
d Particles were too plentiful to count *per* whole grid square. Those on the small target screen were counted and an estimate made *per* whole square.
e The majority (80-90 percent) of rods trapped were one-third to one-fifth the unit length of TMV (300 nm). Note that CP-transgenic tobacco leaf sap also contains very low levels of endogenous TMV-like particles of about this size (see Wilson, 1989).

The results presented in Tables 5 and 6 offer an alternative explanation for the ability of hybrid virions of TMV C_c RNA, packaged in TMV U1 CP, to infect U1 CP-transgenic tobacco mesophyll protoplasts with the same efficiency and kinetics as naked TMV RNA (U1 or C_c) or native C_c particles (Osbourn *et al.*, 1989).

If the 5'-leader of TMV C_c RNA is substantially different from U1, then the earlier observation (Osbourn *et al.*, 1989) becomes analogous to the result in Table 6. During *in vitro* translation experiments, the hybrid virions were, if anything, more stable to pH pretreatment than either native parent virus; and by inoculation of tobacco or cowpea leaves the purity of each parent was confirmed. The hybrid virions only infected cowpea leaves. We had speculated (Osbourn *et al.*,

1989) that in Xanthi protoplasts, the common host, the transgenic U1 CP was unable to interfere with later stages in the replication cycle of C_c TMV RNA beyond uncoating - even though disassembly may be inhibited by approximately 90 percent through compatible U1 CP:CP interactions (extrapolating from the GUS pseudovirion data). The more recent data in Tables 5 and 6 indicate that this presumed level of inhibition of disassembly may be over-optimistic, particularly if the C_cRNA leader is "foreign" to U1 CP. Nevertheless, the significance of multi-level interference by TMV CP on the infection cycle of TMV U1 remains supported by the fact that low levels of TMV U1 RNA (2μg/ml) could not infect CP-transgenic tobaccos (Osbourn, 1990) but led to systemic infection of control Xanthi plants.

Table 5. Inhibition (Percent) of Cotranslational Disassembly of Pseudovirus Particles with Different 5'-Leader Sequences *In Vitro*.

	Inhibition (Percent) of Total Protein Synthesis[a]			
Leader	**Expt. 1**	**Expt. 2**	**Expt. 3**	**Average Percent**
True Leader (68 nt)	55	56	60	57 (2)[e]
12 nt-Leader-7 nt[b]	49	51	52	51 (1)
31 nt Polylinker[c]	13	7	30	17 (8)
53 nt Viral Leader[d]	35	38	30	34 (3)
Native TMV	51	46	49	49 (4)
Uncoated TMV RNA	5	10	8	8 (2)

a Measured as percent reduction in average TCA-insoluble cpm of L-^{35}S-methionine in paired 3μl samples from duplicate 25μl standard rabbit reticulocyte incubations.
b SP6 promoter construct (pSP64) with a *Hin*dIII - *Sal*I cassette of the TMV leader, pJII140 described in Sleat *et al.*, 1987.
c Control pSP64 construct without the TMV leader, only the polylinker region (pJII139).
d Unrelated insect viral gene leader.
e Figures in brackets are +/- 1 S.D.

Some recent results from TMV assembly experiments *in vivo*, in *E. coli*, support the intuitive argument that once they have been initiated, striposomes cannot be arrested, reversed, or disassembled (and the viral RNA repackaged 3'-to-5') by free CP subunits or aggregates *in planta*. The transgenic CP must operate to prevent ANY uncoating occurring at all, or else reverse the very earliest release of CP subunits, before the ribosomes arrive. Figure 10 summarizes our findings and the model.

GUS transcripts containing a 3'-proximal OAS and a leader capable of recruiting 30S and 70S prokaryotic ribosomes, initiate co-transcriptional translation before they can be packaged *in situ* by TMV CP being made in abundance from a separate plasmid in the same *E. coli* cell. In consequence, almost all TMV-like rods prepared from cleared lysates of IPTG-induced 3-18 hour cultures of cells, showed a polysome-like structure attached to the CONVEX end of short (25-40 nm) ribonucleoprotein helices (Bloomer and Butler, 1986). Northern blots of RNA recovered from these structures after EDTA and RNase treatment confirmed the presence of a protected TMV OAS. Thus, even with the correct configuration of RNA in the growing helix (Bloomer and Butler, 1986), biologically active *in vivo* aggregates (20S) of TMV CP cannot strip 70S ribosomes from the 5'-tail of RNA as it tries to pass through the lumen of the particle (Bloomer and Butler, 1986). Instead, the polysome blocks further 3'-to-5'-assembly of the nucleoprotein helix. In striposomes, the RNA is in the wrong configuration for efficient 3'-to-5'-assembly by 20S CP aggregates. It, therefore, seems even less likely that free CP, *in vivo,* could dislodge an already translocating ribosome and repackage the exposed RNA.

Inhibition of Virus Disassembly/Reassembly at Other Stages of the Infection Cycle: Although many plant viruses move from cell-to-cell in a non-virion form (as naked RNA or vRNP) using virus-coded movement protein(s) (MP) to modify the gating limit of plasmodesmata (Hull, 1989; Atabekov and Taliansky, 1990; Ding *et al.*, 1992), some seem to move only as particles or at least to require functional CP (e.g., nepoviruses, comoviruses, carmoviruses, bromoviruses, and luteoviruses) even over such short distances (Heaton and Morris, 1992). Some viruses are capable of moving, albeit slowly, throughout plants as naked RNA (tobraviruses); but most seem to require functional CP and assembly-competent RNA to cause systemic infection. Mutagenesis and complementation studies have supported the general notion that virions are the infectious entity transported through the vasculature of the plant (Saito *et al.*, 1990; Osbourn *et al.*, 1990; Holt and Beachy, 1991). In this case, uncoating must occur once again at secondary sites where virus particles leave the phloem. Exactly how, where, or in which order these exit and disassembly events occur is unknown; but if cotranslational disassembly is involved it confirms that the process is independent of

Table 6. Expression[a] of Chimaeric Leader-GUS mRNAs or Pseudovirus Particles in Control or Coat Protein Transgenic Protoplasts.

	Control(U1 CP⁻)			Coat Protein (U1 CP⁺)		
Leader	mRNA	Ptcles[b]	Ratio[f]	mRNA	Ptcles	Ratio
$(C,U,A)_{68}$[c]	443	323	0.7	133	383	2.9
53 nt Viral L[d]	436	316	0.7	380	376	1.0
AlMV RNA4 L[e]	360	350	1.0	396	403	1.0
TMV L (C's-to-G's)	380	403	1.1	553	33	0.06
$(CAA)_{23}$	293	473	1.6	436	140	0.3
$(AUC)_{23}$	236	246	1.0	396	0	0
$(ACAAUUAC)_6$	196	230	1.2	353	0	0

a Measured as pmoles 4-methyl-umbelliferone produced/min/mg protein.
b Same mRNA with 3'-OAS packaged in TMV U1 CP.
c Random leader sequence of only C, U, and A bases, starts with a G, ends with CCAUGG (*Nco*I site for GUS gene).
d Unrelated insect viral leader (53 nt long).
e Leader of alfalfa mosaic virus sg CP mRNA4.
f Ratio of expression of packaged RNA/naked RNA in either class of protoplast, as a measure for any selective inhibition.

physical wounding, apart from primarily infected cells where the virus first enters the plant. In the case of systemic sites of spread, CP-mediated inhibition of particle disassembly could function once more, as evidenced by the data of Wisniewski and colleagues (Wisniewski *et al.*, 1990). However, recent data (Ding *et al.*, 1992) showing that TMV MP accumulates only in secondary plasmodesmata and not in plasmodesmata between vascular cells, and that the MP does not function between phloem parenchyma and bundle-sheath cells, makes the process of virus and/or viral RNA egress at systemic sites all the more puzzling. Further experiments in this area are required.

Transgenic plants which express intact, defective, or portions of non-structural viral genes have gained prominence for their resistance to infection by the cognate virus or viral RNA, even at high challenge doses (Golemboski *et al.*, 1990; MacFarlane and Davies, 1992; reviewed in Wilson, 1993). Recently, a similar phenotype has been reported for transgenic plants which express an untranslatable potyvirus CP mRNA (Lindbo and Dougherty, 1992). In all these cases, the mechanism of resistance cannot involve simple interference with virus particle disassembly/reassembly and is thought to be due to the general disruption of RNA replication (Lindbo and Dougherty, 1992; Carr and Zaitlin, 1991).

Several examples of CP-mediated protection against potyviruses or luteoviruses have shown resistance to virus delivered by the natural aphid vector. However, in recent work with a nematode-transmitted tobravirus (TRV) it appeared that CP-mediated resistance does not operate as effectively in roots or in leaves as when the same virus was inoculated mechanically (Ploeg, Brown, and Bol, personal communication). This may be attributed to some ingredient in nematode saliva (unlikely!) or to a conformational or chemical change in the challenge virus which occurs between acquisition and transmission, such that its surface structure is no longer recognized by, or compatible with, the endogenously expressed CP. Even so, the low level of RNA delivered, albeit in an effectively "foreign" CP, should still be inhibited by the homologous transgenic CP, if we can extrapolate from the TMV RNA result mentioned above. Experiments with TRV 56 kDa "replicase" transgenic plants (MacFarlane and Davies, 1992) or virus particles in which the exposed and flexible C-terminal arm of the CP (Legorboru and Robinson, personal communication) has been removed by proteolytic digestion, should help clarify this phenomenon and may give further clues to the essential requirements for CP-mediated protection.

Early Events in Monoclonal Antibody (MAb) Transgenic Plants: The ability to express murine monoclonal antibody (MAb) genes *in planta* and, after suitable cross-pollination, for doubly-transgenic plants expressing both the heavy and light chains to produce correctly assembled, chaperoned, and exported MAb's at levels up to about 1.3 percent of total soluble leaf protein, was first reported by Hiatt *et al.* (1989). The capacity to produce MAb's cheaply and in large quantities by extracting fields of transformed tobacco, followed by affinity chromatography, has many appealing features for clinical, biochemical, and structural studies. Furthermore, the plants themselves could be made resistant to a pathogen by virtue of expressing a MAb to an epitope on a critical structural or non-structural protein of the disease agent.

Several groups are currently attempting this approach, for convenience, with genes for MAb's to a variety of plant viral CP's. It remains to be seen whether or not this strategy will be effective in achieving an adequate inhibition of parental virus particle disassembly at the point and instant of inoculation and/or if secondary effects on the viral replication cycle which may require CP can also be inhibited. Data from existing systems suggest that these secondary events may be at least as critical in creating the virus

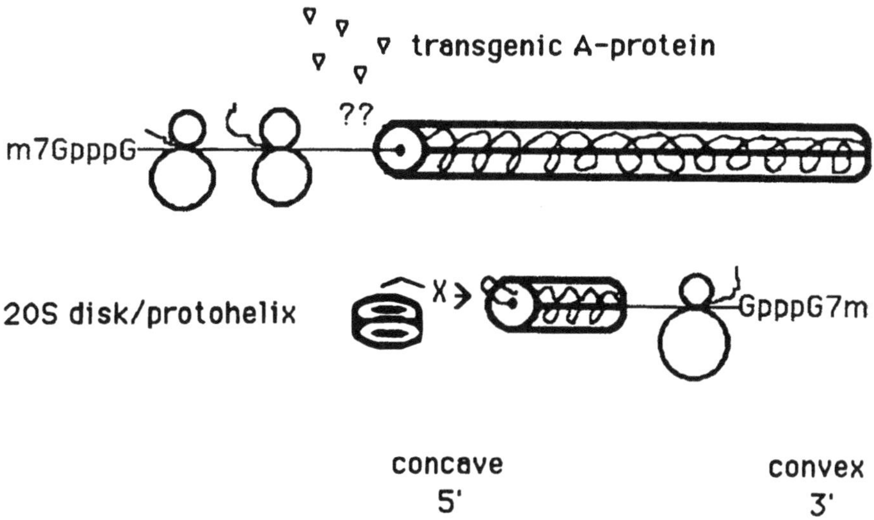

Figure 10. Schematic Diagram Showing the Relative Configurations of the Nucleoprotein Helix, the Decoding Ribosomes, and the Form of TMV CP Presumed to be Active in RNA Packaging, in Eukaryotic Striposomes (Upper Picture), or Partly Assembled Polysomes with a 3'-OAS (Lower Picture). The morphologically distinguishable ends (concave/convex of the CP helix are drawn in the same orientation; but, during assembly, the 5'-capped end (m^7GpppG) of the ssRNA is folded back down the lumen of the helix and is pulled through as the structure extends right-to-left, as drawn.

resistant phenotype as reduced uncoating of the parental virions. This may be attributed either to the presence of CP at the wrong time or place or to the presence of CP transcripts, even if untranslatable. Thus, the ability of CP MAb's to inhibit later replicational events may be questioned.

Very recently, van Regenmortel and colleagues have confirmed that some anti-TMV CP MAb's are effective inhibitors of the cotranslational disassembly of pH8-pretreated virions at MAb:TMV ratios between 50:1 and 200:1 (Saunal *et al.*, 1993). Pretreated virions were incubated with each MAb at 4°C overnight before *in vitro* translation. Experiments to reproduce this phenomenon in TMV challenged transgenic plants expressing one of those MAb's shown to function *in vitro* are currently underway (Hiatt, van Regenmortel, and Wilson). It appears that single chain (scF_v) or complete IgG's, not F_{ab} fragments, may be necessary for interaction with the plant homologue of the molecular chaperone protein BiP and, hence, for efficient IgG assembly *in planta*. Failure to assemble correctly and be exported from the cell must result in a high turnover rate, since we were unable to detect any L or truncated H chains on western blots of sap from 40 seedlings selected at random from approximately 120 first generation progeny of a cross between L-, or truncated H-chain, single-transformed parent tobaccos (Hiatt, Cowan, and Wilson; unpublished work).

The *in vitro* work (Saunal *et al.*, 1993) shows that MAb's to unrelated viral CP's have no effect on TMV expression and that both anti-cryptotope and anti-metatope MAb's are effective but that not all clones of either class function equally well. The former result was unexpected as cryptotopes are present on subunit surfaces that are buried when the CP is polymerized into helical rods. The latter observation was expected, as all published cases where binding of anti-metatope MAb's had been studied showed them to bind to one or a few CP's at the 5'-end of the virion, the site where cotranslational disassembly initiates (Dore *et al.*, 1988; Dore *et al.*, 1990). The fact that some anti-metatope MAb's failed to inhibit is more surprising. Anti-neotope MAb's were only partially effective at inhibiting cotranslational disassembly and at ratios above 200:1. Not surprising, perhaps, when about 800 such epitopes exist and can bind as many MAb's to the surface of a single virus particle (Azimzadeh and van Regenmortel, 1991).

Transgenic plants expressing MAb's against virus-coded replicase proteins, proteases, or cell-to-cell movement proteins will also be interesting to study for resistance to infection by virions or naked viral RNA.

Concluding Comments

From the perspective of a plant molecular virologist, the processes leading to nucleocapsid disassembly remain largely an enigma. In this article, I have attempted to review what little we know and what clues we have. Future studies with transformed plants resistant to virus infection may be the closest approach to having mutant "hosts" to help further elucidate the early structural events and later replicational events which normally proceed unchecked in an otherwise isogenic control plant. Several natural resistance genes (e.g., *Tm-1, Tm-2, -2*^{2}) and mutant viruses which overcome the resistance have illuminated some structural features of viral proteins essential for normal interactions with plant components during the replication cycle. Transgenic plants expressing viral proteins and mutagenized infectious viral RNA's have already advanced our knowledge of molecular pathology and symptomatology (Dawson, 1992).

In no case, to date, can plant resistance be attributed to the complete inhibition of viral nucleocapsid disassembly. Studies indicate that even in the poorest subliminal host, a virus (or at least TMV!) can uncoat and synthesize at least one virus-coded component of the RNA replicase (Wilson and Shaw, 1987). Statistically, any attempt to protect crop plants which relies upon the total inhibition of early structural events for every invading virus particle seems doomed to fail. Secondary levels of interference involving non-structural proteins or genome replication are more likely to succeed. Combining multiple sites of inhibition could be built into crops in an additive, combinatorial way - "mixing-and-matching" strategies of protection, by sequential "cross-pollinations" (e.g., CP plus satellite or DI RNA's; CP plus "54 kDa"-type; antisense/ribozyme plus CP; defective MP plus CP; or any other combination).

Acknowledgements

I am indebted to the many friends and colleagues who have struggled in this difficult area of research and who have kindly provided recent information and/or illustrations to enhance this contribution. The list is long; but, with apologies to those I may have omitted, particular thanks should go to: Roger Beachy, Derek Brown, Lou Heaton, Ari Helenius, John Hillman, Jack Morris, Wolfgang Mundry, Jane Osbourn, Kitty Plaskitt, Ton Ploeg, Helene Saunal, John Shaw, Ila Singh, David Sleat, Marc van Regenmortel, Dick Verduin, Peter Watkins, Lynne Watkins, and Xiao-Jie Wu.

References

Adrian, M., Timmins, P.A., and Witz, J. 1992. J. Gen. Virol. 73:2079-2083.
Argos, P., Tsukihara, T., and Rossmann, M.G. 1980. J. Molec. Evol. 15:169-179.
Atabekov, J.G. and Taliansky, M.E. 1990. Adv. Virus Res. 38:201-248.
Azimzadeh, A. and van Regenmortel, M.H.V. 1991. J. Immunol. Meth. 141:199-208.
Bawden, F.C. and Pirie, N.W. 1957. J. Gen. Microbiol. 17:80-95.

Beachy, R.N., Loesch-Fries, S., and Tumer, N.E. 1990. Annu. Rev. Phytopathol. 28:451-474.

Beachy, R.N., Zaitlin, M., Bruening, G., and Israel, H.W. 1976. Virology 73:498-507.

Bloomer, A.C. and Butler, P.J.G. 1986. In: *The Plant Viruses*. Vol. 2. M.H.V. van Regenmortel and H. Fraenkel-Conrat (eds.). Plenum Publishing, New York.

Brisco, M.J., Haniff, C., Hull, R., Wilson, T.M.A., and Sattelle, D.B. 1986. Virology 148:218-220.

Brisco, M.J., Hull, R., and Wilson, T.M.A. 1985. Virology 143:382-398.

Brisco, M.J., Hull, R., and Wilson, T.M.A. 1986. Virology 148:210-217.

Brown, D.T. and Edwards, J. 1992. Seminars in Virol. 3:519-527.

Carr, J.P. and Zaitlin, M. 1991. Molec. Plant-Microbe Interact. 4:579-585.

Choi, H-K., Tong, L., Minor, W., Dumas, P., Boege, U., and Rossmann, M. 1991. Nature 354:37-43.

Clark, W.G, Register, J.C., Nejidat, A., Eichholtz, D.A., Sanders, P.R., Fraley, R.T., and Beachy, R.N. 1990. Virology 179:640-647.

Clowes, F.A.L. and Juniper, B.E. 1968. Botanical Monographs 8:39-67. (Blackwell Scientific Publications)

D'Arcy, C.J. and De Zoeten, G.A. 1979. Phytopathol. 69:1194-1198.

Datema, K.P., Spruijt, R.B., Verduin, B.J.M., and Hemminga, M.A. 1987. Biochemistry 26:6217-6223.

Dawson, W.O. 1992. Virology186:359-367.

Ding, B., Haundenshield, J.S., Hull, R.J., Wolf, S., Beachy, R.N., and Lucas, W.J. 1992. The Plant Cell 4:915-928.

Dore, I., Ruhlmann, P., Oudet, P., Cahoon, M., Caspar, D.L.D., and van Regenmortel, M.H.V. 1990. Virology 176:25-29.

Dore, I., Weiss, E., Altschuh, D., and van Regenmortel, M.H.V. 1988. Virology 162:279-289.

Durham, A.C.H. 1978. Biomedicine 28:307-314.

Durham, A.C.H., Hendry, D.A., and von Wechmar, M.B. 1977. Virology 77:524-533.

Golemboski, D.B., Lomonossoff, G.P., and Zaitlin, M. 1990. Proc. Natl. Acad. Sci. U.S.A. 87:6311-6315.

Harms, C.T. 1992. Crop Protection 11:291-306.

Harrison, B.D. 1992. Proc. Royal Soc. Edinburgh 99B(3/4):61-77.

Heaton, L.A. and Morris, T.J. 1992. Seminars in Virol. 3:433-439.

Hiatt, A., Cafferkey, R., and Bowdish, K. 1989. Nature 342:76-78.

Holt, C.A. and Beachy, R.N. 1991. Virology 181:109-117.

Hull, R. 1989. Annu. Rev. Phytopathol. 24:213-240.

Kiho, Y., Abe, T., and Ohashi, Y. 1979b. Microbiol. Immunol. 23:1067-1076.

Kiho, Y., Shimomura, T., Abe, T., and Nozu, Y. 1979a. Microbiol. Immunol. 23:735-748.

Kozak, M. 1989. J. Cell Biol. 108:229-241.

Kukla, B.A., Guilley, H., Jonard, G., Richards, K.E., and Mundry, K.W. 1979. Eur. J. Biochem. 98:61-66.

Lackmann, M., Ueckermann, C., Engelmann, K., and Koch, G. 1987. Arch. Virol. 95:1-16.
Laidlaw, W.M.R. 1987. EPPO Bull. 17:81-89.
Lindbo, J.A. and Dougherty, W.G. 1992. Virology 189:725-733.
Lok, S. and Abou-Haidar, M. 1986. Virology 153:289-296.
MacFarlane, S.A. and Davies, J.W. 1992. Proc. Natl. Acad. Sci. U.S.A. 89:5829-5833.
Matthews, R.E.F. and Witz, J. 1985. Virology 144:318-327.
Mundry, K.W., Watkins, P.A.C., Ashfield, T., Plaskitt, K.A., Eisele-Walter, S., and Wilson, T.M.A. 1991. J. Gen. Virol. 72:769-777.
Namba, K., Pattanayek, R., and Stubbs, G. 1989. J. Molec. Biol. 208:307-325.
Osbourn, J.K. 1990. Ph.D. Thesis. University of East Anglia.
Osbourn, J.K., Sarkar, S., and Wilson, T.M.A. 1990. Virology 179:921-925.
Osbourn, J.K., Watts, J.W., Beachy, R.N., and Wilson, T.M.A. 1989. Virology 172:370-373.
Plaskitt, K.A., Watkins, P.A.C., Sleat, D.E., Gallie, D.R., Shaw, J.G., and Wilson, T.M.A. 1988. Molec. Plant-Microbe Interact. 1:10-16.
Powell Abel, P., Nelson, R.S., De, B., Hoffmann, N., Rogers, S.G., Fraley, R.T., and Beachy, R.N. 1986. Science 232:738-743.
Pustowoit, B., Pustowoit, W., and Schuster, G. 1984. Biochem. Phys. Pflanzen. 179:507-516.
Racaniello, V.R. 1992. Seminars in Virol. 3:473-481.
Register, J.C. and Nelson, R.S. 1992. Seminars in Virol. 3:441-451.
Roenhorst, J.W. 1986. Ph.D. Thesis. Wageningen Agricultural University, The Netherlands.
Roenhorst, J.W., van Lent, J.W.M., and Verduin, B.J.M. 1988. Virology 164:91-98.
Roenhorst, J.W., Verduin, B.J.M., and Goldbach, R.W. 1989. Virology 168:138-146.
Saito, T., Yamanaka, K., and Okada, Y. 1990. Virology 176:329-336.
Saunal, H., Witz, J., and van Regenmortel, M.H.V. 1993. J. Gen. Virol. 74:897-900.
Shaw, J.G., Plaskitt, K.A., and Wilson, T.M.A. 1986. Virology 148:326-336.
Shields, S.A., Brisco, M.J., Wilson, T.M.A., and Hull, R. 1989. Virology 171:602-606.
Singh, I. and Helenius, A. 1992. Seminars in Virol. 3:511-518.
Sleat, D.E., Gallie, D.R., Jefferson, R.A., Bevan, M.W., Turner, P.C., and Wilson, T.M.A. 1987. Gene 60:217-225.
Sleat, D.E., Turner, P.C., Finch, J.T., Butler, P.J.G., and Wilson, T.M.A. 1986. Virology 155:299-308.
Smirnyagina, E.V., Morozov, S.Y., Rodionova, N.P., Miroshnichenko, N.A., Solovyev, A.G., Fedorkin, O.N., and Atabekov, J.G. 1991. Biochimie. 73:587-598.
Solis, I. and Garcia-Arenal, F. 1990. Virology 177:553-558.
Tomlinson, J.A., Faithfull, E., Flemett, T.H., and Beards, G. 1982. Nature 300:637-638.

Turnbull-Ross, A.D., Reavy, B., Mayo, M.A., and Murant, A.F. 1992. J. Gen. Virol. 73:3203-3211.
Turner, D.R., McGuigan, C.J., and Butler, P.J.G. 1989. J. Molec. Biol. 209:407-422.
Turner, P.C., Watkins, P.A.C., Zaitlin, M., and Wilson, T.M.A. 1987. Virology 160:515-517.
Verduin, B.J.M. 1992. Seminars in Virol. 3:423-431.
Wei, N., Heaton, L.A., Morris, T.J., and Harrison, S.C. 1990. J. Molec. Biol. 214:85-95.
Wilson, T.M.A. 1984a. Virology 137:255-265.
Wilson, T.M.A. 1984b. Virology 138:353-356.
Wilson, T.M.A. 1986. Virology 152:277-279.
Wilson, T.M.A. 1988. Oxford Surveys Plant Molec. Cell Biol. 5:89-144.
Wilson, T.M.A. 1989. BioEssays 10(6):179-186.
Wilson, T.M.A. 1993. Proc. Natl. Acad. Sci. U.S.A. 90:3134-3141.
Wilson, T.M.A. and Shaw, J.G. 1987. In: *Positive-Strand RNA Viruses.* M.A. Brinton and R.R. Rueckert (eds.). UCLA Symp. Molec. Cell. Biol. 54:159-181 A.R. Liss New York.
Wilson, T.M.A. and Watkins, P.A.C. 1985. Virology 145:346-349.
Wilson, T.M.A. and Watkins, P.A.C. 1986. Virology 149:132-135.
Wisniewski, L.A., Powell, P.A., Nelson, R.S., and Beachy, R.N. 1990. The Plant Cell 2:559-567.
Wu, X-J., Beachy, R.N., Wilson, T.M.A., and Shaw, J.G. 1990. Virology 179:893-895.
Yarwood, C.E. 1957. Adv. Virus Res. 4:243-278.

Function and Interaction of Bromovirus Proteins in RNA Replication and Infection Spread

P. Ahlquist, C. C. Kao, W. De Jong, R. Quadt, R. Hershberger and K. Mise
Institute for Molecular Virology and Department of Plant Pathology
University of Wisconsin - Madison
Madison, Wisconsin 53706

Bromovirus RNA replication requires the expression and function of two large, virus-encoded proteins designated 1a and 2a. 1a contains domains showing sequence conservation with proteins having helicase and capping functions, and 2a contains a polymerase-like domain. We have now found that 1a and 2a form a specific complex *in vitro* and have mapped the interacting domains in both proteins. *In vivo* genetic data implicate the 1a-2a interaction in RNA replication and suggest that it allows the coordinated action of the putative helicase, polymerase, and capping domains in a unified RNA replication complex. Bromoviruses also encode a 3a "movement" protein that is essential for cell-to-cell spread and a coat protein that enhances local spread and is required for long distance spread. Despite prior indications of intergenic compatibility requirements, the 3a movement gene of a bromovirus has now been functionally replaced by the nonhomologous movement gene of a tobamovirus. Because of the extensive differences between the icosahedral bromoviruses and rod-shaped tobamoviruses, this implies that at least the tobamovirus movement protein functions independently of specific interactions with other viral factors, including the coat protein. Both the arrangement of interacting domains in 1a and 2a and the successful function of the tobamovirus movement protein after transfer into a bromovirus genome support the concept of modular evolution in RNA viruses.

Introduction

The bromoviruses are a group of plant viruses with isometric, approximately 30 nm diameter virions that encapsidate a messenger-sense RNA genome (Lane, 1981). The two best characterized members of the group are the type member, brome mosaic virus (BMV), which infects cereals and other grasses, and cowpea chlorotic mottle virus (CCMV), which infects certain legumes.

As shown in Figure 1, the bromovirus genome is divided among three genomic RNA's, which are designated RNA1 (3.2 kb), RNA2 (2.9 kb), and RNA3 (2.1 kb). Each genomic RNA serves as the mRNA for a separate noncapsid protein, designated 1a, 2a, and 3a, respectively. In addition, RNA3 encodes the coat protein gene, which is located 3′- to the 3a gene and is, therefore, not directly translated from RNA3. Expression of coat protein occurs via a subgenomic mRNA, RNA4, which is transcribed by internal initiation on the negative-strand RNA replication intermediate of RNA3.

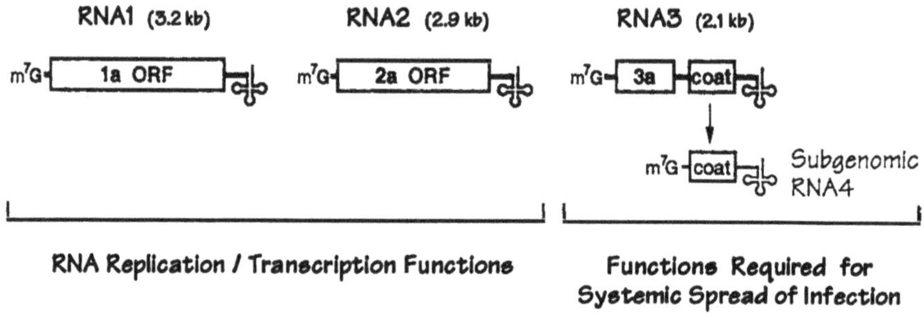

Figure 1. Schematic Diagram of the Tripartite Bromovirus Genome. The four open reading frames are shown by the labelled boxes while noncoding regions are shown by the flanking lines. The 5'-caps (m^7G) and conserved 3'-tRNA-like sequences (cloverleaf structure) are indicated on each RNA. See main text for additional discussion.

Infectious transcripts can be synthesized from cloned bromovirus cDNA (Ahlquist *et al,* 1984; Allison *et al.*, 1988), allowing molecular genetic studies to be performed. Mutational studies show that the 3a and coat protein genes of RNA3 are dispensable for RNA replication and mRNA transcription in isolated protoplasts but are required for infection spread in whole plants. The roles of the 3a and coat protein genes in infection spread are discussed in more detail in later sections.

Similar mutational studies show that both the 1a and 2a proteins are required for genomic RNA replication and subgenomic mRNA transcription (Ahlquist, 1992; Kiberstis *et al.*, 1981; French *et al.*, 1986; Kroner *et al.*, 1989; Kroner *et al.*, 1990; Traynor and Ahlquist, 1990; Traynor *et al.*, 1991). Sequence analysis shows that two large domains in 1a and a central core domain in 2a share sequence conservation with noncapsid proteins encoded by an array of otherwise dissimilar plant and animal RNA viruses, including the animal alphaviruses, such as Sindbis virus (Figure 2) (Haseloff *et al.*, 1984; Ahlquist *et al.*, 1985; Goldbach *et al.*, 1991). This collection of viruses has come to be called the alphavirus-like superfamily. The N-terminal 1a domain is related to an alphavirus-encoded methyltransferase domain believed to operate in viral RNA capping, while the C-terminal 1a domain is related to cellular and DNA virus helicases. The central conserved region in 2a is related to the RNA-dependent RNA polymerases of poliovirus and bacteriophage Qβ. Thus, for all three domains, possible functions have been identified that are consistent with the established role of these domains in viral RNA synthesis (Ahlquist, 1992).

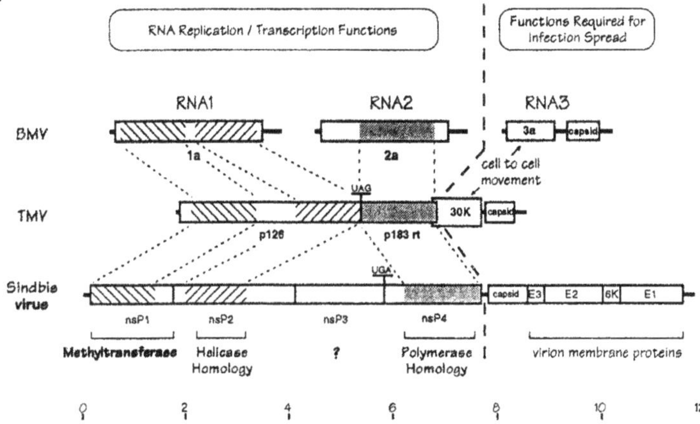

Figure 2. Comparison of the RNA Genomes of BMV, TMV, and Sindbis Virus, Three Representative Members of the Alphavirus-Like Superfamily of Positive-Strand RNA Viruses. Open reading frames are represented by boxed regions, which are labelled with the name of the encoded protein products. Known or putative functions for the gene products are given by annotations where appropriate. The variously crosshatched or shaded regions in the left half of the diagram map the three regions of amino acid sequence conserved between the proteins encoded by each virus. The TMV 183 kDa read-through protein (see main text) is synthesized by occasional translational read-through of the leaky UAG codon indicated at the end of the p126 open reading frame, resulting in continued translation through the downstream region labelled p183rt.

Interaction of Bromovirus RNA Replication Proteins 1a and 2a

Early molecular genetic experiments based on RNA reassortment (Allison *et al.*, 1988) and more recent experiments based on gene exchanges between BMV and CCMV (Dinant, Janda, Kroner, and Ahlquist, unpublished results) show that viral RNA replication requires the satisfaction of some form of compatibility requirement(s) between the 1a and 2a proteins. This compatibility requirement shows that 1a and 2a do not function in a completely independent fashion in viral RNA synthesis but does not reveal whether the interaction between 1a and 2a is direct or indirect. The possibility of a direct 1a-2a interaction was suggested by several considerations, including the fact that tobacco mosaic virus (TMV) encodes a 183 kDa read-through protein that covalently fuses protein domains related to all three intervirally conserved domains in 1a and 2a (Figure 2).

Motivated by these observations, experiments were undertaken to further characterize the basis for communication and compatibility between 1a and 2a (Kao *et al.*, 1992). Immunoprecipitation experiments showed that, when 1a and 2a were co-translated *in vitro*, both proteins could be precipitated by specific antisera to either 1a or 2a. Similar results were obtained when 1a and 2a were separately translated and mixed, although the efficiency of co-precipitation was greater with co-translation. Independent confirmation of the ability of 1a and 2a to selectively interact was provided by a filter binding or "far-western blot" assay. For this test, unlabelled 2a protein was expressed in an insect cell/baculovirus vector system; and total protein extracts from the insect cells were electrophoresed on a denaturing gel and transferred to a membrane. Incubation of the membrane with *in vitro* translated, radiolabelled 1a protein resulted in a specific radioactive signal coinciding with the 2a protein band in extracts from 2a-expressing cells but gave no band with extracts from control, uninfected cells (Kao *et al.*, 1992).

Mapping of the Interacting Domains in 1a and 2a

The protein segments responsible for 1a and 2a interaction were then mapped by *in vitro* translation and immunoprecipitation of 1a and 2a proteins altered by truncation, site-specific mutagenesis, or fusion to heterologous proteins (Kao and Ahlquist, 1992). The combined results of multiple experiments indicated that amino acids 25-140 of the 2a protein were necessary and sufficient to direct interaction with the 1a protein (Figure 3). This segment is located outside of and N-terminal to the conserved polymerase-like core domain of 2a. In the 1a protein, the C-proximal helicase-like domain (Figure 3) was found to interact with 2a even more efficiently than the intact 1a protein. Small deletions at either end of this domain, however, blocked the interaction with 2a.

These *in vitro* mapping results agreed well with prior molecular genetic studies of 1a and 2a function *in vivo*. Of particular interest are the results obtained with BMV/CCMV 2a gene hybrids, which pointed to the importance of sequences in the N-terminal portion of 2a in determining the ability of 2a to function compatibly with the BMV 1a protein (Traynor and Ahlquist, 1990). Similarly, the effects of 2a N-terminal deletions on RNA replication *in vivo* (Traynor *et al.*, 1991) closely parallel the progressively debilitating effects of the same deletions on 2a interaction with 1a *in vitro* (Kao and Ahlquist, 1992).

Model for Coordinated RNA Replication Functions in a 1a-2a Complex

An attractive mechanistic possibility is that the interaction of the 1a and 2a proteins facilitates bromovirus RNA replication and transcription by integrating the separate enzymatic functions of these proteins into a unified, well-coordinated machine (Ahlquist, 1992; Traynor and Ahlquist, 1990; Kao *et al.*, 1992). In consequence, viral RNA synthesis could proceed by a continuous series of directly connected steps, rather than by an interrupted series of discontinuous steps requiring the independent or repeated association of separate factors with the viral RNA templates.

With the demonstration of 1a-2a interaction and suggestive evidence for potential polymerase, helicase, and capping domains in 1a and 2a (see above), it is now possible to suggest how specific functions of 1a and 2a might be integrated in a replication complex. Such models should also take into account the probable

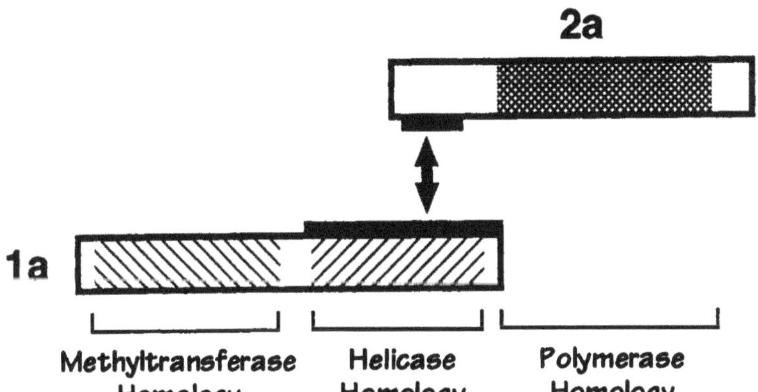

Figure 3. Location of Interacting Regions in the BMV 1a and 2a Proteins. The rectangles represent the primary structure of the 1a and 2a proteins, and the bold lines over 1a and under 2a map regions containing sequences that are necessary and sufficient for formation of a 1a-2a complex. Note that the resulting linear ordering of domains conforms directly to the order of the corresponding domains in the TMV 183 kDa protein (see Figure 2 and main text).

role of host proteins in viral RNA replication and the possibility that host proteins may also interact with 1a and 2a as components of a replication complex. Due to the incomplete nature of current information, such models can only be viewed as tentative but can nevertheless provide a starting point for developing better concepts and experiments concerning the replication process.

For example, suitable coupling of the helicase-like C-terminal domain of 1a to the putative 2a polymerase might provide crucial unwinding activities at one or several steps of RNA synthesis. Alternatively, one or more of these steps might be mediated by a host-encoded helicase. Helicase action is almost certainly required to unwind the highly structured ends of the bromovirus RNA's to allow the polymerase to initiate RNA synthesis. A helicase-polymerase complex should also greatly facilitate elongation by allowing removal of secondary structure from internal regions of the template as RNA synthesis proceeds. Finally, a helicase could be involved in releasing the newly synthesized RNA strand from the template. Inclusion of the possible capping activities of the N-terminal 1a domain in a 1a-2a complex might position these activities to cap new viral RNA's immediately after their 5'-ends are synthesized. This possibility is consistent with the lack of complementation between mutants in the N- and C-terminal domains of 1a, indicating that functional copies of both domains must be present in a single protein or protein complex for successful RNA replication (Kroner *et al.*, 1990).

Function of a Tobamovirus Movement Protein in a Bromovirus Hybrid

The effective host range of many plant viruses is limited at the level of infection spread from primarily inoculated cells, making the process of infection spread a matter of considerable practical interest. As noted in the **Introduction**, both of the genes on RNA3 are required for bromovirus infection spread (DeJong and Ahlquist, 1991). Specifically, site-directed mutations show that the 3a noncapsid protein is required for cell-to-cell spread of infection (DeJong and Ahlquist, 1992; Allison *et al.*, 1990), while the coat protein is required for long range spread of infection (Allison *et al.*, 1990; Sacher and Ahlquist, 1989), apparently by a vascular route (Mise and Ahlquist, unpublished results).

As discussed above, recent results show that bromovirus RNA replication depends on specific interaction between two viral gene products, as well as selective interactions between the replication complex and its specific bromovirus template RNA's. By contrast, previous experiments did not reveal conclusively whether successful infection spread depends on similar specific interactions between the noncapsid 3a movement protein and other viral proteins or between the 3a protein and viral RNA. In some combinations, helper virus co-infection can facilitate the spread of a misadapted or mutant virus in an otherwise nonpermissive host; but such co-infection experiments do not define the number or nature of helper virus genes required for such a role. At least one recent transgenic plant study suggested that the combined action of multiple helper virus genes might be required for such complementation (Taliansky *et al.*, 1992).

Moreover, in many cases co-inoculation does not assist the spread of a defective or misadapted virus, suggesting incompatibility between helper virus movement functions and the dependent virus. Notably, co-inoculating cowpea plants with sunn-hemp mosaic virus (SHMV), a legume-adapted tobamovirus, failed to complement even the local, cell-to-cell movement defect(s) of a CCMV 3a deletion mutant (DeJong and Ahlquist, 1992). Since wild-type CCMV and SHMV are each well adapted for spread in cowpea plants, this implied that the SHMV movement functions were either unable to interact effectively with the CCMV mutant to achieve movement, or else that co-inoculation did not result in expression of the potentially complementing SHMV function(s) throughout all times and/or sites needed to assist the CCMV mutant.

To distinguish between these possibilities, a hybrid virus was constructed in which the 3a gene of CCMV was replaced with the 30 kDa gene of SHMV. The bromovirus 3a and tobamovirus 30 kDa genes and their encoded proteins (Figure 2) do not share any statistically significant sequence similarity, but they are formally analogous in that both are noncapsid proteins that are dispensable for RNA replication but required for cell-to-cell spread of infection.

Testing in cowpea plants showed that this hybrid virus was remarkably competent for cell-to-cell spread and consistently achieved a level of virus accumulation in inoculated leaves that rivaled that of wt CCMV. By contrast, CCMV mutants with deletion or frameshift mutations in the 3a gene, or a derivative of the hybrid virus with a frameshift mutation in the SHMV 30 kDa gene, failed to spread detectably in inoculated tissue. Thus, the 30 kDa gene is able to functionally replace the 3a gene in the context of an otherwise CCMV genome. Because of the fundamental differences in CCMV vs. SHMV genome organization (tripartite vs. single component), virion morphology (isometric vs. rod-shaped), and other properties, this result shows that the cell-to-cell movement functions of the tobamovirus 30 kDa protein function independently of sequence-specific interactions with other viral components or sequences.

Northern blot examination of viral RNA accumulation in protoplasts and inoculated plants revealed that the hybrid RNA3 bearing the SHMV 30 kDa gene accumulated much more poorly than wt CCMV RNA1 and RNA2 in the same infection or wt RNA3 in parallel infections by wt CCMV. Despite this apparent handicap, however, the CCMV/SHMV hybrid not only achieved local cell-to-cell spread in inoculated leaves but also achieved systemic infection in a large fraction of the plants in which it is inoculated. This indicates that the indispensable contribution of bromovirus capsid protein to infection movement does not require specific interaction with the natural 3a movement protein. Both these observations and other results suggest that the 3a and capsid proteins might mediate movement across distinct tissue barriers to infection spread (DeJong and Ahlquist, 1992).

Implications for RNA Virus Evolution

As shown in Figure 2, the members of the alphavirus-like superfamily share a related cluster of protein domains involved in RNA replication, while diverging substantially in capsid genes and other genes not involved in RNA replication. This relationship led immediately (Haseloff *et al.*, 1984; Ahlquist *et al.*, 1985) to the suggestion that these viruses have evolved in a largely modular fashion; i.e., by the rearrangement of conserved functional modules within and between the RNA replication genes and by the reassortment and recruitment (from cellular or viral sources) of different accessory genes specific to their diverse host and vector adaptations.

Additional results from many avenues of RNA virus study have continued to support this concept of modular evolution. Observations of the prevalence and diversity of recombination events in positive-strand RNA viruses have provided precedents for most classes of events required to explain the structure of genomes in the alphavirus-like superfamily by divergent evolutionary models (Allison *et al.*, 1990; Bujarski and Kaesberg, 1986; Allison *et al.*, 1989). Consistent with divergent evolution, Figures 2 and 3 reveal that the arrangement of functional and interacting domains in bromovirus proteins 1a and 2a shows an exact correspondence to the linear order of homologous domains in the TMV 183 kDa readthrough protein. This correspondence suggests that the 1a-2a complex may be both a functional analog and an evolutionary relative of the TMV 183 kDa protein (Ahlquist, 1992; Traynor and Ahlquist, 1990; Kao and Ahlquist, 1992).

Observations on naturally evolved RNA virus genomes have been complemented by artificial variants such as the CCMV/SHMV hybrid. Among pairs of viruses whose genes have been successfully exchanged to date, CCMV and SHMV appear to be the most dissimilar. The viability of the CCMV/SHMV hybrid thus confirms that foreign genes can indeed be recombined from eclectic sources to make a functional virus that is able to replicate and spread in a mature, differentiated host. Such exchanges may provide a paradigm for the natural recruitment of new genes by a virus under selection pressure to adapt to a new host (DeJong and Ahlquist, 1992). Other studies have shown that a remarkable tolerance exists for rearrangement and reassortment of bromovirus regulatory sequences (French and Ahlquist, 1987; French and Ahlquist, 1988; Pacha *et al.*, 1990; Pacha and Ahlquist, 1991; Pacha and Ahlquist, 1992). Taken together, these results reveal that the alphavirus-like viruses possess an extraordinary genetic plasticity that has greatly facilitated their survival; their divergence across plant, insect, and vertebrate hosts; and their study by recombinant DNA technology.

Acknowledgments

This research was supported by NIH grant GM35072 and NSF grant DMB-9004385.

References

Ahlquist, P. 1992. Bromovirus RNA replication and transcription. Curr. Opin. in Genetics and Develop. 2:71-76.

Ahlquist, P., French, R., Janda, M., and Loesch-Fries, L.S. 1984. Multicomponent RNA plant virus infection derived from cloned viral cDNA. Proc. Natl. Acad. Sci. U.S.A. 81:7066-7070.

Ahlquist, P., Strauss, E.G., Rice, M., Strauss, J.H., Haseloff, J., and Zimmern, D. 1985. Sindbis virus proteins nsP1 and nsP2 contain homology to nonstructural proteins from several RNA plant viruses. J. Virol. 53:536-542.

Allison, R., Janda, M., and Ahlquist, P. 1988. Infectious *in vitro* transcripts from cowpea chlorotic mottle virus cDNA clones and exchange of individual components with brome mosaic virus. J. Virol. 62:3581-3588.

Allison, R., Janda, M., and Ahlquist, P. 1989. Sequence of cowpea chlorotic mottle virus RNA's 2 and 3 and evidence of a recombination event during bromoviral evolution. Virology 172:321-330.

Allison, R., Thompson, C., and Ahlquist, P. 1990. Regeneration of a functional RNA virus genome by recombination between deletion mutants and requirement for cowpea chlorotic mottle virus 3a and coat genes for systemic infection. Proc. Natl. Acad. Sci. U.S.A. 87:1820-1824.

Bujarski, J. and Kaesberg, P. 1986. Genetic recombination between RNA components of a multipartite plant virus. Nature 321:528-531.

DeJong, W. and Ahlquist, P. 1991. Bromovirus host specificity and systemic infection. Sem. in Virology 2:97-105.

DeJong, W. and Ahlquist, P. 1992. A hybrid plant virus made by transferring the noncapsid movement protein from a rod-shaped to an icosahedral virus is competent for systemic infection. Proc. Natl. Acad. Sci. U.S.A. 89:6808-6812.

French, R. and Ahlquist, P. 1987. Intercistronic as well as terminal sequences are required for efficient amplification of brome mosaic virus RNA3. J. Virol. 61:1457-1465.

French, R. and Ahlquist, P. 1988. Characterization and engineering of sequences controlling synthesis of brome mosaic virus subgenomic RNA. J. Virol. 62:2411-2420.

French, R., Janda, M., and Ahlquist, P. 1986. Bacterial gene inserted in an engineered RNA virus: Efficient expression in monocotyledenous plant cells. Science 231:1294-1297.

Goldbach, R., LeGall, O., and Wellink, J. 1991. Alpha-like viruses in plants. Semin. Virol. 2:19-25.

Haseloff, J., Goelet, P., Zimmern, D., Ahlquist, P., Dasgupta, R., and Kaesberg, P. 1984. Striking similarities in amino acid sequence among nonstructural proteins encoded by RNA viruses that have dissimilar genomic organization. Proc. Natl. Acad. Sci. U.S.A. 81:4358:4362.

Kao, C. and Ahlquist, P. 1992. Identification of the domains required for direct interaction of the helicase-like and polymerase-like RNA replication proteins of brome mosaic virus. J. Virol. 66:(in press).

Kao, C., Quadt, R., Hershberger, R., and Ahlquist, P. 1992. Brome mosaic virus RNA replication proteins 1a and 2a form a complex *in vitro*. J. Virol. 66:6322-6329.

Kiberstis, P.A., Loesch-Fries, L.S., and Hall, T.C. 1981. Viral protein synthesis in barley protoplasts inoculated with native and fractionated brome mosaic virus RNA. Virology 112:804-808.

Kroner, P., Richards, D., Traynor, P., and Ahlquist, P. 1989. Defined mutations in a small region of the brome mosaic virus 2a gene cause diverse temperature-sensitive RNA replication phenotypes. J. Virol. 63:5302-5309.

Kroner, P.A., Young, B.M., and Ahlquist, P. 1990. Analysis of the role of brome mosaic virus 1a protein domains in RNA replication, using linker insertion mutagenesis. J. Virol. 64:6110-6120.

Lane, L.C. 1981. Bromoviruses. In: *Handbook of Plant Virus Infections and Comparative Diagnosis*. E. Kurstak (ed). Elsevier/North Holland Press.

Pacha, R.F., Allison, R.F., and Ahlquist, P. 1990. *cis*-acting sequences required for *in vivo* amplification of genomic RNA3 are organized differently in related bromoviruses. Virology 174:436-443.

Pacha, R.F. and Ahlquist, P. 1991. Use of bromovirus RNA3 hybrids to study template specificity in viral RNA amplification. J. Virol. 65:3693-3703.

Pacha, R.F. and Ahlquist, P. 1992. Substantial portions of the 5'- and intercistronic noncoding regions of cowpea chlorotic mottle virus RNA3 are dispensable for systemic infection but influence viral competitiveness and infection pathology. Virology 187:298-307.

Sacher, R. and Ahlquist, P. 1989. Effects of deletions in the N-terminal basic arm of brome mosaic virus coat protein on RNA packaging and systemic infection. J. Virol. 63:4545-4552

Taliansky, M.E., Malyshenko, S.I., Kaplan, I.B., Kondakova, O.A., and Atabekov, J.G. 1992. Production of the tobacco mosaic virus (TMV) transport protein in transgenic plants is essential but insufficient for complementing foreign virus transport: A need for the full length TMV genome or some other TMV-encoded product. J. Gen. Virol. 73:471-474.

Traynor, P. and Ahlquist, P. 1990. Use of bromovirus RNA hybrids to map *cis*- and trans-acting functions in a conserved RNA replication gene. J. Virol. 64:69-77.

Traynor, P., Young, B.M., and Ahlquist, P. 1991. Deletion analysis of brome mosaic virus 2a protein: Effects on RNA replication and systemic spread. J. Virol. 65:2807-2815.

The Replication and Gene Expression of Turnip Yellow Mosaic Virus

T.W. Dreher[1,2,3], C.S. Bozarth[1], K.L. Bransom[1,2], C.-H. Tsai[1,2], and J.J. Weiland[2,4]
Department of Agricultural Chemistry[1], Program in Genetics[2], and Center for Gene Research and Biotechnology[3]
Oregon State University
Agricultural & Life Sciences Room 1007
Corvallis, Oregon 97331-7301

Barley Unit, Northern Crop Sciences Laboratory[4]
Agricultural Research Service, USDA
Fargo, North Dakota 58105-5677

Turnip yellow mosaic virus (TYMV) is a positive strand RNA virus with a monopartite genome 6.3 kb long. Analysis of the gene expression *in vitro* reveals the efficient expression of each of the three major open reading frames. These encode a 69 kDa protein necessary for virus movement from cell-to-cell, the 20 kDa coat protein, and a 206 kDa protein that encodes activities necessary for RNA replication. The 206 kDa protein is autocatalytically cleaved by the action of a protease domain to yield an N-terminal 150 kDa protein and a C-terminal 70 kDa protein, which are presumed to provide capping and helicase activities and polymerase activities, respectively, for the replication of the viral RNA. In mixed inoculation experiments, RNA's encoding active forms of the 150 kDa and 70 kDa proteins were observed to replicate in preference to coinoculated RNA's expressing a mutant form of one of these proteins. A model proposing the formation of a 150 kDa/70 kDa protein complex that preferentially binds to potential RNA templates *in cis* is presented to explain the *cis*-preferential replication observed for TYMV. The proposed complex is presumed to participate in the interaction of the replication complex with the 3'-noncoding region to facilitate minus strand synthesis. Two structural elements in the 3'-noncoding region have been recognized to play a role in viral RNA replication: the tRNA-like structure, which can be efficiently esterified with valine, and an adjacent pseudoknot. Studies with substitution mutants have demonstrated a correlation between the ability of the viral RNA to be valylated *in vitro* and its ability to replicate *in vivo*. Thus, it is likely that the valylation of the viral RNA templates is an important step in their efficient replication.

Introduction

Turnip yellow mosaic virus (TYMV) is a small (29 nm) icosahedral virus that infects members of the Cruciferae. TYMV is the type member of the tymoviruses, for which the genomes of five members have now been sequenced: TYMV (Morch *et al.*, 1988; Keese *et al.*, 1989; Dreher and Bransom, 1992), eggplant mosaic virus (Osorio-Keese *et al.*, 1989), kennedya yellow mosaic virus (Ding *et al.*, 1990), ononis yellow mosaic virus (Ding *et al.*, 1989), and erysimum latent virus (Srifah, *et al.*, 1992). Tymoviral genomes have characteristically high cytosine (35-40 percent) and low guanine contents (15-17 percent). The TYMV genome consists of a single RNA strand of positive polarity 6.3 kb long and possesses a 5'-^{7}mGpppG cap structure (Briand *et al.*, 1978) and a tRNA-like structure at the 3'-end (Pinck *et al.*, 1970; Rietveld *et al.*, 1983). TYMV infection in Chinese cabbage induces characteristic cytopathology involving chloroplast clumping and the appearance of invaginations of the chloroplast outer membrane that seem to be the site of RNA replication (Matthews, 1991).

The TYMV Genome

Coding Regions: The 6318 nucleotide-long TYMV genome (Morch *et al.*, 1988; Keese *et al.*, 1989; Dreher and Bransom, 1992) encodes three major open reading frames (ORF's), diagrammed in Figure 1:

ORF 69, which initiates at nucleotide 88 and encodes a 69 kDa protein;

ORF 206, which initiates at nucleotide 95 and overlaps with ORF 69 to encode a 206 kDa protein; and

CP-ORF, situated in the 3'-region of the genome, which encodes the 20 kDa coat protein that is expressed from an encapsidated subgenomic RNA.

The TYMV genome is very efficiently utilized for protein-coding sequences; ORF 69 almost completely overlaps ORF 206, and there is only a very short intergenic region between ORF 206 and CP-ORF. Further, read-through of the ORF 206 amber termination codon extends the ORF to overlap 74 percent of CP-ORF. While the overall genome plan is similar for all tymoviruses sequenced to date (Srifah *et al.*, 1992), the existence of a significant read-through domain associated with the longest ORF is not conserved.

Computer-assisted alignments of amino acid sequences predicted by the genome sequences have identified three domains within ORF

206 that have sequence similarities with other positive strand RNA viruses (indicated in Figure 1b). The most N-terminal of these domains, encoded by amino acids 58-219, has been suggested to encode a methyltransferase activity involved in 5-cap synthesis (Rozanov *et al.*, 1992), due to a distant sequence relationship with nsP1 of Sindbis virus, for which a methyltransferase function has been established experimentally (Mi and Stollar, 1991). A second domain encoded by amino acids 973-1204 shares similarity with essential domains of other positive strand RNA viruses and a broader family of nucleic acid helicases (Habili and Symons, 1989; Gorbalenya and Koonin, 1989); this domain contains the putative nucleotide binding element Gly-Xxx-Gly-Lys-(Ser/Thr) (marked GxGKT in Figure 1b). The most C-terminal of the conserved domains, between amino acids 1496 and 1728 (Habili and Symons, 1989; Koonin, 1991), is related to verified RNA-dependent RNA polymerases such as the 3D

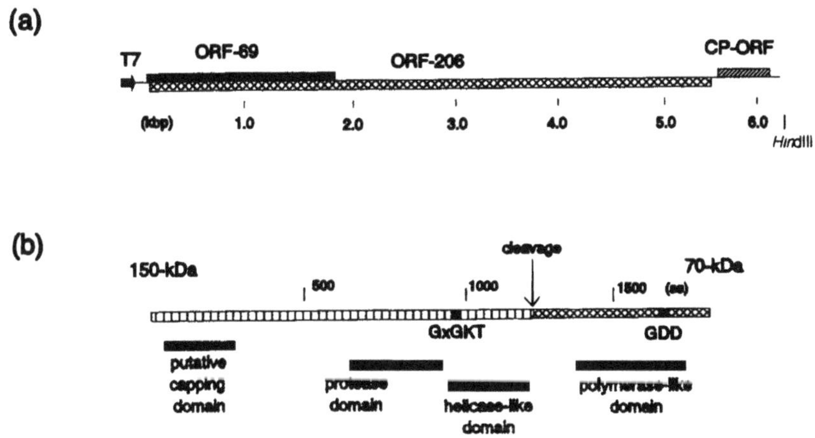

Figure 1. The TYMV Genome and Its Expression. A: Diagram of the cDNA insert of pTYMC, which can be transcribed with T7 RNA polymerase after linearization at the *Hind*III site, yielding infectious genomic transcripts. The major open reading frames encoded by TYMV are shown. B: The expression of ORF 206. The 150 kDa and 70 kDa products arise by cleavage of the precursor 206 kDa protein at the indicated site, catalyzed by the upstream protease domain. The other sequence elements and domains indicated share similarities with their analogues in the genomes of other positive strand RNA viruses.

polymerase of poliovirus (Jablonski *et al.*, 1991); this domain contains the Gly-Asp-Asp signature sequence of RNA-dependent RNA polymerases (Argos, 1988) (marked GDD in Figure 1b).

Phylogenetic reconstructions based on genomic sequence comparisons have classified the tymoviruses among the alpha-like viruses (Goldbach, 1987), most closely related to the potexviruses, carlaviruses, and apple chlorotic leafspot virus (Ding *et al.*, 1990; Rozanov *et al.*, 1992).

The 3'-Noncoding Region: The 3'-noncoding region of TYMV RNA contains a tRNA-like structure that functionally mimics $tRNA^{Val}$. Thus, the virion RNA, which terminates in 3 -CC is a substrate for the host enzyme (CTP,ATP):tRNA nucleotidyltransferase, resulting in 3'-CCA termini that are active substrates for aminoacylation with valine by valyl-tRNA synthetase. TYMV RNA is an efficient substrate *in vitro* for yeast (Giegé *et al.*, 1978) or wheat germ (Dreher *et al.*, 1988) valyl-tRNA synthetase. A portion of the pool of TYMV RNA in infected Chinese cabbage cells has been shown to exist in the valylated state (Joshi *et al.*, 1982), although the virion RNA, which lacks the 3'-A residue necessary for aminoacylation, is apparently packaged immediately after release from the replication complexes and is not accessible to the host adenylation and valylation enzymes. Valylated TYMV RNA has been shown to bind the elongation factors EF-Tu from *Escherichia coli* (Joshi *et al.*, 1984) and eEF1α from wheat germ (Joshi *et al.*, 1986).

The TYMV tRNA-like structure has only limited sequence similarity to higher plant $tRNA^{Val}$ but does possess a valine -CAC- anticodon and an -ACAC- sequence in the anticodon loop that is present in all eukaryotic $tRNA^{Val}$ species (Sprinzl *et al.*, 1991). The tRNA-like structure has clear similarities with tRNA's on a secondary and tertiary structure level, however, as emphasized in the L-conformation of Figure 2. An important feature of the viral structure is the involvement of a pseudoknot in building the amino acid acceptor stem (Rietveld *et al.*, 1983). Further properties of the TYMV tRNA-like structure are discussed below.

Another structural feature present in the 3'-noncoding region is a pseudoknot immediately upstream of the tRNA-like structure, overlapping with the CP-ORF termination codon. Pseudoknots are a common feature of the 3'-noncoding regions of positive strand RNA viruses (Pleij *et al.*, 1987).

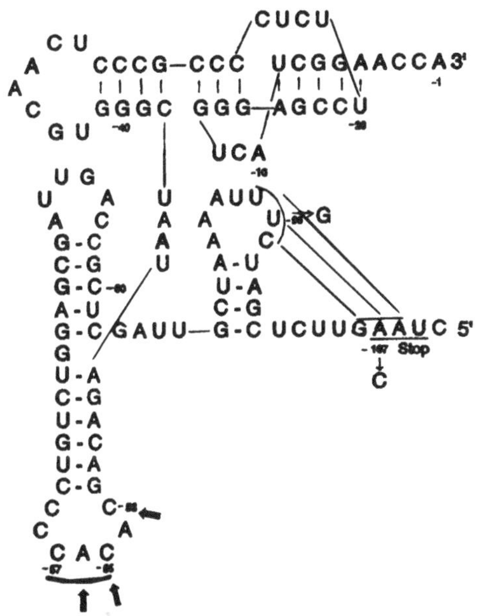

Figure 2. Model for the Secondary Structure of the 3'-Noncoding Region of TYMV RNA. A pseudoknot (nt -87 to -108; note that nucleotides are numbered from the 3'-end) is positioned immediately upstream of the L-shaped tRNA-like structure, which resembles tRNA^{Val}. The anticodon (-57 to -55) is underscored, and the three nucleotides that constitute the valine identity determinants recognized by wheat germ valyl-tRNA synthetase are marked with arrows. The normal termination codon of CP-ORF is underlined and marked 'Stop.' The substitutions at nucleotides -96 and -107 that are discussed in the text are marked and can be seen to interrupt the base-pairing (long diagonals) that stabilize the pseudoknotted conformation.

TYMC, A cDNA-Based TYMV Clone

Our studies on the biology of TYMV are conducted with the TYMC (Corvallis) strain of TYMV-type (Weiland and Dreher, 1989). This strain is maintained in the form of full-length genomic cDNA cloned in a pUC8 plasmid and is amplified in *E. coli.* Infectious, capped genomic RNA can be obtained by *in vitro* transcription with T7 RNA polymerase, using as template pTYMC that has been linearized at the 3'-end of the genome with *Hin*d III restriction endonuclease (Weiland and Dreher, 1989) (Figure 1). The resultant transcripts have the correct 5'-termini but carry a few additional nucleotides (-AGCU) at the 3'-end that originate from the *Hind* III site. The host cytoplasm is thought to provide the exonuclease and 3'-adenylation activities needed to generate correct 3'-CCA termini, as occurs with tRNA's (Deutscher, 1982). The establishment of a cDNA-cloned source of infectious viral RNA permits manipulation of the genome by directed mutagenesis.

TYMC RNA is infectious to Chinese cabbage, turnip, or *Arabidopsis thaliana*. Most of our experiments have been conducted with Chinese cabbage (cv. Spring A1) as host, although our more recent experiments have used turnips (cv. Just Right), which supports higher levels of TYMC replication.

TYMV Gene Expression

Proteins originating from the major ORF's can be detected in either rabbit reticulocyte (Figure 3) or wheat germ extracts, but ORF 69 expression can be overlooked in studies utilizing ^{35}S[methionine] labelling (Weiland and Dreher, 1989) since ORF 69 encodes only three methionine residues. The ORF 69 product migrates in SDS-PAGE at a relative MW of about 75 kDa, presumably due to the high proportion of basic amino acids (pI = 12.1). When *in vitro*-synthesized proteins are labelled with ^{3}H[leucine] and the products are immunoprecipitated with ORF-specific antisera, it is evident that the expression levels of ORF 69 and ORF 206 products are approximately equal in rabbit reticulocyte lysates.

Experiments have shown that the closely spaced initiation codons are not selected for initiation by the normal 5'→3' scanning mechanism, in which downstream AUG codons act as initiators only as a result of "leaky scanning" through an upstream AUG codon (Kozak, 1989). Nucleotide substitutions in the ORF 69 initiation codon had no detectable effect on ORF 206 expression in rabbit reticulocyte lysates, while substitution in the ORF 206 initiation codon substantially enhanced ORF 69 expression (Weiland and Dreher, 1989). These results contrast with the typical behavior of ORF's expressed by leaky scanning, in which removal of upstream initiation codons results in enhanced expression, while removal of downstream initiation codons has little or no effect (Kozak, 1989). Similar results have been obtained with influenza B virus RNA segment six, which has two closely spaced initiation codons that are equally expressed (Williams and Lamb, 1989). In that system, an uncharacterized element in the mRNA leader appears to regulate the balanced usage of both initiation codons. The order and relative spacing of the initiation codons of ORF 206 and ORF 69 and their analogues are conserved among the tymoviruses, suggesting that the close spacing of the initiation codons plays a role in regulating the balanced expression of the overlapping ORF's.

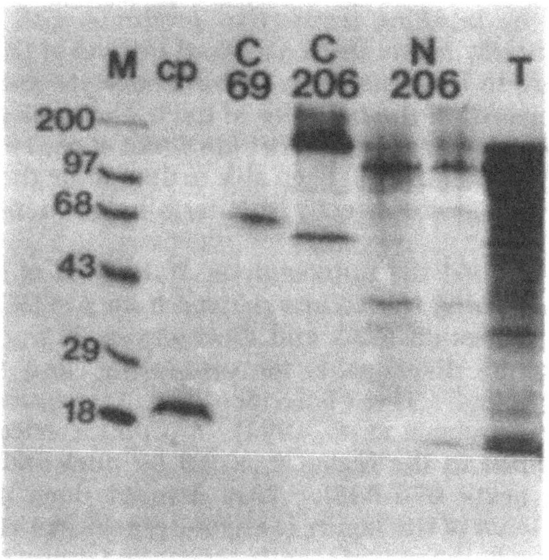

Figure 3. Protein Synthesis *in vitro* Directed by TYMV Virion RNA. TYMV RNA was translated in rabbit reticulocyte lysate in the presence of [^{35}S]methionine. The products were immunoprecipitated with ORF-specific antisera as indicated, prior to separation by SDS-PAGE: immunoprecipation with anti-TYMV antiserum (cp), and antisera raised against synthetic peptides from ORF 69 (C-69), the C-terminus of ORF 206 (C-206), and the N-terminal region of ORF 206 (N-206). T, total translation products; M, MW markers. Lane C-206 shows the 206 kDa protein and smaller amounts of the 70 kDa protein, which appears to be unstable. Lanes N-206 show the 206 kDa and 150 kDa proteins and smaller amounts of non-specific products.

Proteolytic Maturation in ORF 206 Expression: ORF 206 expression involves the proteolytic maturation of the 206 kDa translation product. We have studied this event *in vitro* using rabbit reticulocyte lysates, detecting products by immunoprecipitation using antisera corresponding to the N-terminal and C-terminal regions encoded by ORF 206 (Bransom *et al.*, 1991). The most abundant product arising from ORF 206 is typically a 150 kDa protein that can be immunoprecipitated with the N-terminal region-specific antiserum (Figure 3). A corresponding C-terminal cleavage product does not accumulate in stoichiometric amounts but can be detected as a 70 kDa protein after immunoprecipitation with antiserum recognizing the C-terminal region of ORF 206. The 70 kDa protein can also be distinguished by labelling translation products with ^{35}S[cysteine], which preferentially labels the C-terminal regions of ORF 206. This protein appears to be unstable in both reticulocyte and wheat germ lysates, reminiscent of the behavior of nsP4, the analogous putative polymerase encoded by Sindbis virus (de Groot *et al.*, 1991). Because of this instability, we have not been able to determine the 150 kDa/70 kDa protein cleavage site directly by N-terminal sequencing of the 70 kDa protein.

We have studied the autocatalytic proteolysis of the ORF 206 product further using transcripts derived from pTYMC. Sequences upstream of nucleotide 1755 and downstream of nucleotide 4085 were found to be dispensable for processing, and two domains necessary for proteolysis have been identified in the intervening region of the genome (Bransom *et al.*, 1991). A protease-encoding domain has been mapped to the region encoded by nucleotides 2160 and 2750 (amino acids 670-885). This domain does not have the sequence elements of the family of chymotrypsin-like viral proteases and may belong to the more recently recognized group of papain-like viral enzymes that utilize cysteine and histidine residues in their catalytic centers (Gorbalenya *et al.*, 1991). By substitution mutagenesis, we have shown that cysteine^{783} and histidine^{869} are essential for proteolysis (Bransom and Dreher, unpublished). These residues are conserved among the various tymoviruses and are consistent with the active site residues of a papain-like protease.

The second domain necessary for proteolysis has been mapped to nucleotides 3851-3886 encoding the amino acid sequence 1253-Gly-Pro-Lys-Leu-Asn-Gly-Ala-Thr-Pro-Ser-Ala-Ser-1264, which presumably includes the cleavage site. Substitution of amino acids 1256-Leu-Asn-Gly-1258 with Ala-Ala-Ala abolished processing, suggesting that cleavage occurs between or adjacent to these residues (Bransom *et al.*, 1991).

The placement of the protease domain and probable 150 kDa/70 kDa cleavage site on the genome plan shown in Figure 1 exemplifies the relationship between the genome plans of TYMV and other viruses belonging to the alphavirus-like supergroup. The protease domain is distinct from the other conserved regions, being located immediately upstream of the helicase-like domain. The proteolytic maturation during ORF 206 expression produces a 150 kDa protein whose C-terminus is shortly downstream of the helicase-like domain and which is, thus, analogous to the 126 kDa protein encoded by tobacco mosaic virus (TMV) and the 1a protein encoded by tricornaviruses such as brome mosaic virus (Goldbach, 1987). The TYMV 70 kDa protein is analogous to the 2a polymerase protein encoded by the tricornaviruses and nsP4 encoded by Sindbis virus.

Are There Significant Alternative Pathways for ORF 206 Expression? Although we believe that the major expression pathway of ORF 206 has been identified, three issues are at present unresolved:

1. Proteolytic cleavage of the full-length 206 kDa protein does not go to completion *in vitro*, even after overnight incubations. It is, thus, possible that the 206 kDa protein is present in infected cells and that it plays a role during infection.

2. The amber termination codon at the end of ORF 206 can be suppressed to a low degree during translation *in vitro*, and the rate of suppression can be enhanced by supplementing translation mixtures with bovine liver tRNA. The resultant 221 kDa protein is a minor but significant product of such translations. It is not known whether suppression occurs *in vivo*. On the one hand, *in vivo* read-through seems possible, since the sequence immediately surrounding the amber codon is similar to that present at the end of the p126 coding region of tobacco mosaic virus, where amber suppression plays an important physiological role (Beier *et al.*, 1984; Skuzeski, 1990). On the other hand, the existence, length, and sequence of a read-through domain are not conserved among the tymoviruses that have been sequenced (Morch *et al.*, 1988; Keese *et al.*, 1989; Dreher and Bransom, 1992; Osorio-Keese *et al.*, 1989; Ding *et al.*, 1990; Ding *et al.*, 1989; Srifah *et al.*, 1992), suggesting that read-through may not play a significant role in TYMV gene expression. The 221 kDa read-through product would presumably be proteolytically processed, to yield the 150 kDa protein and a 85 kDa C-terminal protein.

3. A 115 kDa protein has been reported as a viral subunit of purified TYMV replicase (Mouchès *et al.*, 1984). The appearance of such a viral protein is not explained by our current understanding of TYMV gene expression, which is based on *in vitro* studies. Further studies are needed to determine whether ORF 206 can be expressed by an alternative pathway *in vivo* or whether the 115 kDa protein represents a non-specific degradation product.

Roles of the TYMV Gene Products

The Essential Replication Proteins Encoded by ORF 206: As discussed above, ORF 206 encodes proteins with domains that are expected to play a role in catalyzing the capping, helicase, and polymerase processes of viral RNA replication. We have verified that ORF 206 expression is essential for RNA replication (Weiland and Dreher, 1989) and that the expression of active helicase-like and polymerase-like domains is required for replication (Weiland and Dreher, 1993). Nucleotide substitutions in the initiation codon of ORF 206 result in inviable RNA's that are incapable of replication in Chinese cabbage protoplasts (Weiland and Dreher, 1989). By contrast, substitutions in the initiation codon of ORF 69, or insertion of termination codons in the 5'-region of ORF 69 (mutations that do not alter the coding of ORF 206), have no significant effect on viral RNA replication and coat protein expression in Chinese cabbage protoplasts (Bozarth *et al.*, 1992). Likewise, a TYMV genome from which two-thirds of the CP-ORF has been deleted is viable (Weiland and Dreher, 1993). Thus, ORF 206 is the only ORF essential for the replication of TYMV RNA.

We have created a number of defective TYMV genomes by introducing nucleotide substitutions, frameshift mutations, or deletions into ORF 206 (Weiland and Dreher, 1993). Interference in ORF 206 expression uniformly results in inviability. Two informative defective, non-viable mutants studied were TY-K982S and TY-G1663R RNA's (Figure 4), which encode a mutant helicase-like domain and polymerase-like domain, respectively. The substitution mutation in the putative nucleotide binding fold of the TY-K982S helicase-like domain (Gly-Xxx-Gly-Lys-Thr to Gly-Xxx-Gly-*Ser*-Thr) is analogous to the Gly-Xxx-Gly-Lys-Thr to Gly-Xxx-Gly-*Arg*-Thr substitution reported to inactivate the DNA helicase activity of the RAD3 protein of yeast (Sung *et al.*, 1988). The Gly-Asp-Asp to *Arg*-Asp-Asp substitution of the polymerase-like domain encoded by the 70 kDa coding region of

TY-G1663R RNA is analogous to the substitutions of the corresponding Gly residue of the poliovirus 3D polymerase with various amino acids, which resulted in large losses of polymerase activity (Jablonski *et al.*, 1991). The results are, thus, consistent with the assignments of helicase and polymerase activities to the 150 kDa and 70 kDa proteins of TYMV, respectively.

The Role of ORF 69 in Virus Movement: We have used substitution mutations that inactivate the ORF 69 initiation codon and place termination codons within the 5'-region of ORF 69 to study the role of this ORF in TYMV infections (Bozarth *et al.*, 1992). Such mutant genomes replicate normally in protoplasts but fail to infect Chinese cabbage plants systemically. Local lesions also fail to form on the inoculated leaves. ORF 69 expression thus appears to be involved in permitting the cell-to-cell spread of the virus, and perhaps functions analogously to the TMV 30 kDa movement protein (Deom *et al.*, 1992). ORF 69 encodes a high proportion of basic amino acids that may confer nucleic acid binding properties on the ORF 69 product, as has been demonstrated for the TMV 30 kDa and cauliflower mosaic virus gene 1 movement proteins (Deom *et al.*, 1992).

Studies on a variant of TYMC that produces more severe symptoms on Chinese cabbage plants, comprising severe chlorosis with uniformly yellow-green regions of leaves, further illustrate the importance of ORF 69 in the establishment of a successful systemic infection. A plant inoculated with TYMC developed the more severe symptoms that were subsequently shown to be due to the spontaneous mutation U1888→C in TYMC RNA (Tsai and Dreher, 1993). This mutation results in the substitution of histidine for tyrosine at codon 601 of ORF 69 but does not alter the coding by ORF 206. In protoplasts, the replication of TYMC-C1888 is similar to that of TYMC, but TYMC-C1888 accumulates to levels about 4-fold those of TYMC in Chinese cabbage plants. The increased yields of virus are apparently a result of improved movement throughout the plant. Interestingly, the yield of mutant virus is not enhanced relative to TYMC in turnip plants.

ORF 69 expression has been detected in infected plants using Western blots probed with ORF 69-specific antiserum (Bozarth *et al.*, 1992). Infections originating from cloned TYMC RNA showed a c. 80 kDa protein, while plants infected with uncloned TYMV (type strain) RNA contained a doublet with a MW of about 75-80K. Expression was maximal in young, expanding, systemically-infected leaves and was associated with the insoluble cell wall fraction. The doublet apparently arises from the existence of sequence variants and cannot be attributed to partial N-glycosylation: based on the normal infectivity of a TYMC variant with a substitution in a potential N-glycosylation site encoded by ORF 69, post-translational modification with N-glycosyl groups is not important for infectivity and probably does not occur (Bozarth *et al.*, 1992).

The Role of the Coat Protein in Virus Movement: Deletion of two-thirds of the CP-ORF results in a genome (TYMC-Δ5708-6062 RNA; Figure 4) that is able to replicate in protoplasts and induce normal chlorotic local lesions on the inoculated leaves but which is unable to induce systemic symptoms (Weiland and Dreher, unpublished). Very low levels of viral RNA were detected in upper leaves of inoculated turnip plants. The coat protein is, thus, important in permitting long distance movement of the virus but is dispensable for cell-to-cell spread.

Analysis of the replication of TYMC-Δ5708-6062 RNA in turnip protoplasts revealed that (+)-strand genomic and subgenomic RNA's accumulated to levels about 10 percent those of wild-type TYMC infections, although normal levels of (-) strand genomic RNA were present (Weiland and Dreher, 1993). Altered (+):(-) strand ratios have been commonly observed after interfering with coat protein expression by positive strand RNA plant viruses (Nassuth and Bol, 1983), presumably due to the more rapid degradation of RNA's that are normally protected by encapsidation. A direct role of the coat protein in regulating the balance between (+) and (-) strand synthesis cannot be excluded, however.

Replication of TYMV RNA

The Role of the 3'-Noncoding Region and Its Valylation: We have investigated the role of the valylation of the TYMV tRNA-like structure by conducting a mutagenic survey designed to create RNA variants incapable of aminoacylation on the one hand and capable of aminoacylation with an amino acid other than valine on the other hand. Our studies have focussed on the anticodon loop of TYMV RNA, in which the nucleotides *-56*-ACAC-*-53* are identical to the analogous nucleotides of all eukaryotic valylatable tRNA's (Sprinzl *et al.*, 1991) or tRNA-like structures (van Belkum *et al.*, 1987). Substitutions of these nucleotides were used to map the valine identity determinants of 264-nucleotide-long 3'-fragments of TYMV RNA that were made *in vitro* by transcription with T7 RNA polymerase from a specially designed vector, pT7YSma (Dreher *et al.*, 1988).

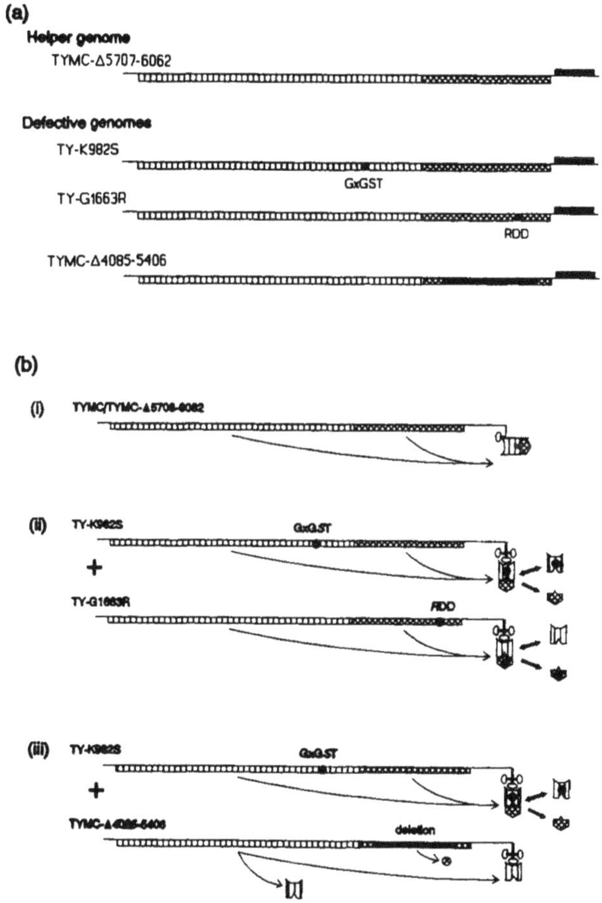

Figure 4. ***Cis*-Preferential Replication of TYMV RNA's. A: Genome plans of some of the RNA's used to study the replication of TYMV RNA *in trans*. TYMC-Δ5707-6062 and TYMC-Δ4085-5406 RNA's contain deletions (black boxes) in the CP-ORF and 70 kDa coding region, respectively. TY-K982S and TY-G1663R RNA's encode mutant 150 kDa and 70 kDa proteins, respectively, as a result of single amino acid substitutions. B: Representation of the proposed interactions between the 150 kDa and 70 kDa proteins that result in *cis*-preferential replication. 1. Formation of a stable, productive complex *in cis* on replication-proficient RNA's. 2. Two defective RNA's that mutually complement inefficiently (replication = 1 percent relative to TYMC) due to the trapping of mutant proteins in relatively stable complexes that form *in cis*. 3. Two defective RNA's whose very efficient mutual complemention (replication = 50 percent relative to TYMC) is proposed to result from the inability of a complex to form *in cis* on one of the RNA's, due to the large deletion in the 70 kDa protein.**

Substitutions of major determinants of valine identity (nucleotides recognized by valyl-tRNA synthetase in the productive selection of its substrates) were expected to result in non-charging, and perhaps mischarging, mutant RNA's.

The valine identity of TYMV RNA as recognized by wheat germ valyl-tRNA synthetase *in vitro* was found to reside predominantly in three anticodon loop nucleotides (Dreher *et al.*, 1992). The most important nucleotide is A-56 in the middle of the anticodon, followed by C-55, the 3'-anticodon nucleotide. C-53 was found to contribute more weakly but, nevertheless, significantly to valine identity. The catalytic specificity constant V_{max}/K_M was decreased at least 10^3-fold by single or combined substitutions in these three nucleotides, resulting in RNA's with very low valine charging activity.

The same mutations studied in *in vitro* aminoacylation studies were also placed into the genomic cDNA (pTYMC) to permit a study of the effect of altered valylation on viral replication (Tsai and Dreher, 1991). For mutations in the anticodon loop, a strong correlation has been observed between the *in vitro* valylatability and *in vivo* RNA replication in Chinese cabbage protoplasts. A low basal level of RNA replication, detected as (-) strand synthesis in inoculated protoplasts, was observed for mutants with V_{max}/K_M values for *in vitro* valylation that were $<10^{-2}$ relative to wild-type. These mutants were unable to infect whole plants. By contrast, mutants such as those with substitutions in the wobble position (-57) that retained efficient valylatability, replicated to levels similar to wild-type in both protoplasts and whole plants. Mutants with intermediate levels of valylatability (C-54 and A-53 mutants) replicated to intermediate levels *in vivo* (Table 1).

The correlation between replication and valylatability was further strengthened by the characterization of a mutant that arose spontaneously in a plant inoculated with TYMC-A55 RNA, which does not support systemic infection in plants (Tsai and Dreher, 1992). Presumably as a result of polymerase error during replication, the plant became systemically infected with progeny having a U-57/A-55 genotype. This genotype was reconstructed in TYMC-U57/A55 RNA, which was shown to replicate to levels comparable to TYMC *in vivo* (Tsai and Dreher, 1992). Further, this mutant RNA is capable of relatively efficient valylation: although the V_{max}/K_M is 0.075 relative to wild-type, the double mutant RNA can readily be valylated to completion (Dreher *et al.*, 1992). The C-57→U mutation has thus clearly suppressed the effect of the C-55→A mutation in both the valylation and replication properties. We conclude that valylation of the viral RNA by host valyl-tRNA synthetase is a requisite for the efficient replication of TYMV RNA.

The mechanistic basis of the requirement for valylation is not known at present, nor is it known whether efficient charging with an amino acid other than valine is compatible with efficient replication. It is likely that the valylation status of a molecule of TYMV RNA influences its potential as a template for (-) strand synthesis, the first stage of viral RNA replication. The effect may be exerted at the level of (-) strand promoter selection. Although the (-) strand promoter of TYMV RNA has not been characterized in detail, it appears to reside at least partly within the tRNA-like structure (Gargouri-Bouzid *et al.*, 1991; Georgel and Dreher, unpublished). It is, thus, conceivable that the 3'-esterified valine acts as an additional, enhancing element of the promoter.

Table 1. Correlation Between Valylation and Replication of TYMV RNA's.

Mutant	Anticodon loop sequence	Relative efficiency of *in vitro* valylation	Relative replication in protoplasts
wild-type	59-CCCACAC-53	100	100
TY-U57	. . U	114	150
-G56	. . . G . . .	0.02	0.2
-A55	. . . . A . .	0.46	2.0
-C54	. . . . . C .	22	24
-A53	. . . . . . A	9.3	12
-U57/A55	. . U.A . .	7.5	85

***In vitro* valylation efficiencies (V_{max}/K_M) were determined on 264-nucleotide-long 3'-fragments of TYMV RNA using wheat germ valyl-tRNA synthetase (data from Dreher *et al.*, 1992). The replication levels report the accumulations of coat protein in Chinese cabbage protoplasts 48 hours after inoculation (data from Tsai and Dreher, 1991; 1992).**

The Role of a Pseudoknot Upstream of the tRNA-Like Structure: Nucleotides between the CP-ORF termination codon and the tRNA-like structure can be folded into a pseudoknot (Figure 2), a structural element commonly found in viral RNA's and which has also been implicated in translational frameshifting in RNA viruses (Pleij, 1990). Because of their distinctive structure, pseudoknots may function as protein recognition elements involved in the regulation of translation and/or RNA replication.

We observed that mutation of U-96→G in the genomic RNA, which destabilizes the base-pairing needed for the formation of one of the helical segments of the pseudoknot upstream of the TYMV tRNA-like structure, resulted in a poor replication rate and inability to infect plants systemically (Tsai and Dreher, 1992). When protoplasts infected with TYMC-G96 were used to inoculate Chinese cabbage plants, some plants did develop symptoms; but these infections were in all instances due to progeny with altered sequences. Most plants contained progeny with simple G-96→U reversions; but in a few cases, second-site suppressing mutations were responsible for the ability to develop systemic symptoms (Tsai and Dreher, 1992). In one such case, the suppressing mutation was found to be the A-107→C substitution, which recovered the ability to form a pseudoknot similar to that present in the wild-type RNA (Figure 2). A second consequence of this mutation was the abolition of the normal CP-ORF termination codon, such that the extended CP-ORF terminated five codons downstream. TYMC-C107/G96 is able to produce normal systemic symptoms in plants; but the virions are less stable than normal, resulting in the degradation of virion RNA in mature, infected leaves as virions dissociate. We, thus, conclude that the extended coat protein plays no role in rescuing the poor replication caused by the G-96 mutation but that the rescue is due to the fact that the U-107→C mutation recovers the ability to form a pseudoknot. The pseudoknot may represent a binding element recognized by the replication machinery in assembling a (-) strand replication complex. A similarly positioned pseudoknot has been shown to be necessary for the replication of TMV RNA (Takamatsu *et al.*, 1990).

The Replication of TYMV RNA is cis-Preferential: In the course of our studies with TYMV, it was noticed that TYMC helper RNA was unable to efficiently support the replication *in trans* of a variety of deletion-bearing TYMC genomes. Since this apparent *cis*-preferential replication contrasted with the behavior of many positive strand RNA viral systems, which are able to replicate molecules such as defective interfering RNA's very efficiently *in trans*, we have studied the phenomenon further with TYMV. *Trans*-replication has been studied in two formats (Weiland and Dreher, 1993) in which it was attempted to rescue the replication of defective RNA's unable to replicate autonomously by coinoculation with (Figure 4):

1. TYMC-Δ5707-6062 RNA, which contains a deletion in the CP-ORF, as helper genome;

2. a second defective RNA to permit replication to occur as a result of mutual complementation.

The use of TYMC-Δ5707-6062 RNA as helper RNA permitted the specific detection in Northern blots of potentially co-replicating defective RNA's (by using probes corresponding to the deleted sequences), so that even inefficient levels of *trans*-replication could be studied and quantitated. Most defective RNA's, regardless of whether the substitution or in-frame deletion mutations were located in the 150 kDa or 70 kDa coding region, were inefficiently coreplicated in the presence of helper RNA, with accumulations of defective RNA ranging from 1-20 percent relative to helper RNA. Defective RNA TYMC-Δ4085-5406, however, which contains a deletion of most of the 70 kDa coding region, accumulated to levels similar to those of helper RNA. Thus, while TYMV replication is typified by a preferential replication of those RNA's encoding active essential proteins, it is possible to overcome this *cis*-limitation. One may thus expect TYMV to support the replication of defective interfering RNA's, as so many other positive strand RNA viruses do, but such RNA's have not been reported; and we have not been able to force their appearance by repeated passaging in plants.

For coinoculation experiments involving two defective RNA's, mutual complementation was considered possible between an RNA with a lethal mutation in the 150 kDa coding region and another with a mutation in the 70 kDa coding region. The mutations were selected so as to avoid interfering with the proteolysis that is necessary in generating these essential proteins (Figure 4). All combinations tested were found to result in the mutual rescue of replication, but most combinations supported only inefficient *trans*-replication. TYMC-Δ4085-5406 RNA was again the most efficient *trans*-replication substrate tested. Indeed, the inoculum combination TY-K983S + TYMC-Δ4085-5406 RNA replicated to levels about half those of TYMC infections in turnip protoplasts and even replicated throughout inoculated turnip plants to produce systemic symptoms. The genome of this novel bipartite form of TYMV was stable for five serial passages in plants, providing an illustration of the ease with which reorganizations in RNA viral genomes can evolve. The genetic stability of the bipartite genome suggests that recombination (which could readily regenerate a wild-type TYMC genome) is less active for TYMV than for other RNA viruses. We have shown that the *trans*-replication of coinoculated defective RNA's resulted from complementation rather than recombination and have not observed any RNA recombination events.

It was generally observed in coinoculation experiments that RNA's encoding defective 150 kDa proteins were less readily replicated *in trans* than those encoding defective 70 kDa proteins. Thus, in TY-K983S + TYMC-Δ4085-5406 infections, TYMC-Δ4085-5406 RNA accumulated to levels 10-fold higher than TY-K982S RNA. This suggests that the 70 kDa protein can be more readily supplied *in trans* than the 150 kDa protein, although *cis*-limitation is not limited to RNA's encoding defective 150 kDa protein. We have developed a model to explain the observed phenomena that is based on the formation *in cis* of a relatively stable complex between the 150 kDa and 70 kDa proteins and the viral RNA. Thus, it is suggested that the newly synthesized proteins are channelled to their site of action at the (-) strand promoter. This strategy would be expected to ensure an efficient beginning of the replication cycle, which occurs when viral proteins and RNA's are present at very low concentrations that do not favor efficient reaction rates. As a result of the formation of the proposed complex, the replication of RNA's encoding active 150 kDa and 70 kDa proteins would be favored, since slow complex dissociation would limit the supply of active proteins *in trans*. For RNA's encoding a defective essential protein, the formation of the proposed complex would result in *cis*-inhibition due to the trapping of an inactive subunit. This *cis*-limitation could be avoided, however, by avoiding the formation of a complex *in cis*; we propose that the deletion present in the 70 kDa coding region of TYMC-Δ4085-5406 RNA removes a domain necessary for the formation of the complex, permitting the more rapid exchange of protein subunits in order to assemble an active replication complex on the defective RNA. While it appears that the 70 kDa protein can be efficiently supplied *in trans*, the 150 kDa protein is more *cis*-limited, perhaps due to an affinity for RNA that results in preferential binding to its own messenger (genomic) RNA.

Future Perspectives

TYMV is a well-characterized virus that is an excellent system for studying the detailed mechanisms of viral RNA replication. The roles in this process of the essential viral proteins, the nature of their interactions that result in *cis*-preferential replication, and their assembly (presumably with one or more host factors) into an active replicational complex on the viral transcriptional promoters are important topics of future research. Understanding these processes will provide insight into the role of the tRNA mimicry and the valylation of the TYMV tRNA-like structure. Given the broad conservation across the positive strand RNA viruses of domains involved in RNA replication (Rozanov *et al.*, 1992; Habili and Symons, 1989; Gorbalenya and Koonin, 1989; Koonin, 1991; Goldbach, 1987), it is to be expected that a thorough understanding of the replication of well-characterized viruses such as TYMV will contribute to an understanding of the replication strategies of many viruses that are not so amenable to study.

Acknowledgements

This research was funded by grants from USDA-CSRS, NSF, and Oregon State University.

References

Argos, P. 1988. A sequence motif in many polymerases. Nucl. Acids Res. 16:9909-9916.

Beier, H., Barciszewska, M., Krupp, G., Mitnacht, R., and Gross, H.J. 1984. UAG read-through during TMV RNA translation: Isolation and sequence of two tRNA's^{Tyr} with suppressor activity from tobacco plants. EMBO J. 3:351-355.

Bozarth, C.S., Weiland, J.J., and Dreher, T.W. 1992. Expression of ORF 69 of turnip yellow mosaic virus is necessary for viral spread in plants. Virology 187:124-130.

Bransom, K.L., Weiland, J.J., and Dreher, T.W. 1991. Proteolytic maturation of the 206 kDa nonstructural protein encoded by turnip yellow mosaic virus RNA. Virology 184:351-358.

Briand, J.-P., Keith, G., and Guilley, H. 1978. Nucleotide sequence at the 5'-extremity of turnip yellow mosaic virus genome RNA. Proc. Natl. Acad. Sci. U.S.A. 75:3168-3172.

de Groot, R.J., Rümenapf, T., Kuhn, R.J., Strauss, E.G., and Strauss, J.H. 1991. Sindbis virus RNA polymerase is degraded by the N-end rule pathway. Proc. Natl. Acad. Sci. U.S.A. 88:8967-8971.

Deom, C.M., Lapidot, M., and Beachy, R.N. 1992. Plant virus movement proteins. Cell 69:221-224.

Deutscher, M.P. 1982. tRNA nucleotidyltransferase. The Enzymes 15:183-215.

Ding, S.-W., Keese, P., and Gibbs, A. 1989. Nucleotide sequence of the ononis yellow mosaic tymovirus genome. Virology 172:547-554.

Ding, S.-W., Keese, P., and Gibbs, A. 1990. The nucleotide sequence of the genomic RNA of kennedya yellow mosaic tymovirus-Jervis Bay isolate: Relationships with potex- and carlaviruses. J. Gen. Virol. 71:925-931.

Dreher, T.W. and Bransom, K.L. 1992. Genomic RNA sequence of turnip yellow mosaic virus isolate TYMC, a cDNA-based clone with verified infectivity. Plant Mol. Biol. 18:403-406.

Dreher, T.W., Florentz, C., and Giegé, R. 1988. Valylation of tRNA-like transcripts from cloned cDNA of turnip yellow mosaic virus RNA demonstrate that the L-shaped region at the 3'-end of the viral RNA is not sufficient for optimal aminoacylation. Biochimie 70:1719-1727.

Dreher, T.W., Tsai, C.-H., Florentz, C., and Giegé, R. 1992. Specific valylation of turnip yellow mosaic virus RNA by wheat germ valyl-tRNA synthetase is determined by three anticodon loop nucleotides. Biochemistry 31:9183-9189.

Gargouri-Bouzid, R., David, C., and Haenni, A.-L. 1991. The 3'-promoter region involved in RNA synthesis directed by the turnip yellow mosaic virus genome *in vitro*. FEBS Lett. 294:56-58.

Giegé, R., Briand, J.-P., Mengual, R., Ebel, J.-P., and Hirth, L. 1978. Valylation of the two RNA components of turnip yellow mosaic virus and specificity of the tRNA aminoacylation reaction. Eur. J. Biochem. 84:251-256.

Goldbach, R. 1987. Genome similarities between plant and animal RNA viruses. Microbiol. Sci. 4:197-202.

Gorbalenya, A.E. and Koonin, E.V. 1989. Viral proteins containing the purine NTP-binding sequence pattern. Nucl. Acids Res. 17:8413-8440.

Gorbalenya, A.E., Koonin, E.V., and Lai, M.M.-C. 1991. Putative papain-related thiol proteases of positive-strand RNA viruses. FEBS Lett. 288:201-205.

Habili, N. and Symons, R.H. 1989. Evolutionary relationship between luteoviruses and other RNA plant viruses based on sequence motifs in their putative RNA polymerases and nucleic acid helicases. Nucl. Acids Res. 17:9543-9555.

Jablonski, S.A., Luo, M., and Morrow, C.D. 1991. Enzymatic activity of poliovirus RNA polymerase mutants with single amino acid changes in the conserved YGDD amino acid motif. J. Virol. 65:4565-4572.

Joshi, R.L., Faulhammer, H., Chapeville, F., Sprinzl, M., and Haenni, A.-L. 1984. Aminoacyl RNA domain of turnip yellow mosaic virus Val-RNA interacting with elongation factor EF-Tu. Nucl. Acids Res. 12:7467-7478.

Joshi, R.L., Ravel, J.M., and Haenni, A.-L. 1986. Interaction of turnip yellow mosaic virus Val-RNA with eukaryotic elongation factor EF-1α: Search for a function. EMBO J. 5:1143-1147.

Joshi, S., Chapeville, F., and Haenni, A.-L. 1982. Turnip yellow mosaic virus RNA is aminoacylated *in vivo* in Chinese cabbage leaves. EMBO J. 1:935-938.

Keese, P., Mackenzie, A., and Gibbs, A. 1989. Nucleotide sequence of the genome of an Australian isolate of turnip yellow mosaic tymovirus. Virology 172:536-546.

Koonin, E.V. 1991. The phylogeny of RNA-dependent RNA polymerases of positive-strand RNA viruses. J. Gen. Virol. 72:2197-2206.

Kozak, M. 1989. The scanning model for translation: An update. J. Cell Biol. 108:229-241.

Matthews, R.E.F. 1991. *Plant Virology*. Third Edition, pp. 235-238. Academic Press, San Diego, California.

Mi, S. and Stollar, V. 1991. Expression of Sindbis virus nsP1 and methyltransferase activity in *E. coli*. Virology 184:423-427.

Morch, M.-D., Boyer, J.-C., and Haenni, A.-L. 1988. Overlapping open reading frames revealed by complete nucleotide sequencing of turnip yellow mosaic virus genomic RNA. Nucl. Acids Res. 15:6157-6173.

Mouchès, C., Candresse, T., and Bové, J.M. 1984. Turnip yellow mosaic virus RNA-replicase contains host and virus-encoded subunits. Virology 134:78-90.

Nassuth, A. and Bol, J.F. 1983. Altered balance of plus- and minus-strand RNA's induced by RNA's 1 and 2 of alfalfa mosaic virus in the absence of RNA 3. Virology 124:75-85.

Osorio-Keese, M.E., Keese, P., and Gibbs, A. 1989. Nucleotide sequence of the genome of eggplant mosaic tymovirus. Virology 172:547-554.

Pinck, M., Yot, P., Chapeville, F., and Duranton, H.M. 1970. Enzymatic binding of valine to the 3'-end of TYMV RNA. Nature 226:954-956.

Pleij, C.W.A. 1990. Pseudoknots: A new motif in the RNA game. Trends Biochem. Sci. 15:143-147.

Pleij, C.W.A., Abrahams, J.P., van Belkum, A., Rietveld, K., and Bosch, A. 1987. The spatial folding of the 3'-noncoding region of aminoacylatable plant viral RNA's. In *Positive strand RNA Viruses*. M.A. Brinton and R.R. Rueckert (eds.), pp. 299-316. Alan R. Liss, Inc. New York.

Rietveld, K., Pleij, C.W.A., and Bosch, L. 1983. Three-dimensional models of the tRNA-like 3'-termini of some plant viral RNA's. EMBO J. 2:1079-1085.

Rozanov, M.N., Koonin, E.V., and Gorbalenya, A.E. 1992. Conservation of the putative methyltransferase domain: A hallmark of the Sindbis-like' supergroup of positive strand RNA viruses. J. Gen. Virol. 73:2129-2134.

Skuzeski, J.M., Nichols, L.M., and Gesteland, R.F. 1990. Analysis of leaky viral translation termination codons *in vivo* by transient expression of improved β-glucuronidase vectors. Plant Mol. Biol. 15:65-79.

Sprinzl, M., Dank, N., Nock, S., and Schön, A. 1991. Compilation of tRNA and tRNA gene sequences. Nucl. Acids Res. 19:Supplement:2127-2171.

Srifah, P., Keese, P., Weiller, G., and Gibbs, A. 1992. Comparisons of the genomic sequences of erysimum latent virus and other tymoviruses: A search for the molecular basis of their host specificities. J. Gen. Virol. 73:1437-1447.

Sung, P., Higgins, D., Prakash, L., and Prakash, S. 1988. Mutation of lysine-48 to arginine in the yeast RAD3 protein abolishes its ATPase and DNA helicase activities but not the ability to bind ATP. EMBO J. 7:3263-3269.

Takamatsu, N., Watanabe, Y., Meshi, T., and Okada, Y. 1990. Mutational analysis of the pseudoknot region in the 3'-noncoding region of tobacco mosaic virus RNA. J. Virol. 64:3686-3693.

Tsai, C.-H. and Dreher, T.W. 1991. Turnip yellow mosaic virus RNA's with anticodon loop substitutions that result in decreased valylation fail to replicate efficiently. J. Virol. 65:3060-3067 and 1992 Authors' Correction, *ibid* 66:3977.

Tsai, C.-H. and Dreher, T.W. 1992. Second-site suppressor mutations assist in studying the function of the 3'-noncoding region of turnip yellow mosaic virus RNA. J. Virol. 66:5190-5199.

Tsai, C.-H. and Dreher, T.W. 1993. Increased viral yield and symptom severity result from a single amino acid substitution in the turnip yellow mosaic virus movement protein. Molec. Plant Microbe Interact. 6:268-273.

van Belkum, A., Bingkun, J., Rietveld, K., Pleij, C.W.A., and Bosch, L. 1987. Structural similarities among valine accepting tRNA-like structures in tymoviral RNA's and elongator tRNA's. Biochemistry 26:1144-1151.

Weiland, J.J. and Dreher, T.W. 1989. Infectious TYMV RNA from cloned cDNA: Effects *in vitro* and *in vivo* of point substitutions in the initiation codons of two extensively overlapping ORF's. Nucl. Acids Res. 17:4675-4687.

Weiland, J.J. and Dreher, T.W. 1993. *Cis*-preferential replication of the turnip yellow mosaic virus RNA genome. Proc. Natl. Acad. Sci. U.S.A. 90:6095-6099.

Williams, M.A. and Lamb, R.A. 1989. Effect of mutations and deletions in a bicistronic mRNA on the synthesis of influenza B virus NB and NA glycoproteins. J. Virol. 63:28-35.

Expression, Processing, and Transport of Tobacco Etch Potyviral Proteins

J. C. Carrington, V. V. Dolja, X. H. Li, M.A. Restrepo-Hartwig, and J. Verchot
Department of Biology
Texas A&M University
College Station, Texas 77843

The tobacco etch potyvirus (TEV) genome encodes at least nine proteins that arise from a single large polyprotein. Three virus-encoded proteinases catalyze cleavage of the polyprotein through a combination of *cis*- and *trans*-processing reactions. Although protein processing and RNA replication are thought to occur in the cytoplasm of infected cells, two replication-associated proteins (NIa and NIb) accumulate predominantly in the nucleus. Both proteins contain localization signals that are sufficient to transport heterologous proteins to the nucleus. Evidence suggests that nuclear translocation of NIa is controlled by interaction with other TEV proteins. Thus, the functional concentrations of NIa in the cytoplasm may be subject to regulation by novel (nontranslational, nontranscriptional) mechanisms. To explore further the events associated with potyviral replication as well as virus movement, we have constructed a "tagged" virus that expresses the gene for β-glucuronidase.

Introduction

The potyviruses, with several hundred individual members, form the largest group of plant viruses (Milne, 1988). They cause diseases of varying severity in most cultivated crop species throughout the world. Transmission of the vast majority of potyviruses is mediated by invertebrate vectors, most commonly aphids.

The potyvirus particle is characterized by a flexuous rod-shaped morphology and is composed of a positive-strand RNA of nearly 10,000 nucleotides and over 2,000 copies of a coat protein subunit. Genomic RNA is attached covalently at the 5'-terminus to a virus-encoded protein, termed VPg, rather than a m^7GpppG cap structure that is typical of eukaryotic mRNA's. The complete nucleotide sequences of several potyvirus genomes, including those for the well-characterized tobacco etch virus (TEV), tobacco vein mottling virus, and potato virus Y (the type member), have been elucidated (Riechmann *et al.*, 1992; Dougherty and Carrington, 1988). Based on genome organizational properties and amino acid sequence similarities, the potyviruses clearly belong to the picornavirus-like superfamily of positive-strand RNA viruses. The potyviral genome contains only one open reading frame that encodes all viral proteins through synthesis and processing of a polyprotein (Figure 1).

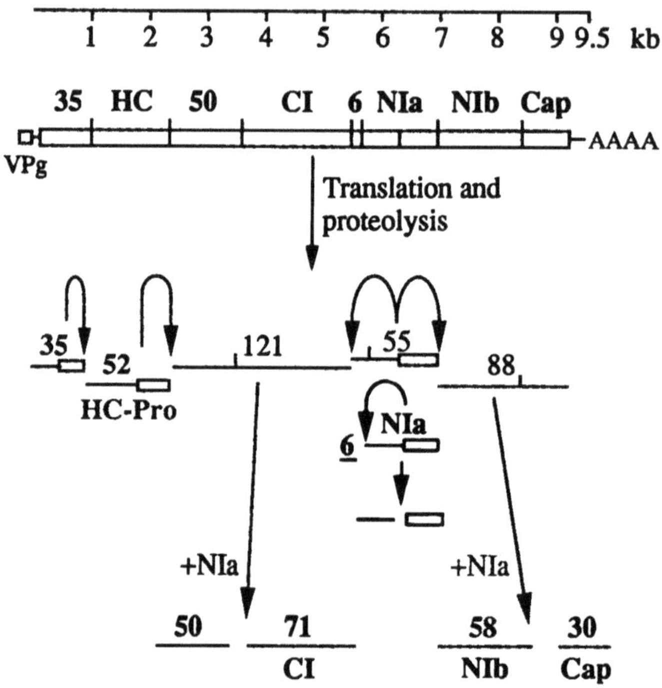

Figure 1. Genetic Organization of TEV and Proteolytic Processing of the Viral Polyprotein. The genome map at top shows the open reading frame (long rectangular box) and positions of coding sequences for proteolytic cleavage sites (vertical lines). The processing map below indicates the three proteinase domains (open rectangular boxes), as well as the sites recognized by each proteinase (arrows). Modified from Restrepo-Hartwig and Carrington (1992).

As a result of the large quantity of coat protein required for assembly of each virion, non-structural proteins apparently are manufactured in significant excess over that needed for RNA replication. Regulation of viral gene products, therefore, must depend on mechanisms other than differential synthesis via translation or transcription control. We have investigated the functions and activities of several TEV-encoded non-structural proteins, and the potential roles of proteolytic processing and differential subcellular localization in regulating their availability.

Proteolytic Processing of the TEV Polyprotein

Three TEV-encoded proteins contain the proteinase activities required for complete processing of the viral polyprotein. Each enzyme is characterized by high specificity toward selective sites within the polyprotein, and at least two (probably all three) possess one or more additional activities involved in the replicative cycle. Because of the prominent role of proteolytic processing in expression of the viral genome, detailed investigations have been performed to probe the nature of each proteinase and the molecular bases for their extreme specficities.

The NIa proteinase catalyzes proteolysis of the RNA replication-associated and capsid proteins in the C-terminal region of the polyprotein (Carrington and Dougherty, 1987). It is evolutionarily related to the 3C proteinase of picornaviruses and the chymotrypsin-like family of cellular serine proteinases (Gorbalenya *et al.*, 1989; Bazan and Fletterick, 1988). Unlike the cellular enzymes, however, the NIa and 3C proteinases contain a cysteine rather than serine as the nucleophilic residue within the active site. NIa excises from the viral polyprotein by an autoproteolytic or *cis* mechanism and, in a subset of molecules, undergoes further autoprocessing between the VPg and proteinase domains (Dougherty and Parks, 1991). Additional *cis*- and *trans*-proteolytic reactions involving NIa proteinase occur at three sites. The basis for specificity of NIa involves recognition of a heptapeptide motif, containing both strictly conserved and nonconserved residues, surrounding each cleavage site (Dougherty *et al.*, 1989). Importantly, the amino acid residues in the nonconserved positions were found to influence the rates of cleavage at two *trans*-proteolytic sites (Dougherty *et al.*, 1989), suggesting that individual sites may be programmed to process with different efficiencies. The predominant form of this proteinase within infected cells localizes to the nucleus (Restrepo *et al.*, 1990) as the 49 kDa NIa protein, which is actually a polyprotein composed of a proteinase domain (C-terminal) and VPg domain (N-terminal). Due to the presence of most other TEV proteins in the cytoplasm, however, we presume that NIa-mediated proteolysis occurs prior to translocation.

Proteolytic processing by the helper component-proteinase (HC-Pro) is restricted to a single autoprocessing event at its C-terminus (Carrington *et al.*, 1989). The presence of essential cysteine and histidine residues at the putative catalytic reaction center, as well as sequence similarities with other proteinases, suggest that HC-Pro is a highly divergent, papain-like cysteine proteinase (Oh and Carrington, 1989). Like the NIa proteinase, HC-Pro recognizes an extended amino acid sequence motif (five residues) around the site of cleavage (Carrington and Herndon, 1992), ensuring high processing specificity. Also like NIa, HC-Pro is a multifunctional protein with the proteolytic domain confined to the C-terminal half. An N-terminal domain performs the helper component function during aphid transmission (Thornbury *et al.*, 1990), presumably by mediating an interaction between the virus particle and the aphid.

The third TEV proteolytic enzyme lies near the N-terminus of the viral polyprotein and is referred to as the 35 kDa proteinase (Verchot *et al.*, 1991). Specific His, Asp, and Ser residues were identified recently as putative components of the catalytic triad, similar to the trypsin/chymotrypsin family of serine proteinases (Verchot *et al.*, 1992). The spatial distribution of these residues within the 35 kDa proteinase resembles most closely the arrangement within the Sindbis virus capsid proteinase (Figure 2), which was demonstrated by crystallographic analysis to be related to the large cellular serine proteinases (Choi *et al.*, 1991).

As with other TEV proteinases, the C-terminal half of the 35 kDa proteinase functions as the proteolytic domain that preferentially cleaves by an autolytic mechanism at its C-terminus. Cleavage occurs within a Tyr304-Ser305 dipeptide as shown by microsequence analysis of the N-terminus of HC-Pro encoded by a related potyvirus (Mavankal and Rhoads, 1991). Elimination of Tyr304 was shown to inhibit proteolysis *in vitro*, suggesting a critical function of this residue in cleavage site recognition (Verchot *et al.*, 1992).

The 35 kDa proteinase may be a multifunctional protein similar to NIa and HC-Pro. Although deletion of the N-terminal half of the 35 kDa proteinase does not affect proteolysis, viral infectivity appears to be reduced after inoculation of tobacco plants with RNA transcripts

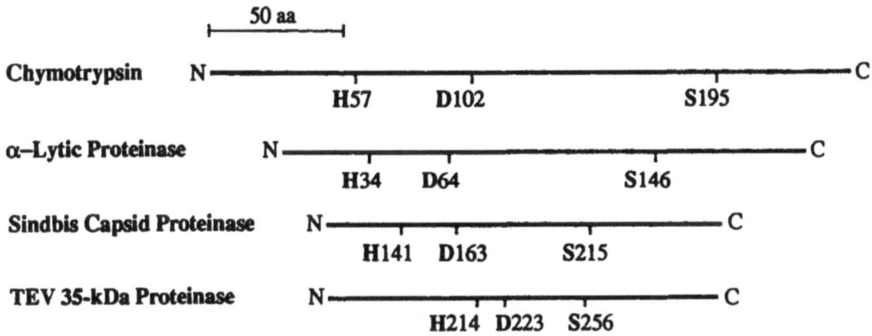

Figure 2. Comparison of the Distribution of Active Site Residues Within Four Serine Proteinases. The numbering of amino acid (aa) residues starts at the N-terminus of each protein. Note that both the Sindbis capsid protein and TEV 35 kDa protein contain non-proteinase N-terminal domains; and, therefore, the amino acid numbering appears high.

lacking this domain (unpublished data). In addition, proteinase-debilitating mutations at codons for active site residues reduce or eliminate viral infectivity (unpublished data). These results in combination suggest that a mature 35 kDa proteinase and/or HC-Pro is required for successful viral infectivity and that the 35 kDa protein may play additional roles that have yet to be defined in the viral life cycle.

Nuclear Transport of TEV Proteins

Tobacco etch virus polyprotein processing and RNA replication most likely occur in the cytoplasm or in association with a cytoplasmic membrane. Subpopulations of the NIa polyprotein, as well as a proteolytic fragment consisting of the N-terminal half of NIa, are found covalently attached to genomic RNA in the cytoplasm (Murphy *et al.*, 1990). Although virus replication and assembly presumably are cytoplasmic events, the majority of NIa and NIb (RNA-dependent RNA polymerase) molecules partition to the nucleus (Restrepo *et al.*, 1990; Baunoch *et al.*, 1988). As NIa and NIb provide several essential functions during virus multiplication, we analyzed the process of nuclear transport of both proteins. In particular, we investigated the nuclear transport signal sequences within NIa and NIb and possible mechanisms whereby translocation is regulated.

Nuclear localization signals (NLS's) were mapped in NIa and NIb by expression of gene fusions with β-glucuronidase (GUS, encoded by the *Escherichia coli uidA* gene) in tobacco protoplasts and transformed tobacco plants. Nuclear transport activity of fusion proteins was analyzed by a fluorometric GUS assay of nuclear and non-nuclear fractionated extracts and by an *in situ* histochemical assay using colorimetric substrates. The former method permits a quantitative analysis of translocation, while the latter provides a qualitative assessment. When expressed as a nonfusion protein, GUS (68 kDa) localized to the cytoplasm of plant cells. Fusion proteins consisting of GUS and either NIa or NIb, on the other hand, localized to the nucleus (Restrepo *et al.*, 1990), indicating that NIa and NIb possess independent nuclear localization signals (NLS's). Deletion and point mutational analyses of GUS/NIa fusions indicated that two basic sequences, one between amino acid residues 1-11 and the other between residues 40-72, are necessary for optimal translocation activity of NIa in protoplast transient assays (Carrington *et al.*, 1991), although only amino acid residues 40-49 are required for nuclear transport in transgenic plants (unpublished data). Discrete NLS sequences within NIb were more difficult to map, since deletion of several different segments of NIb debilitated nuclear transport. In addition, each of several clustered subsitutions introduced throughout the protein blocked translocation (unpublished data). This may indicate that NIb requires a stringent three-dimensional conformation, in addition to one or more NLS's, for nuclear translocation.

The polyprotein expression strategy of potyviruses precludes regulation of viral protein production by differential transcription or translation mechanisms. In the case of TEV, controlled subcellular transport of NIa and NIb to the nucleus may provide a means to limit the abundance of these proteins in the cytoplasm. Such a mechanism would effectively down regulate NIa and NIb, the over-accumulation of which may result in cytotoxicity or interference with completion of the replicative cycle. The roles of proteolytic processing and protein-protein interaction in the control of translocation of NIa and NIb are now under investigation.

Autoproteolytic cleavage by NIa at its N-terminus separates NIa from a 6 kDa polypeptide of unknown function. If cleavage occurs first, however, at the N-terminus of the 6 kDa protein, a 6/NIa polyprotein would form transiently. This polyprotein was shown to autoprocess with a half-time of approximately 20-30 minutes after synthesis *in vitro* (Restrepo-Hartwig and Carrington, 1992). The effect of the 6 kDa protein on nuclear translocation of NIa was tested by fusion of 6/NIa polyproteins to GUS and expression in transgenic plants. Two 6/NIa/GUS fusion proteins were generated. One contained a wild-type NIa proteinase that could catalyze removal of the 6 kDa protein from the N-terminus of NIa after synthesis. The other contained a proteinase-debilitating substitution of Ala in place of Cys339 in NIa, rendering the 6/NIa/GUS polyprotein proteolytically inactive. *In situ* localization of GUS activity revealed that NIa-containing polyproteins capable of undergoing proteolysis between the 6 kDa protein and NIa were transported to the nucleus, whereas polyproteins in which cleavage was prevented remained in the cytoplasm (Restrepo-Hartwig and Carrington, 1992) (Figure 3).

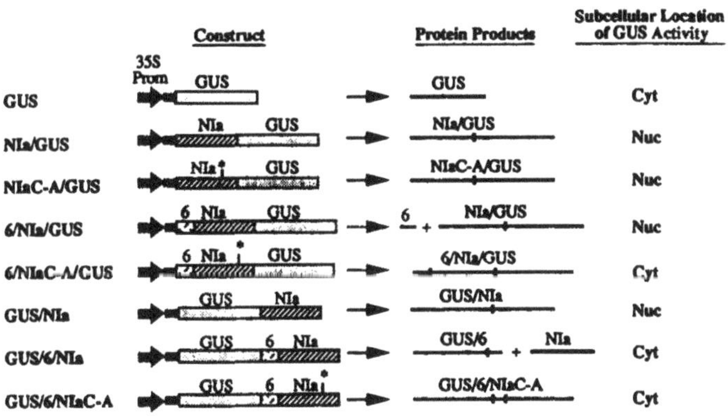

Figure 3. Nuclear Transport Activity of GUS Fusion Proteins in Transgenic Tobacco Plants. The gene fusion constructs adjacent to the cauliflower mosaic virus 35S promoter (thick arrow) are illustrated at the left of the thin arrows. The polyprotein or proteolytically processed products are shown at the right. In NIa-containing polyproteins with the Cys339 to Ala substitution (represented by the asterisk at the left), cleavage between the 6 kDa protein and NIa was inhibited. The subcelluar location of GUS activity in transgenic plants is presented at the far right.

These data suggest that the 6 kDa protein impedes NIa nuclear translocation and that this effect requires *cis*-linkage of the two proteins. Interference by the 6 kDa protein may be mediated by physical masking of the NIa NLS, or by binding to a cytoplasmic anchor (such as a membrane). Differential utilization of proteolytic processing sites at either end of the 6 kDa protein, therefore, may play a role in regulating NIa transport.

Engineering the TEV Genome

Full-length cDNA clones representing the TEV genome were constructed in vectors containing a bacteriophage SP6 promoter. Transcripts synthesized *in vitro* were infectious when applied to *Nicotiana tabacum* plants (Dolja *et al.*, 1992). The TEV genome was modified by the addition of the coding sequence for GUS, which was inserted in frame between the 35 kDa proteinase and HC-Pro coding regions such that the 35 kDa proteinase cleavage site remained functional. Proteolytic processing of the polyprotein encoded by the resulting TEV-GUS genome yielded a fusion protein consisting of GUS and HC-Pro. This recombinant virus was fully infectious in whole plants and protoplasts (Dolja *et al.*, 1992), suggesting that normal replication, cell-to-cell movement, and packaging functions were not disturbed.

Because of the sensitivity of assays for GUS activity in plant cells and cell extracts, TEV-GUS represents a powerful tool with which to further analyze potyviral functions. In whole plant time-course experiments, GUS levels served as an indirect quantitative measure of virus activity in primarily and systemically infected tissues (Figure 4).

Furthermore, *in situ* histochemical anlaysis using colorimetric GUS substrates permitted detection of initially infected, single epidermal cells on inoculated leaves, as well as inititally infected phloem-associated cells after long distance transport to stems, petioles, leaves, and roots (Dolja *et al.*, 1992). This recombinant virus, therefore, has utility to investigate factors involved in potyvirual short- and long-distance movement. In protoplasts, TEV-GUS and mutagenized derivatives should provide a unique approach to examine RNA replication and translation functions. Time-course experiments in protoplasts inoculated with TEV-GUS RNA revealed a lag period of approximately 12-16 hours followed by a viral growth phase of at least 60 hours (Figure 4, inset). Viral proteins or genome sequences involved in positive-strand RNA synthesis (which should be

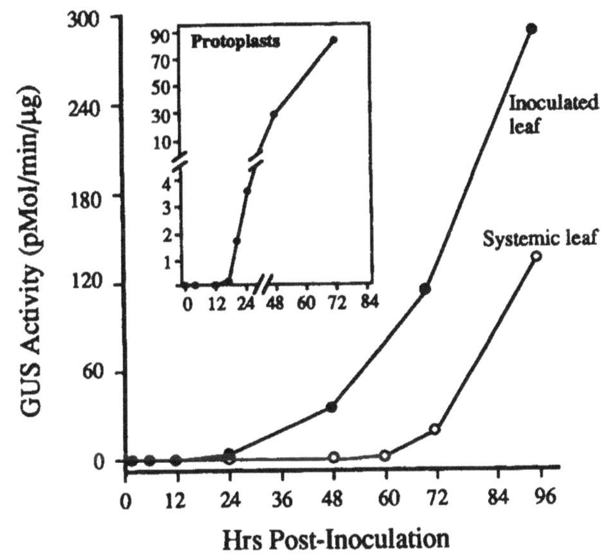

Figure 4. GUS Activity in Whole Plants and Protoplasts Infected by TEV-GUS Recombinant Virus. Extracts were prepared from inoculated or systemic leaves of tobacco at intervals up to 96 hours, and GUS activity was quantified using a fluorometric assay (Jefferson *et al.*, 1987). (Inset) Tobacco protoplasts were inoculated with RNA purified from TEV-GUS (Dolja *et al.*, 1992), and GUS activity was determined at intervals up to 72 hours.

reflected in the GUS accumulation profiles) can now be anlayzed genetically in potyviruses using a highly sensitive and rapid assay. Of particular interest to be addressed by this approach are the roles of differential proteolyic processing and subcellular transport of viral proteins during thc rcplicativc cycle.

Acknowledgements

This work was supported by grants from the National Institute of Allergy and Infectious Disease (AI27832), the National Science Foundation (IBN-9158559), the United States Department of Agriculture (91-37303-6435), and a Collaborative Plant Biology Program grant from the Samuel Roberts Noble Foundation.

References

Baunoch, D., Das, P., and Hari, V. 1988. Intracellular localization of TEV capsid and inlcusion proteins by immunogold labeling. J. Ultrastruc. Mol. Struct. Res. 99:203.

Bazan, J.F. and Fletterick, R.J. 1988. Viral cysteine proteases are homologous to the trypsin-like family of serine proteases: Structural and functional implications. Proc. Natl. Acad. Sci. U.S.A. 85:7872.

Carrington, J.C., Cary, S.M., Parks, T.D., and Dougherty, W.G. 1989. Second proteinase encoded by a plant potyvirus genome. EMBO J. 8:365.

Carrington, J.C. and Dougherty, W.G. 1987. Small nuclear inclusion protein encoded by a plant potyvirus genome. J. Virol. 61:2540.

Carrington, J.C., Freed, D.D., and Leinicke, A.J. 1991. Bipartite signal sequence mediates nuclear translocation of the plant potyviral NIa protein. The Plant Cell 3:953.

Carrington, J.C. and Herndon, K.L. 1992. Characterization of the potyviral HC-Pro autoproteolytic cleavage site. Virology 187:308.

Choi, H.-K., Tong, L., Minor, W., Dumas, P., Beoge, U., Rossmann, M.G., and Wengler, G. 1991. Structure of Sindbis virus core protein reveals a chymotrypsin-like serine proteinase and the organization of the virion. Nature (London) 354:37.

Dolja, V.V., McBride, H.J., and Carrington, J.C. 1992. Tagging of plant potyvirus replication and movement by insertion of β-glucuronidase into the viral polyprotein. Proc. Natl. Acad. Sci. U.S.A. 89:10208.

Dougherty, W.G. and Carrington, J.C. 1988. Expression and function of potyviral gene products. Annu. Rev. Phytopathol. 26:123.

Dougherty, W.G., Cary, S.M., and Parks, T.D. 1989. Molecular genetic analysis of a plant virus polyprotein cleavage site: A model. Virology 171:356.

Dougherty, W.G. and Parks, T.D. 1989. Molecular genetic and biochemical evidence for the involvement of the heptapeptide cleavage sequence in determining the reaction profile at two tobacco etch virus cleavage sites in cell-free assays. Virology 172:145.

Dougherty, W.G. and Parks, T.D. 1991. Post-translational processing of the tobacco etch virus 49 kDa small nuclear inclusion polyprotein: Identification of an internal cleavage site and delimitation of VPg and proteinase domains. Virology 183:449.

Gorbalenya, A.E., Donchenko, A.P., Blinov, V.M., and Koonin, E.V. 1989. Cysteine proteases of positive strand RNA viruses and chymotrypsin-like serine proteases. FEBS Lett. 243:103.

Jefferson, R.A., Kavanagh, T.A., and Bevan, M.W. 1987. GUS fusions: β-glucuronidase as a sensitive and versatile gene fusion marker in higher plants. EMBO J. 6:3901.

Mavankal, G. and Rhoads, R.E. 1991. *In vitro* cleavage at or near the N-terminus of the helper component protein in the tobacco vein mottling virus polyprotein. Virology 185:721.

Milne, R.G. 1988. In: *The Plant Viruses, Vol. 4: The Filamentous Plant Viruses.* R. G. Milne (ed.). Plenum Publishing Coroporation, New York.

Murphy, J.F., Rhoads, R.E., Hunt, A.G., and Shaw, J.G. 1990. The VPg of tobacco etch virus RNA is the 49 kDa proteinase or the N-terminal 24 kDa part of the proteinase. Virology 178:285.

Oh, C.-S. and Carrington, J.C. 1989. Identification of essential residues in potyvirus proteinase HC-Pro by site-directed mutagensis. Virology 173:692.

Restrepo, M.A., Freed, D.D., and Carrington, J.C. 1990. Nuclear transport of plant potyviral proteins. The Plant Cell 2:987.

Restrepo-Hartwig, M.A. and Carrington, J.C. 1992. Regulation of nucleartransport of a plant potyvirus protein by autoproteolysis. J. Virol. 66:5662.

Riechmann, J.L., Laín, S., and García, J.A. 1992. Highlights and prospects of potyvirus molecular biology. J. Gen. Virol. 73:1.

Thornbury, D.W., Patterson, C.A., Dessens, J.T., and Pirone, T.P. 1990. Comparative sequences of the helper component (HC) region of potato virus Y and a HC-defective strain, potato virus C. Virology 178:573.

Verchot, J., Herndon, K.L., and Carrington, J.C. 1992. Mutational analysis of the tobacco etch potyviral 35 kDa proteinase: Identification of essential residues and requirements for autoproteolysis. Virology 190:298.

Verchot, J., Koonin, E.V., and Carrington, J.C. 1991. The 35 kDa protein from the N-terminus of a potyviral polyprotein functions as a third virus-encoded proteinase. Virology 185:527.

Identification of Sequences Involved in Beet Necrotic Yellow Vein Virus RNA Replication and Encapsidation

D. Gilmer, E. Balmori*, C. Schmitt, H. Guilley, G. Jonard, and K. Richards
Institut de Biologie Moléculaire des Plantes du CNRS et de l'Université
Louis Pasteur
12 Rue du Général Zimmer
Strasbourg 67084 (Cedex)
France
**Permanent address: Center of Genetic Engineering and Biotechnology*
P.O. Box 6162
Havana, Cuba

Biologically active transcripts corresponding to BNYVV RNA's 2, 3, and 4 have been used to identify sequences involved in RNA replication and encapsidation. Mutagenesis of RNA's 3 and 4, which are not required for infection of leaves, reveals that the long 5'-proximal *cis*-acting domain of each contains several discrete subdomains separated by nonessential sequences. For RNA 3, important secondary structural features have been identified and a sequence required for RNA encapsidation located. A model for secondary structure has been developed. The 5'-proximal coat protein cistron of RNA 2 is followed by an in-frame 54 kd ORF which is expressed *via* translational read-through to produce a coat protein-54 kd fusion protein. Mutagenesis studies show that, in addition to coat protein, a *trans*-acting element within the read-through domain is important for efficient viral assembly. Possible mechanisms for its action will be considered, and the situation with other viruses having similar genetic organization discussed.

Introduction

Beet necrotic yellow vein virus (BNYVV) is a multicomponent plus-sense RNA virus which is field-transmitted by the soil-borne fungus *Polymyxa betae* and is associated with sugarbeet rhizomania (Jupin *et al.*, 1991; Richards and Tamada, 1992). Virions are nucleoprotein helices containing multiple copies of the 21 kilodalton (kd) coat protein and four distinct, separately packaged RNA species which are referred to as RNA's 1 to 4 in decreasing order of size (RNA 1: 6.8 kb;

RNA 2: 4.7 kb; RNA 3: 1.8 kb; RNA 4: 1.5 kb) (Bouzoubaa *et al.*, 1985; Bouzoubaa *et al.*, 1986; Bouzoubaa *et al.*, 1987). A fifth RNA species, RNA 5 (1.45 kb), is present in some but not all Japanese isolates of BNYVV but has not yet been detected in European or American isolates and is not required for production of rhizomania disease symptoms (Tamada *et al.*, 1989); it will not be discussed further here.

The genetic map of RNA's 1-4, based on sequence analysis of cloned cDNA (Bouzoubaa *et al.*, 1985; Bouzoubaa *et al.*, 1986; Bouzoubaa *et al.*, 1987), is presented in Figure 1. Close sequence homology among the RNA's is limited to the 5'-terminal 8-9 nucleotides (nts) and the last ~70 nts preceding the 3'-poly(A) tail. RNA 1 encodes a long nonstructural protein tentatively identified as the viral RNA-dependent RNA polymerase (Bouzoubaa *et al.*, 1987). RNA 2 encodes the 21 kd viral coat protein and a coat protein-54 kd fusion product (P75) expressed by translational suppression of the amber termination codon of the coat protein cistron (Bouzoubaa *et al.*, 1986; Ziegler *et al.*, 1985; Niesbach-Klösgen *et al.*, 1990). The 3'-proximal half of RNA 2 carries a group of three tightly packed or overlapping genes (the 'Triple Block') that are closely similar to a cluster of three genes in many other rod-shaped plant viruses (Morozov *et al.*, 1989). The BNYVV Triple Block genes are required for cell-to-cell movement (Gilmer *et al.*, 1992a). The species P14, encoded by the most 3'-proximal gene in RNA 2 (Figure 1), is not essential for replication or cell-to-cell movement; but in its absence virus accumulation in both leaves and protoplasts is diminished (Gilmer *et al.*, 1992a). P14 may regulate synthesis or expression of other viral gene products.

RNA's 3 and 4 each carry one long open reading frame (encoding the species P25 for RNA 3 and P31 for RNA 4) separated in each case from the 5'-extremity by a lengthy noncoding sequence containing several presumably nonfuctional AUG's (Bouzoubaa *et al.*, 1985). In neither case is the context of the initiation codon marking the beginning of the long open reading frame particularly favorable; but both P25 and P31 have been detected in infected tissue by serological techniques, as have all the other BNYVV proteins predicted from the sequence except for P237 and P15 (Niesbach-Klösgen *et al.*, 1990).

The behavior of RNA's 1 and 2 differs from that of RNA's 3 and 4 depending upon the host for virus propagation. In the roots of naturally infected sugar beets displaying symptoms of rhizomania, RNA's 1-4 are always present (RNA 5 may or may not be present) (Tamada *et al.*, 1989; Koenig *et al.*, 1986); but only RNA's 1 and 2 are

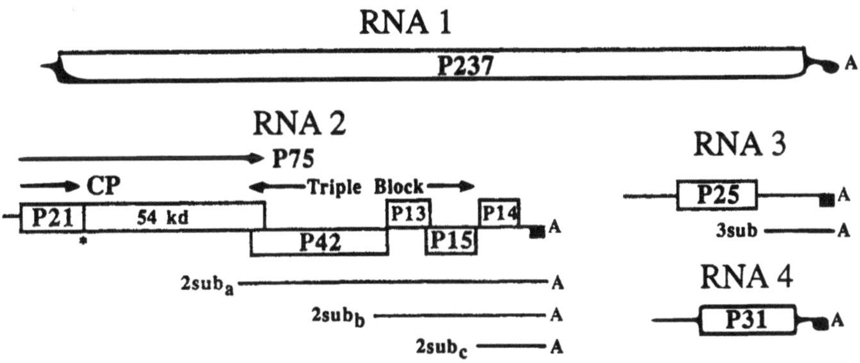

Figure 1. **Genetic Map of BNYVV RNA's. Open reading frames referred to in the text are represented by hollow rectangles and the 3'-terminal domains of homology among the four RNA's by filled rectangles. The coat protein (CP) and the read-through protein (P75) are represented by arrows. The suppressible amber termination codon at the end of the coat protein cistron is indicated by an asterisk. Subgenomic RNA's derived from RNA 2 and 3 (Gilmer *et al.*, 1992a; Bouzoubaa *et al.*, 1991) are indicated below their maps.**

required for infection when the virus is transmitted by mechanical inoculation to leaves of local lesion hosts such as *Tetragonia expansa* or *Chenopodium quinoa* (Tamada *et al.*, 1989; Koenig *et al.*, 1986; Burgermeister *et al.*, 1986; Kuszala *et al.*, 1986; Quillet *et al.*, 1989). Thus, RNA's 1 and 2 apparently encode all the basic 'housekeeping' functions required for multiplication in any host, whereas RNA's 3 and 4 feature specifically in the natural infection process. (It should be noted, however, that RNA 3 can intensify symptoms on leaves (Tamada *et al.*, 1989; Jupin *et al.*, 1992).) In particular, the presence of RNA 4 has been shown to enhance the efficiency of fungus-mediated infection of roots (Tamada and Abe, 1989) (as will be discussed more fully below, the RNA 2-encoded P75 also intervenes in this process) while the presence of RNA 3 is associated with enhanced virus proliferation within the root system and the appearance of the rhizomania symptom phenotype (Tamada *et al.*, 1989; Koenig and Burgermeister, 1989; Tamada *et al.*, 1990).

Mapping *cis*-Acting Sequence Elements on RNA 3 and 4

3'-Proximal cis-Acting Sequences: Because RNA's 3 and 4 do not intervene *in trans* in the virus multiplication cycle in leaves, they are useful experimental objects for identification of sequence features recognized *in cis* by the viral replication machinery. In view of the commonly held model for replication of viruses such as BNYVV in which plus-strand RNA serves as template for synthesis of minus-strand RNA which in turn serves as template for plus-strands (Quadt and Jaspars, 1989), the sequence elements that are recognized by the viral polymerase to initiate plus- and minus-strand replication are expected to lie near the extremities of the viral RNA. Circumstantial evidence for such a location has come from characterization of some of the deleted forms of RNA's 3 and 4 which can appear spontaneously during propagation of BNYVV isolates on leaves (Bouzoubaa *et al.*, 1991). The deletions in these shortened but viable forms of RNA 3 and 4 never encompass the 5'- or 3'-extremity of the RNA's, indicating that any *cis*-active elements are probably confined to these regions. Direct evidence for this hypothesis has come from deletion mutagenesis of RNA 3 and 4 (Jupin *et al.*, 1990; Gilmer *et al.*, 1992b). Bacteriophage T7 RNA polymerase run-off transcripts bearing deletions extending from near the center of the RNA 3 or 4 molecule toward the 3'-extremity can replicate on leaves (in all such experiments wild-type RNA's 1 and 2 were coinoculated to furnish *trans*-acting replication factors) provided that the deletions did not encroach upon the conserved ~70 nt domain immediately preceding the 3'-poly(A) tail. This domain (Domain A) of BNYVV RNA's 1-4 can be folded into a conserved double hairpin secondary structure in which variant residues fall within single-stranded regions of the structure or are accommodated within stems by covariation (Jupin *et al.*, 1990). Domain A almost certainly contains the 'core promoter' sequence for initiation of minus-strand RNA synthesis.

5'-Proximal cis-Acting Sequences in RNA 3: It was expected that the 3'-proximal *cis*-acting domain would coincide with regions of sequence similarity among the four viral RNA's as all four species are undoubtedly recognized by the same viral RNA replication enzymes. Rather surprisingly, however, the situation is quite different at the 5'-terminus. Here, the sequence matchup among RNA's 1-4 is limited to the first 8-9 residues immediately following the cap. Yet deletion mutagensis of RNA 3 and 4 transcripts has shown that *cis*-acting sequence elements occupy the first ~300 residues of RNA 3 and the first ~400 residues of RNA 4 (Jupin *et al.*, 1990; Gilmer *et al.*, 1992b).

Fine-scale mapping of the ~300 nt 5'-proximal *cis*-acting domain of RNA 3 has revealed that it consists of a mosaic of essential and nonessential subdomains (Gilmer *et al.*, 1992b). Box I, the 10 nt *cis*-essential element most distal from the 5'-terminus, can base pair with the first 10 residues at the 5'-extremity, referred to as Box I' (Figure 2A). A second 8 nt *cis*-essential element, Box II, is separated from Box I by a 38 residue nonessential sequence (Figure 2A) and can base pair with a second 5'-proximal element (Box II'; Figure 2A). Box II is preceded by a nonessential subdomain of about 30 residues (Figure 2A) which must contain a sequence signal important for RNA 3 encapsidation since transcripts bearing deletions in this region replicated *in planta* but were not packaged. Additional *cis*-acting sequence elements are certainly situated within the first 150 nts of RNA 3 but have not yet been mapped in detail.

The base complementarity between Boxes I and I' and Boxes II and II' suggested that secondary structure involving base pairing between these relatively distant sequences might be important for function. To test this idea, mutants were created in which unrelated sequences (corresponding to restriction sites) were substituted for one or the other member of each pair of the Boxes to destroy the base pairing potential (Gilmer *et al.*, 1992b). In most cases, such modifications dramatically inhibited RNA 3 accumulation. Introduction of compensatory mutations designed to restore Box I/I' or Box II/II' complementarity restored viability, indicating that secondary structure involving these elements rather than their exact sequence is the critical feature for replicability (Gilmer *et al.*, 1992b).

Secondary Structure at the 5'-Terminus of RNA 3: A possible secondary structure for the first 360 nts of RNA 3 is shown in Figure 2B. In the structure, Boxes I and I' and II and II' are base paired, in line with the mutagenesis results. The structure was generated by the computer program MULFOLD (Jaeger *et al.*, 1989) with minor adjustments based upon measurements of the accessibility of residues in different parts of the 5'-domain to modification by structure-sensitive chemical probes (Gilmer *et al.*, 1993). It should be noted the computer-generated structure took into account only the first 360 residues of RNA 3, whereas the chemical modification experiments were carried out on full-length RNA 3 transcript. The

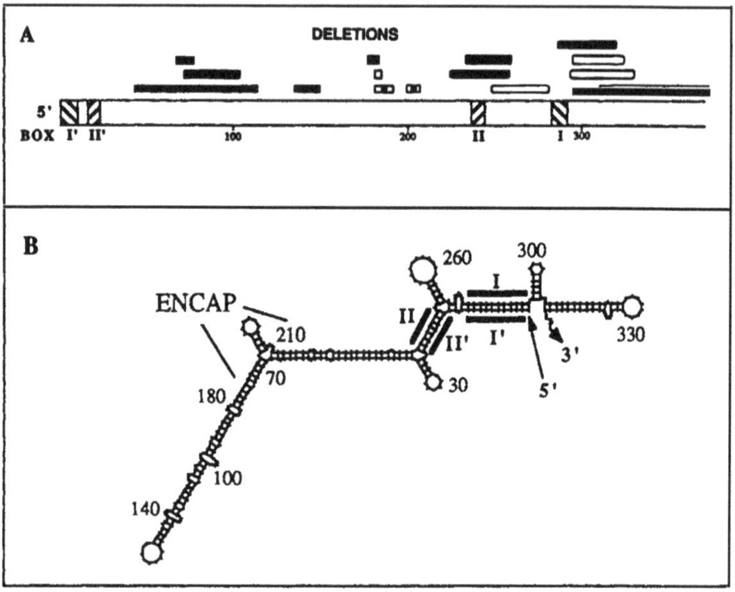

Figure 2. A: Identification of *cis*-Acting Elements in the 5'-Proximal 300 nts of RNA 3. The complementary sequence Boxes referred to in the text are indicated by hatching. The filled bars at the top indicate deletions which abolish RNA 3 accumulation and the empty bars deletions which do not interfere with RNA 3 accumulation. The empty bars containing black circles represent deletions which do not inhibit replication but in which the progeny RNA 3 are not encapsidated. B: Secondary Structure Model for the 5'-Proximal Region of RNA 3. The positions of Boxes I, I', II, and II' and the domain implicated in RNA 3 assembly are indicated.

close ãgreement between the results obtained with the two approaches indicates that long-range interactions involving the 5'-terminal domain and more distant parts of RNA 3 may be neglected to a first approximation.

5'-Proximal cis-Acting Sequences in RNA 4: Site-directed mutagenesis of BNYVV RNA 4 has localized *cis*-essential sequences to the 5'-terminal ~400 nts and the 3'-terminal 70 nts (Gilmer *et al.*, 1992b). Fine scale dissection of the 5'-proximal *cis*-essential region is not yet complete, but the available results indicate that the region consists of a mosaic of essential and nonessential elements as in RNA 3. So far, however, no extended motif of sequence or structural homology with the corresponding region of RNA 3 has been detected. In particular, there is no evidence for a structure involving base pairing between an internal sequence and the 5'-terminus like that between Box I and I' of RNA 3. Thus, the putative promoter sequence that is recognized by the viral replication machinery and which is expected to be common to the RNA 3 and 4 5'-terminal regions may, in fact, consist of several rather short motifs which are relatively distant from one another on the RNA but are juxtaposed in space in the folded form of the molecule.

It was our original assumption that the 5'-proximal *cis*-essential sequences identified in RNA's 3 and 4 are, in fact, the complements of the true promoter elements recognized by the viral replicase in plus-strand RNA synthesis, which would be active on the viral RNA minus-strand. In this regard, it is noteworthy that the interaction between Boxes I and I' and Boxes II and II' in RNA 3 involve only G-C and A-U base pairs (no G-U pairs). Thus, the base pairing scheme involving Boxes I and I' and II and II' of RNA 3 can also exist in the minus strand. Recently, however, evidence has been presented that 5'-proximal *cis*-acting structural elements (presumably involved in initiating plus-strand RNA synthesis of poliovirus (Andino *et al.*, 1990) and brome mosaic virus (Pogue and Hall, 1992) are functional on the plus-strand and not on the minus-strand viral RNA. We suggest that a similar situation holds for BNYVV (Gilmer *et al.*, 1993).

RNA 2-Encoded Proteins and Viral Assembly

Mutations in the Coat Protein cistron and the 54 kd Read-through Domain of RNA 2 do not Interfere with Replication in Leaves: Just as biologically active transcripts of RNA 3 and 4 have proven useful for mapping *cis*-acting replication signals, mutagenesis studies with RNA 2 transcripts have provided information about the role of RNA 2-encoded proteins in the viral infection cycle. Thus, transcripts with mutations interrupting the open reading frames which encode the Triple Block proteins P42, P13, and P15 (Figure 1) were nonviable in infections of whole leaves but active in protoplast infections, indicating that the proteins in question are involved in cell-to-cell movement of the virus (Gilmer *et al.*, 1992a).

Most deletion and frameshift mutations in the 21 kd viral coat protein cistron of RNA 2 did not interfere with infection when transcripts carrying such mutations were coinoculated along with wild-type RNA 1 transcript to leaves (Schmitt *et al.*, 1992). This rule did not hold for deletions encompassing the N-terminal portion of the coat protein cistron (e.g., mutant B218-ΔA9; Figure 3), but such deletions were probably lethal because they removed 5'-proximal *cis*-acting sequences involved in RNA 2 replication. As expected, the viral RNA produced during infections with the viable coat protein mutants was not packaged into virions, as judged by its susceptibility to degradation by endogenous nucleases in tissue extracts (Schmitt *et al.*, 1992). Apparently, functional coat protein and virion assembly are not required for viral RNA replication or cell-to-cell movement. Furthermore, the frameshift mutants B218-N and B218-ΔS (Figure 3) terminate translation prematurely in the coat protein cistron and are not expected to produce a read-through protein. Indeed, no such product could be detected with an antiserum specific for the read-through domain on Western blots of protein from tissue infected with these mutants (Schmitt *et al.*, 1992). Thus, expression of the read-through domain is also nonessential for infection of leaves.

RNA 2 Transcripts Carrying Mutations in the N-Terminal Portion of the 54 kd Read-through Domain are Deficient in Assembly. Experiments with RNA 2 transcripts carrying various deletions or frameshift mutations in the 54 kd read-through domain (Figure 3) have provided confirmation that intact P75 is not required for productive infection of leaves (Schmitt *et al.*, 1992). Unexpectedly, however, mutations in the N-terminal portion of the read-through domain (mutants B218-ΔM1, B218-ΔM2, B218-16_1, and B218-81K; Figure 3) dramatically inhibited production of virions, as judged by the sensitivity of the viral RNA to nuclease in crude extracts (Schmitt *et al.*, 1992) and the low titer of virus particles visible in electron micrographs of such extracts or in thin sections of infected tissue. The low yield of virions in these experiments is not a consequence of deficient synthesis of viral coat protein as coat protein was readily detectable on Western blots of protein extracted from the infected tissue (Schmitt *et al.*, 1992). A deletion in the C-terminal portion of the read-through domain (mutant B218-ΔAc; Figure 3), on the other hand, did not interfere with virion formation (Schmitt *et al.*, 1992).

The foregoing observations implicate P75, and in particular the N-terminal portion of the P75 read-through domain, in the viral assembly pathway. This finding was unexpected because P75 is present in only very low amounts (no more than ~1 copy per 100 virions) in purified virus preparations (authors' unpublished observations). One possible explanation is that P75 might serve as a 'scaffolding' protein which only associates with assembling virions in a transient fashion. Alternatively, P75 might be a minor virion component of virions *in planta* but be lost during virus purification. In either model, it is most likely that P75 would act at the assembly initiation step, the only stage in the encapsidation process which is expected to be amenable to fine control. P75 (or a multimer of P75), for example, could specifically recognize the assembly initiation sequence on the viral RNA in a manner similar to the way in which tobacco mosaic virus (TMV) protein disks react with the assembly origin sequence on TMV RNA (Butler, 1984). Presumably, P75 would interact with the viral RNA *via* its N-terminal coat protein domain while the read-through domain would be responsible for specifically targeting the assembly initiation sequence on the viral RNA. The failure to detect significant amounts of P75 in purified virus preparations is not a major obstacle to such a model as the read-through domain of P75 subunit incorporated into the nucleoprotein helix would almost certainly protrude from the structure and be sensitive to proteolytic attack in the course of virus purification.

After the sequence-specific nucleation step, 21 kd coat protein subunits could then add to the growing end of the nucleoprotein helix in a step-wise and nonsequence-specific fashion. It should be noted, however, that wild-type P75 is apparently not absolutely required for assembly-deficient P75 mutants (Figure 3). Thus, there must be a second less efficient assembly pathway which can operate in parallel. The observation (Schmitt *et al.*, 1992) that full-length virions in low yield are produced by this pathway rather than incomplete particles is also consistent with the idea that assembly initiation is the rate limiting step in which P75 intervenes.

Finally, although unlikely in our view, it is difficult to rule out indirect modes of action for P75 in assembly. For example, wild-type P75 could influence the compartmentalization of coat protein and viral RNA so that, in its absence, they do not encounter one another. A definitive test of these various models will probably require development of an *in vitro* assembly system.

P75 and Virus-Vector Interactions: Recent observations of Tamada and Kusume (1991) have shown that P75 is essential for efficient transmission of BNYVV by *P. betae*. Two spontaneous deletion mutants of RNA 2 which arose during serial passage of virus on *T. expansa* leaves were found to form virions but to be deficient in transmission by the fungus vector (Tamada and Kusume, 1991). Sequence analysis has shown that the deletions map to the C-terminal portion of the 54 kd read-through domain (Tamada, personal

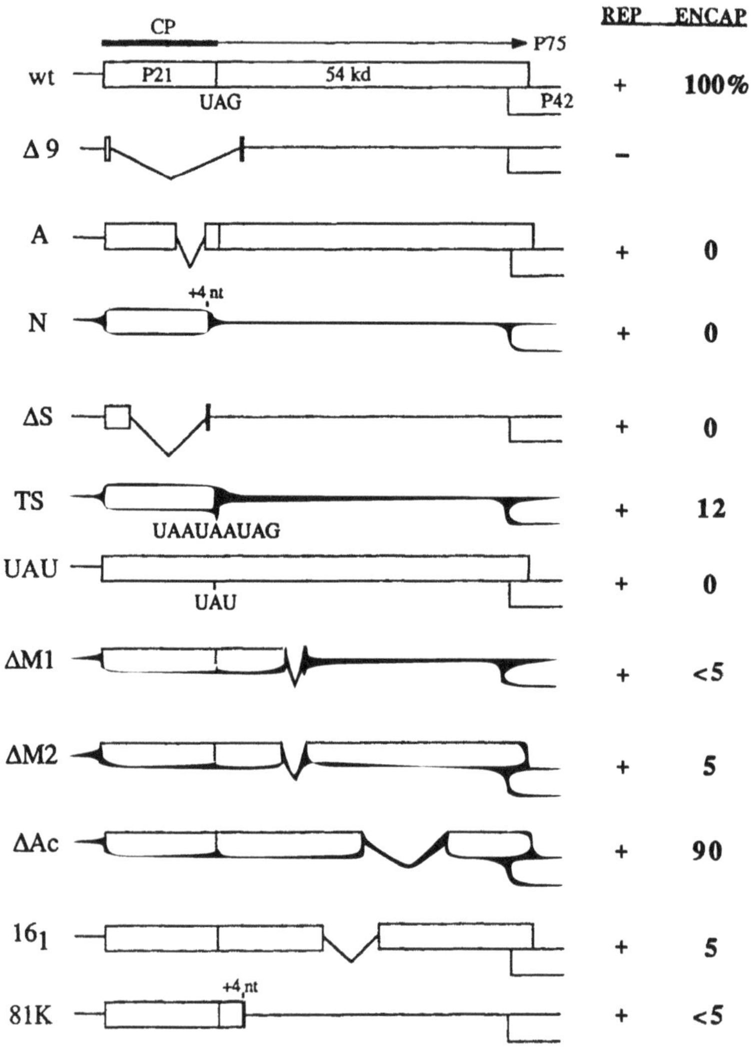

Figure 3. Effect of Mutations in the Coat Protein and the 54 kd Read-through Domain of RNA 2 on the Accumulation of RNA 1 and 2 in Inoculated Tissue and on its Encapsidation. Deletions are represented by angled lines and C-terminal missense sequences resulting from frameshift by filled rectangles. REP: accumulation of viral RNA in infected tissues; ENCAP: degree of protection of progeny viral RNA 1 and 2 from nuclease degradation in crude extracts relative to progeny of wild-type (wt) transcript B218 (100 percent protection).

communication). It will be interesting to determine if our mutant B218-ΔAc, which carries a very similiar deletion, is also nontransmissible. Thus, the evidence suggests that the 54 kd read-through domain of P75 is bifunctional with the C-terminal portion participating in virus-vector interactions while the N-terminal portion intervenes in assembly. Evidently, such bifunctionality could provide a mechanism for insuring that the majority if not all virions in infected tissue will be associated with the vector transmission factor.

Possible Parallels with Other Viruses: BNYVV is not the only soil-borne virus in which a coat protein read-through species exists. RNA 2 of the furovirus soil-borne wheat mosaic virus (SBWMV; vector: *Polymyxa graminis*) carries the cistron for the 19.7 kd viral coat protein at its 5'-extremity followed by a long in-frame open reading frame (Shirako and Wilson, 1993). Translation read-through into this adjacent reading frame produces a 90 kd species with the coat protein sequence at its N-terminus (Hsu and Brakke, 1985). A role of SBWMV P90 in virus-vector interactions and possibly in viral assembly can certainly be envisaged. There is strong evidence from sequence analysis, however, that coat protein read-through does not occur in another putative furovirus, African peanut clump virus (Manohar *et al.*, 1993).

Read-through of coat protein cistron has been reported for luteoviruses, which are transmitted in a persistent manner by aphids instead of fungi (Bahner *et al.*, 1990; Mayo *et al.*, 1989; Miller *et al.*, 1988; Veidt *et al.*, 1988). The read-through protein of potato leafroll luteovirus can be found associated with virions in semipurified virus preparations (Bahner *et al.*, 1990); and it is entirely possible that at least some of the determinants governing aphid transmissibility, previously assumed to be carried exclusively by the coat protein, may, in fact, lie within the read-through domain (Bahner *et al.*, 1990). Experiments with biologically active transcripts of the luteovirus beet western yellow virus (BWYV), however, have shown that the read-through domain of BWYV can tolerate extensive deletions without interfering with virion assembly, as measured by the resistance of viral RNA to degradation in extracts of infected protoplasts (Reutenauer *et al.*, 1993). Thus, with regard to possession of a coat protein read-through extension, the superficial similarity between BNYVV and the luteoviruses appears to conceal significant differences in the manner in which these read-through proteins function in assembly.

References

Andino, G.E., Rieckhof, G.E., and Baltimore, D. 1990. A functional ribonucleoprotein complex forms around the 5'-end of poliovirus RNA. Cell 63:369-380.

Bahner, I., Lamb, J., Mayo, M.A., and Hay, R.T. 1990. Expression of the genome of potato leafroll virus: Read-through of the coat protein termination codon *in vivo*. J. Gen. Virol. 71:2251-2256.

Bouzoubaa, S., Guilley, H., Jonard, G., Richards, K., and Putz, C. 1985. Nucleotide sequence analysis of RNA-3 and -4 of beet necrotic yellow vein virus, isolates F2 and G1. J. Gen Virol. 66:1553-1564.

Bouzoubaa, S., Niesbach-Klösgen, U., Jupin, I., Guilley, H., Richards, K., and Jonard, G. 1991. Shortened forms of beet necrotic yellow vein virus RNA-3 and -4. J. Gen. Virol. 72:259-266.

Bouzoubaa, S., Quillet, L., Guilley, H., Jonard, G., and Richards, K. 1987. Nucleotide sequence of beet necrotic yellow vein virus RNA-1. J. Gen. Virol. 68:615-626.

Bouzoubaa, S., Ziegler, V., Beck, D., Guilley, H., Richards, K., and Jonard, G. 1986. Nucleotide sequence of beet necrotic yellow vein virus RNA-2. J. Gen. Virol. 67:1689-1700.

Burgermeister, W., Koenig, R., Weich, H., Sebald, W., and Lesemann, D.E. 1986. Diversity of RNA's in thirteen isolates of beet necrotic yellow vein virus in *Chenopodium quinoa* detected by means of cloned cDNA's. J. Phytopath. 115:229-242.

Butler, P.J.G. 1984. The current picture of the structure and assembly of tobacco mosaic virus. J. Gen. Virol. 65:253-279.

Gilmer, D., Allmang, C., Ehresmann, C., Guilley, H., Richards, K., Jonard, G., and Ehresmann, B. 1993. The secondary structure of the 5'-noncoding region of beet necrotic yellow vein virus RNA 3: Evidence for a role in viral RNA replication. Nucleic Acids Res. 21:1389-1395.

Gilmer, D., Bouzoubaa, S., Hehn, A., Guilley, H., Richards, K., and Jonard, G. 1992a. Efficient cell-to-cell movement of beet necrotic yellow vein virus requires 3'-proximal genes located on RNA 2. Virology 189:40-47.

Gilmer, D., Richards, K., Jonard, G., and Guilley, H. 1992b. *Cis*-active sequences near the 5'-termini of beet necrotic yellow vein virus RNA's 3 and 4. Virology 190:55-67.

Hsu, Y.H. and Brakke, M.K. 1985. Cell-free translation of soil-borne wheat mosaic virus RNA's. Virology 143:272-279.

Jaeger, J.A., Turner, D.H., and Zuker, M. 1989. Predicting optimal and suboptimal secondary structure for RNA. Methods in Enzymology 183:281-306.

Jupin, I., Guilley, H., Richards, K.E., and Jonard, G. 1992. Two proteins encoded by beet necrotic yellow vein virus RNA 3 influence symptom phenotype on leaves. EMBO J. 11:479-488.

Jupin, I., Richards, K., Jonard, G., Guilley, H., and Pleij, C.W.A. 1990. Mapping sequences required for productive replication of beet necrotic yellow vein virus RNA 3. Virology 178:273-280.

Jupin, I., Tamada, T., and Richards, K.E. 1991. Pathogenesis of beet necrotic yellow vein virus. Seminars in Virology 2:121-129.

Koenig, R. and Burgermeister, W. 1989. Mechanical inoculation of sugar beet roots with isolates of beet necrotic yellow vein virus having different RNA compositions. J. Phytopath. 124:249-255.
Koenig, R., Burgermeister, W., Weich, H., Sebald, W., and Kothe, C. 1986. Uniform RNA patterns of beet necrotic yellow vein virus in sugar beet roots, but not in leaves from several plant species. J. Gen. Virol. 67:2043-2046.
Kuszala, M., Ziegler, V., Bouzoubaa, S., Richards, K., Putz, C., Guilley, H., and Jonard, G. 1986. Beet necrotic yellow vein virus: Different isolates are serologically similar but differ in RNA composition. Ann. Appl. Biol. 109:155-162.
Manohar, S.K., Guilley, H., Dollet, M., Richards, K., and Jonard, G. 1993. Nucleotide sequence and genetic organization of peanut clump virus RNA 2 and partial characterization of deleted forms. Virology 195:33-41.
Mayo, M.A., Robinson, D.J., Jolly, C.A., and Hyman, L. 1989. Nucleotide sequence of potato leafroll luteovirus RNA. J. Gen. Virol. 70:1037-1051.
Miller, W.A., Waterhouse, P.M., and Gerlach, W.L. 1988. Sequence and organization of barley yellow dwarf virus genomic RNA. Nucleic Acids Res. 16:6097-6111.
Morozov, S.Y., Dolja, V.V., and Atabekov, J.G. 1989. Probable reassortment of genomic elements among elongated RNA-containing plant viruses. J. Mol. Evol. 29:52-62.
Niesbach-Klösgen, U., Guilley, H., Jonard, G., and Richards, K. 1990. Immunodetection *in vivo* of beet necrotic yellow vein virus encoded proteins. Virology 178:52-61.
Pogue, G.P. and Hall, T.C. 1992. The requirement for a 5'-stem-loop structure in brome mosaic virus replication supports a new model for viral positive-strand RNA initiation. J. Virol. 66:674-684.
Quadt, R. and Jaspars, E.M.J. 1989. RNA polymerases of plus-strand RNA viruses of plants. Mol. Plant Micr. Inter. 2:219-223.
Quillet, L., Guilley, H., Jonard, G., and Richards, K. 1989. *In vitro* synthesis of biologically active beet necrotic yellow vein virus RNA. Virology 172:293-301.
Reutenauer, A., Ziegler-Graff, V., Lot, H., Scheidecker, D., Guilley, H., Richards, K., and Jonard, G. 1993. Identification of beet western yellows luteovirus genes implicated in viral replication and particle morphogenesis. Virology 195:692-699.
Richards, K.E. and Tamada, T. 1992. Mapping functions on the multipartite genome of beet necrotic yellow vein virus. Annu. Rev. Phytopath. In press.
Schmitt, C., Balmori, E., Jonard, G., Richards, K.E., and Guilley, H. 1992. *In vitro* mutagenesis of biologically active transcripts of beet necrotic yellow vein virus RNA 2: Evidence that a domain of the 75 kDa read-through protein is important for efficient virus assembly. Proc. Natl. Acad. Sci. U.S.A. 89:5715-5719.
Shirako, Y. and Wilson, T.M.A. 1993. Complete nucleotide sequence and organization of the bipartite RNA genome of soil-borne wheat mosaic virus. Virology 195:16-32.
Tamada, T. and Abe, H. 1989. Evidence that beet necrotic yellow vein virus RNA-4 is essential for efficient transmission by the fungus *Polymyxa betae*. J. Gen. Virol. 70:3391-3398.

Tamada, T. and Kusume, T. 1991. Evidence that the 75K read-through protein of beet necrotic yellow vein virus RNA-2 is essential for transmission by the fungus *Polymyxa betae*. J. Gen. Virol. 72:1497-1504.

Tamada, T., Saito, M., Kiguchi, T., and Kusume, T. 1990. Effect of isolates of beet necrotic yellow vein virus with different RNA components on the development of rhizomania symptoms. In: *Proceedings of the First Symposium of the International Working Group on Plant Viruses with Fungal Vectors*. R. Koenig (ed.), pp. 41-44. Eugen Ulmer GmbH, Stuttgart.

Tamada, T., Shirako, Y., Abe, H., Saito, M., Kiguchi, T., and Harada, T. 1989. Production and pathogenicity of isolates of beet necrotic yellow vein virus with different number of RNA components. J. Gen. Virol. 70:3399-3409.

Veidt, I., Lot, H., Leiser, M., Scheidecker, D., Guilley, H., Richards, K., and Jonard, G. 1988. Nucleotide sequence of beet western yellow virus RNA. Nucleic Acids Res. 16:9917-9932.

Ziegler, V., Richards, K., Guilley, H., Jonard, G., and Putz, C. 1985. Cell-free translation of beet necrotic yellow vein virus: Read-through of the coat protein cistron. J. Gen. Virol. 66:2079-2087.

The Role of Replication and Movement Functions in the Pathogenesis of Squash Leaf Curl Virus

S. G. Lazarowitz, L. C. Wu, and D. J. Ingham
Department of Microbiology
University of Illinois
Urbana, Illinois 61801

J. S. Elmer
Plant Molecular Biology Division
The Monsanto Company
St. Louis, Missouri 63198

The bipartite geminiviruses such as squash leaf curl virus (SqLCV) encode a single essential replication protein (AL1) and three proteins essential for systemic movement in the plant. The movement protein AL2 has been implicated as a transactivator of viral gene expression, and genetic studies of SqLCV have established that both replication and movement functions determine viral host range and pathogenic potential. Much of the regulated expression of these viral functions appears to be clustered within the intergenic 200 nt viral common region. Using an *in vivo* replication assay we have localized a viral replication origin for both ssDNA and dsDNA synthesis to a 90 nt fragment within the common region that contains the conserved viral stem-loop structure and 60 nt of 5'-upstream sequence. The AL1-encoded protein recognizes this fragment in a sequence-specific manner, and the stem-loop structure is essential for replication. Genetic analysis of SqLCV-R has shown the movement protein BR1 to be an important determinant of viral host range. Mutations within the BR1 movement protein specifically alter viral host range without affecting disease symptoms.

Introduction

The squash leaf curl viruses SqLCV-E and SqLCV-R are members of the bipartite virus subfamily of plant pathogenic geminiviruses (Lazarowitz and Lazdins, 1991; Lazarowitz, 1991). These bipartite geminiviruses, which include tomato golden mosaic virus (TGMV) and African cassava mosaic virus (ACMV), contain two ~2.7 kb circular single-stranded DNA genomic components designated A and B (Lazarowitz and Lazdins, 1991; Stanley and Gay, 1983; Hamilton *et al.*, 1983; Hamilton *et al.*, 1984), which together encode six genes (Figure 1). The viral genes are named according to their component location (A or B) and position on the virion (rightward) or complementary (leftward) DNA strand.

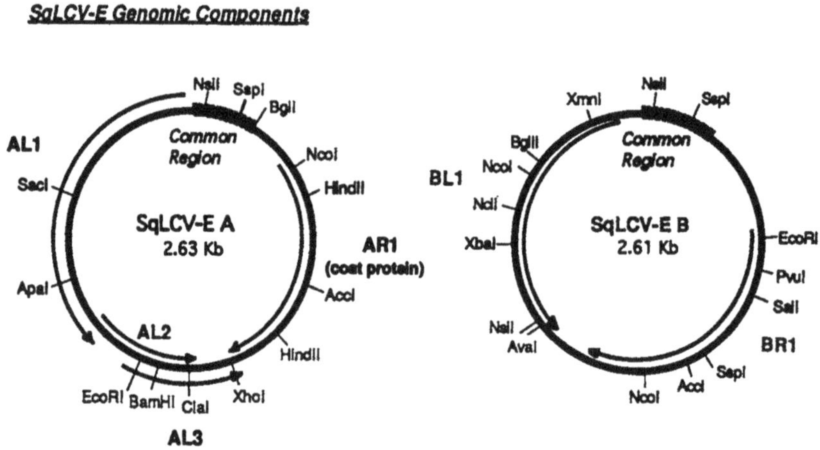

Figure 1. Genomic Map of the Bipartite Geminivirus SqLCV-E. Shown are the limits and direction of transcription of each individual gene (circumferential arrows). Gene designations are by component location (A or B) and direction of transcription (Rightward = virion strand; Leftward = complementary strand). Also indicated is the intergenic viral common region (▨). See text for details.

The viral A component encodes those genes essential for replication, while the B component is required for systemic movement in the plant. The two B-encoded genes (BR1, BL1) and AL2 are needed for viral movement in the plant (Etessami *et al.*, 1988; Brough *et al.*, 1988; Gardiner *et al.*, 1988), the latter appearing to have an indirect role as a transactivator of coat protein and BR1 gene expression (Sunter and Bisaro, 1991; Haley *et al.*, 1992). On the A component, in addition to the coat protein gene, are two genes involved in viral replication, AL1 and AL3. Mutational analyses and expression of viral genes in transgenic plants have shown that the AL1 protein is the only virus-encoded product required for replication, being essential for the replication of both the A component and its cognate B component (Brough *et al.*, 1988; Elmer *et al.*, 1988; Etessami *et al.*, 1991; Hanley-Bowdoin *et al.*, 1990). The AL3 gene product increases the efficiency of viral DNA replication but is not essential (Sunter *et al.*, 1990).

The cognate A and B components comprising the genome of a given bipartite virus are distinguished by an ~200 nt intergenic common region (CR) (Figure 1) (Lazarowitz, 1987; Lazarowitz, in press) the sequence of which is identical in the two genomic components of any single bipartite geminiviruses, but completely different among the separate geminiviruses. The striking exception to this complete sequence divergence among the different geminiviral CR's is a conserved 29-32 nt sequence element which can potentially form a hairpin in which the loop sequence TAATATTAC is present in all viruses. In the geminiviruses infecting dicotyledonous hosts, this conserved element has the consensus sequence $GGCCA^{T}_{A}CCGN^{T}_{A}{}^{A}_{T}TAATATTACCGG^{A}_{T}TGGCC$ (Lazarowitz, 1987; Lazarowitz, in press), although its sequence is quite divergent among those viruses infecting monocotyledonous plants.

Genetic anaysis of SqLCV-E (extended host range) and SqLCV-R (restricted host range) has demonstrated that both viral replication and movement functions are important determinants of viral host range and pathogenesis (Lazarowitz, 1991). Both viruses are highly infectious for *Nicotiana benthamiana*, but differ in their abilities to infect pumpkin, bean, and squash. Differences in the interactions of these two highly homologous viruses has been shown to be the result of differences in their abilities to replicate in these different hosts. A naturally occurring mutant B component, B_R^{td1} of SqLCV-R, has been identified which has lost the ability to infect *N. benthamiana* (hence *td1* for tobacco defective mutant 1), although it is still infectious for pumpkin, bean, and squash (Lazarowitz, 1991).

We have further characterized the mechanism of viral replication and the function of the viral movement proteins on the molecular level to better understand their underlying roles in viral host range and pathogenesis. Using an *in vivo* replication assay, we have identified a viral origin for both ssDNA and dsDNA replication within the CR, demonstrating that sequence-specific recognition by the viral AL1-encoded protein occurs within a small 90 nt fragment and that the viral stem-loop structure is required for replication to occur (Lazarowitz *et al.*, 1992). Analysis of the mutant B_R^{td1} component of SqLCV-R in leaf discs had demonstrated that the virus did replicate in *N. benthamiana* (Lazarowitz, 1991), thus leading to the conclusion that viral movement was defective in this host. The mutation in B_R^{td1} has been mapped to the coding sequence of the viral movement protein BR1. Sequence analysis has identified three altered amino acids in this mutant as compared to the wild-type BR1 protein.

Identification of a Viral Replication Origin

The intergenic location and sequence characteristics of the viral CR, as well as a few mutational studies which targeted the CR and eliminated viral infectivity (Revington *et al.*, 1989; Ugaki *et al.*, 1991), have led to the general belief that much of the regulated expression of viral gene functions appears to be clustered within the 200 nt CR. To define the viral replication origin and demonstrate that the viral-encoded AL1 protein interacted with sequences within the viral CR in a sequence-specific manner to affect viral DNA replication, we devised an *in vivo* leaf disc replication assay that utilized the CR's of SqLCV-E and TGMV and their encoded AL1 proteins (Lazarowitz *et al.*, 1992). Choice of SqLCV-E and TGMV was based on several considerations. First, sequence analysis clearly shows that SqLCV-E and TGMV are as distantly related as are all other bipartite geminiviruses characterized to date (Lazarowitz and Lazdins, 1991). Second, their CR's display no clear sequence homology, except for the canonical geminivirus sequence element and 18 nts at the start of both viral CR's. Third, both viruses have a common host in *N. benthamiana*, thus permitting the development of an *in vivo* replication assay. Finally, the exchange of components between SqLCV-E and TGMV demonstrates that each viral A component selectively replicates its cognate B component CR but is unable to replicate a viral B component with a noncognate CR (Lazarowitz *et al.*, 1992).

Identification of Viral Replication Origin(s) Within the CR: To precisely map the viral replication origin and demonstrate sequence-specific interactions with the AL1 protein, we devised an *in vivo* assay that would identify functional elements within the CR based on their ability to direct replication when present, rather than an assay based on loss of replication when such elements were absent (Lazarowitz *et al.*, 1992). We first directly demonstrated that viral replication origins are contained completely within the CR by assaying the ability of cloned CR sequences to direct the replication of nonviral DNA segments. PCR was used to generate precise cloned copies of each viral CR containing BglII at each end (Lazarowitz *et al.*, 1992). These SqLCV or TGMV CR's were then cloned into a unique BglII site located between direct duplications of the bacterial chloramphenicol transacetylase (CAT) coding sequence contained within an *Agrobacterium* binary vector (Figure 2). Earlier studies of the excision

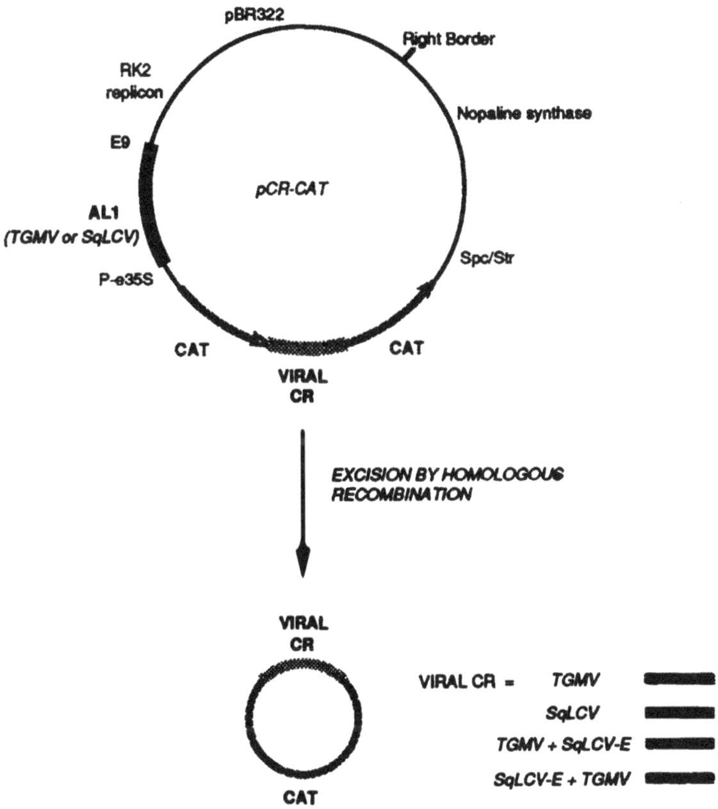

Figure 2. Schematic Diagram of AL1-Dependent Replication Assay. Binary vector *pCR-CAT* contains Common Region construct to be tested (▒▒▒) cloned between two copies of the CAT coding sequence and a cassette (▬▬) expressing the AL1 of TGMV or SqLCV-E under the control of the enhanced 35S CaMV promoter (P-e35S). Upon introduction into a leaf disc, a copy of the CR::CAT mini-circle DNA is excised by homologous recombination within the CAT sequences. If the AL1 being expressed recognizes the CR, the CR::CAT mini-circle is amplified and can be detected on blots. (Reproduced with permission from S.G. Lazarowitz, L.C. Wu, S.G. Rogers, and J.S. Elmer, The Plant Cell, Vol. 4, July 1992, copyright 1992 American Society of Plant Physiologists.)

and replication of maize streak virus had demonstrated that such vectors would work, with circular replicating chimeric DNA's consisting of fused viral and CAT sequences being generated in the plant cells by homologous recombination within the duplicated CAT sequences (Lazarowitz *et al.*, 1989). For the assay as diagrammed in Figure 2, a cassette expressing the appropriate AL1 protein was cloned into the same binary vector as the cloned CR. This assured that each cell receiving the CR-CAT fusions would also express the AL1 protein and made it straightforward to test different combinations of cloned CR constructs and viral AL1 proteins. Hence, following agroinoculation (Grimsley and Bisaro, 1987) of leaf discs, mini-circle chimeric DNA containing a single copy of CAT and the cloned CR would be excised from the vector (Figure 2) and could be amplified and, thus, detected on Southern blots only if it contained a functional origin that was recognized by the AL1 protein being expressed.

Using this assay, each viral CR was recognized and replicated by its cognate AL1 protein (Figure 3A and Table 1), as expected. Minicircle DNA's of the expected sizes were detected using a CAT probe following agroinoculation of *N. benthamiana* leaf discs with either *pMON1603* (containing the intact TGMV CR and TGMV AL1-AL2-AL3 expression cassette) or pSQE33-116 (containing the intact SqLCV-E CR and SqLCV-E AL1-AL2-AL3 expression cassette). These small circular DNA's were not detected if the expression cassette was omitted. As shown in Figure 3, both ssDNA and dsDNA circular forms were replicated. Identification of the slower migrating small DNA as covalently-closed, circular dsDNA was based on its resistance to digestion by *S*1 nuclease and susceptibility to digestion by restriction endonucleases uniquely cutting within the different viral CR's to give distinct fragments from the integrated and excised DNA forms. That the faster migrating small replicated form was ssDNA was demonstrated by its sensitivity to digestion with mung bean nuclease (Figure 3B).

Sequence-Specific Recognition of the Functional Origin(s) by AL1 Protein: Having thus localized viral replication origins for both viral ssDNA and dsDNA replication to sequences completely within the viral CR's, the precise origin sequences were further delineated by testing chimeric CR's containing sequences from each viral CR cloned so as not to disrupt potentially critical structural features. The very nature of the assay assured that we would also more precisely define sequence elements being recognized by the AL1 protein. As shown in Figure 4, chimeric viral CR clones were joined at the unique SspI site present in the conserved geminiviral sequence element located within the loop of the potential hairpin structure (Lazarowitz *et al.*, 1992). These were tested for their ability to be replicated by the AL1 protein of either TGMV or SqLCV-E (Figure 3 and Table 1). Such CR constructs allowed us to precisely define the viral origin within the context of its specificity of interactions with either the TGMV or SqLCV-E AL1 protein. Thus, failure of a particular construct to replicate in the presence of one of the viral AL1 proteins could clearly be interpreted as lack of recognition of the origin by the particular AL1 protein rather than a structural defect within the construct being tested, since the same chimeric CR should be replicated by the alternative viral AL1 protein.

As shown in Figure 3, and summarized in Table 1 and Figure 4, specific recognition by the viral AL1 protein and, hence, the viral replication origin was mapped to the left half of the CR. Thus, the chimeric CR in which sequences to the left of the SspI site were derived from TGMV and those to the right from SqLCV-E was replicated by the TGMV AL1 protein (*pMON1714R*) but not by the AL1 protein from SqLCV-E (derivatives of *pMON1714R*) in which the TGMV AL1-AL2-AL3 cassette was replaced by the SqLCV-E AL1-AL2-AL3 cassette. Similarly, the chimeric CR consisting of sequences from the left half of the SqLCV-E CR joined to the right half of the TGMV CR was replicated by the SqLCV-E AL1 protein (*pMON1716*) and not by the TGMV AL1 protein (derivatives of *pMON1716* in which the SqLCV-E AL1-AL2-AL3 cassette was replaced by the TGMV AL1-AL2-AL3 cassette). Both ssDNA and dsDNA of the expected sizes were found to be replicated (Figure 3). These results strongly suggest that the viral replication origin(s) for both dsDNA-to-ssDNA and dsDNA-to-dsDNA replication are located with a specific fragment contained within the left half (AL1 proximal side) of the viral CR and the potential stem-loop structure.

Because these anlayses had utilized the complete AL1-AL2-AL3 cassettes from either TGMV or SqLCV-E, the specific AL1-dependence of the assay was demonstrated by testing other specific expression cassettes from SqLCV-E. Unexpectedly, SqLCV-E AL1 expression cassette clones did not support the replication of either chimeric viral CR within the level of sensitivity of our assay, even though precisely the same CR clones were replicated by either the SqLCV-E AL1-AL2 or AL1-AL2-AL3 cassettes, or TGMV AL1-AL2-AL3 cassette, depending on the derivation of the left half of the CR. To eliminate the possibility that the AL2 protein was affecting the results of our assay and thereby demonstrate the dependence of the assay solely on the

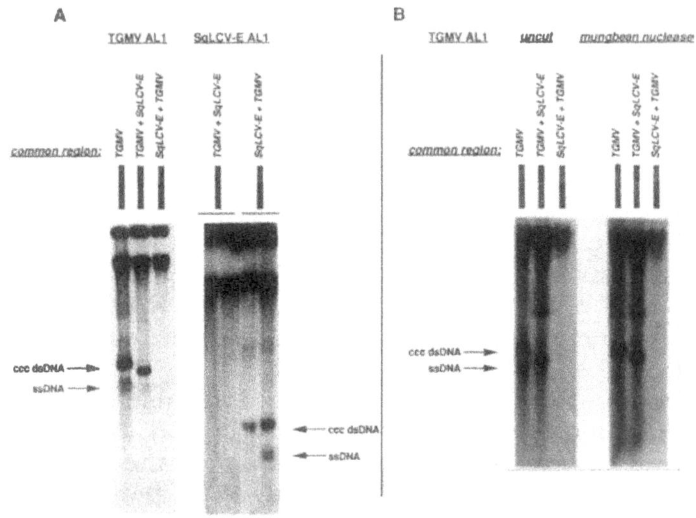

Figure 3. Replication of CR::CAT Constructs in the Presence of AL1 from TGMV or SqLCV-E. DNA gel blots of extracts from agroinoculated leaf discs hybridized with 32P-labeled CAT coding sequence probe. A: Replication in the presence of the AL1 protein from TGMV or SqLCV-E as indicated above blots. Plasmid constructs tested are (left gel) *pMON1603* (TGMV CR), *pMON1714* (TGMV$_{left}$+SqLCV-E$_{right}$ CR ▬▬), *pMON1716* (SqLCV-E$_{left}$+TGMV$_{right}$ CR ▬▬); and (right gel) *pMON1714-14, pMON1714-29* (TGMV$_{left}$+SqLCV-E$_{right}$ CR lanes ▬▬), *pMON1716-243, pMON1716-255* (SqLCV-E$_{left}$+ TGMV$_{right}$ CR lanes ▬▬). B: Sensitivity to digestion with mung bean nuclease. Replication is in the presence of AL1 protein from TGMV. Extracts were incubated without or with addition of mung bean nuclease as indicated. Plasmids are *pMON1603* (TGMV CR), *pMON1714* (TGMV$_{left}$+SqLCV-E$_{right}$ CR ▬▬). *pMON1716* (SqLCV-E$_{left}$+TGMV$_{right}$ CR ▬▬). Marked are the positions of covalently closed circular CR::CAT double-stranded (ccc dsDNA) and single-stranded DNA (ssDNA). (Reproduced with permission from S.G. Lazarowitz, L.C. Wu, S.G. Rogers, and J.S. Elmer, The Plant Cell, Vol. 4, July 1992, copyright 1992 American Society of Plant Physiologists.)

AL1 protein, we assayed the ability of the intact SqLCV-E CR and the chimeric CR's to replicate in the presence of SqLCV-E AL1-2 and AL1-2-3 expression cassettes in which a frameshift mutation eliminated AL2 function. As expected, these AL2-deficient cassettes did not replicate the chimeric CR containing the left half of the TGMV CR. However, both of these AL2 mutant cassettes did support the replication of the chimeric CR containing the left half of the SqLCV-E CR (not shown) or subregions of the SqLCV-E CR (see below). Thus,

Table 1. Specificity of AL1 Protein for Replication of Common Region Constructs

Cloned DNA	Replication in the presence of AL1[a] from	
	TGMV[b]	SqLCV-E[c]
TGMV B (intact)[d]	+[e]	-[e]
SqLCV-E B (intact)[d]	-	+
TGMV-CR/CAT (*pMON1603*)	+	-
SqLCV-E-CR/CAT (*pSQE-33*)	-	+
SqLCV-R-CR/CAT (*pSQR-48*)	-	+
TGMV:SqLCV-E/CAT (*pMON1714R* and derivatives)	+	-
SqLCV-E:TGMV/CAT (*pMON1716* and derivatives)	-	+

[a] Supplied either by co-inoculation with an intact A component (TGMV or SqLCV-E) or inoculation into leaf discs from transgenic *N. benthamiana* expressing TGMV AL1 (Hanley-Bowdoin *et al.*, 1990).

[b] Expression cassette contained TGMV AL1-2-3.

[c] Results the same with SqLCV-E AL1-2, or AL1-2-3, AL1-2-, and AL1-2-3 expression cassettes.

[d] Dimeric construct of intact viral genomic component.

[e] +, replication based on detection of CAT mini-circle DNA of the expected size. -, no replication detected.

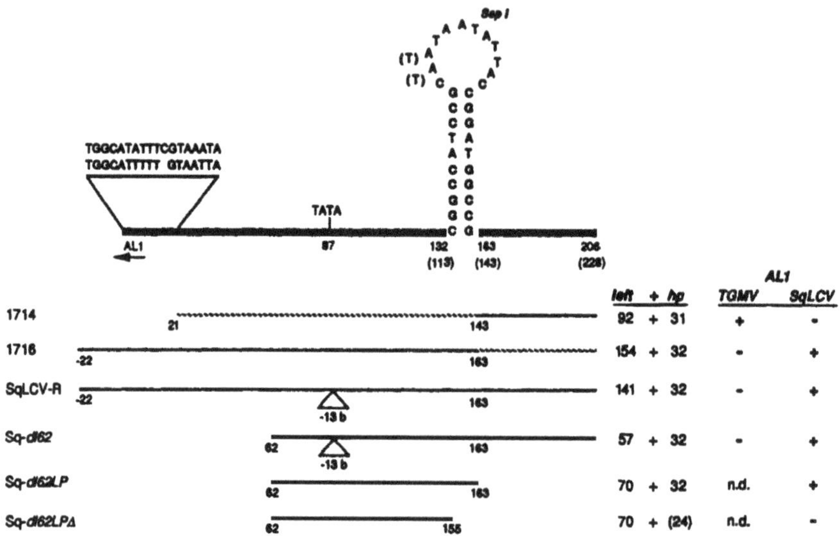

Figure 4. Diagrammatic Representation of the CR's (Virion Strand 5'-to-3') of TGMV and SqLCV E Aligned Beginning at the Second Codon of AL1 (left) and Showing the Near Identical Sequences at the Beginning of the CR (top = TGMV, bottom = SqLCV-E) and Sequence of the Potential Stem-Loop from SqLCV-E with the Changes in TGMV (TT) in the Loop as Indicated. Numbers underneath are the nucleotide positions within the CR of SqLCV-E or TGMV (in parenthesis), with the position of the apparent TATA box for AL1 marked. Aligned beneath are the fragments containing the replication origin(s) based on their sequence specific interaction with AL1, and summary of their replication in the presence of AL1 from TGMV or SqLCV-E. "left+hp" indicates the size of the stem-loop ("hp") and 5'-upstream sequences ("left"). SqLCV CR sequences ———— , TGMV CR sequences ~~~~~~~~ . (Reproduced with permission from S.G. Lazarowitz, L.C. Wu, S.G. Rogers, and J.S. Elmer, The Plant Cell, Vol. 4, July 1992, copyright 1992 American Society of Plant Physiologists.)

these results did demonstrate that the geminivirus AL1 protein interacts with the viral CR in a sequence-specific manner, recognizing the AL1 proximal half of the CR up to and possibly including the stem-loop. While it is quite possible that this sequence-specific interaction could involve host-encoded proteins, it clearly does not require the participation of any other virus-encoded proteins.

Further Delimitation and Structural Features of the Replication Origin(s): The analysis of both intact and chimeric TGMV and SqLCV-E CR's localized the viral replication origin(s) to within a 123 nt sequence within the CR that included the canonical geminivirus sequence element (potential stem-loop structure) and ~90 nt of 5'-upstream sequence (Figure 4). In order to further delimit the viral origin sequences and determine the requirement for the conserved potential stem-loop structure, the replication of specific subregions of the SqLCV-E CR were tested (Lazarowitz *et al.*, 1992). The CR's of SqLCV-E and SqLCV-R differ by a 13 nt deletion in the latter located 1 nt upstream of the apparent TATA element for AL1 (Lazarowitz, 1991). This deletion, although possibly critical in the less efficient replication of SqLCV-R in certain host plants, was predicted not to be an essential element of the viral replication origin. That this is indeed the case was demonstrated by the ability of the SqLCV-E AL1-AL2-AL3 expression cassette to support the replication of the SqLCV-R CR (Lazarowitz *et al.*, 1992). PCR was used to directly introduce this 13 nt deletion into the cloned SqLCV-E CR along with a deletion removing the first 83 nts from the 5'-end of the CR (Figure 4), thus placing the left end of the CR subclone 25 nts from the TATA box for AL1. This mutant SqLCV-E CR (*pSQ-dl62*), with its 5'-end located at nucleotide 62 within the viral CR, was efficiently replicated in our CAT reporter gene assay by the SqLCV-E AL1-AL2-AL3 expression cassette (Figure 5). Taken together with the chimeric CR analyses, these results are consistent with the viral replication origin(s) being within an ~90 nt segment of the viral CR that contains the conserved viral stem-loop and ~60 nt of 5'-upstream sequence.

To eliminate the unlikely possibility that sequences from the right half of the viral CR's were affecting our results, as well as demonstrate the strict dependence of these results on a sequence-specific interaction with the AL1 protein, the region of the SqLCV-E CR located between nucleotides 62 and 163 (*pSQ-dl62LP*) was cloned by PCR. Thus, this subclone contained those SqLCV-E CR sequences located precisely between nucleotide 62 and the 3'-end of the stem-loop (Figure 4). *pSQ-dl62LP* was tested for its ability to direct the replication of the CAT gene in our assay in the presence of the intact SqLCV-E AL1-AL2-AL3 expression cassette (*pSQ-dl62LP-125*), as well as in the presence of the SqLCV AL1-AL2$^-$ and AL1-AL2$^-$-AL3 expression cassettes containing the AL2 frameshift mutation (*pSQ-dl62LP-23* and *pSQ-dl62LP-34*) (Lazarowitz *et al.*, 1992). As shown in Figure 5 and summarized in Figure 4, this CR subclone beginning at nucleotide 62 within the SqLCV-E CR and ending precisely at the

3'-end of the stem-loop was efficiently replicated by all three expression cassettes. These results clearly identify the ~90 nt segment of the viral common region encompassing the conserved viral stem-loop structure and ~60 nts of 5'-upstream sequence as both containing the viral replication origin(s) and being specifically recognized by the viral AL1 protein to effect DNA replication.

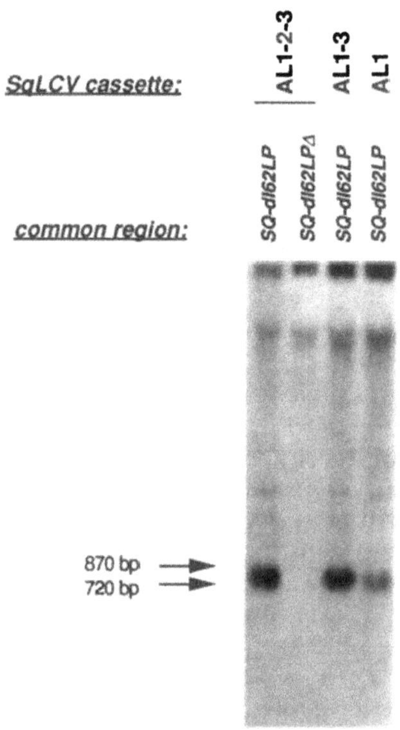

Figure 5. Replication of pSQ-dl62LP and pSQ-dl62LPΔ CR Clones in the Presence of SqLCV Expression Cassettes AL1-2-3 (lanes 1 and 2), AL1-2 -3 (lane 3) or AL1-2$^-$ (lane 4). DNA gel blots of extracts from agroinoculated leaf discs hybridized with 32P-labeled CAT coding sequence probe. Detected are covalently closed dsDNA mini-circle forms. Arrows indicate the position of linear dsDNA size markers. Specific plasmid constructs used are: pSQ-dl62LP-125 (lane 1), pSQ-dl62LPΔ-2432 (lane 2), pSQ-dl62LP-23 (lane 3), and pSQ-dl62LP-34 (lane 4). Size of the SQ-dl62LP::CAT mini-circle is 821 nucleotides, that of the SQ-DL62LPΔ::CAT is 813 nucleotides. (Reproduced with permission from S.G. Lazarowitz, L.C. Wu, S.G. Rogers, and J.S. Elmer, The Plant Cell, Vol. 4, July 1992, copyright 1992 American Society of Plant Physiologists.)

To elucidate the requirement for the actual stem-loop structure in viral DNA replication, PCR was used to clone *pSQ-dl62LP*$_\Delta$ containing SqLCV-E CR sequences located between nucleotides 62 and 155 (Lazarowitz *et al.*, 1992) (Figure 4). While beginning at the same position as *pSQ-dl62LP*, this CR subclone terminates two nucleotides past the loop sequence and, thus, cannot form the potential stem-loop structure. *pSQ-dl62LP*$_\Delta$ was not replicated by the SqLCV-E AL1-AL2-AL3 expression cassette (Figure 5). Thus, deletion of the downstream copy of the inverted repeat from the 90 nt origin-containing fragment destroyed its ability to direct replication, indicating that the potential stem-loop structure is essential for viral DNA replication to occur.

Our findings are consistent with the rolling circle replication studies of BCTV (Stenger *et al.*, 1991) and suggest two possible models concerning the viral replication origin and its recognition by the AL1 protein. Specificity for binding of the AL1-encoded protein may reside in the ~60 nts upstream of the conserved viral stem-loop, with the latter structural element functioning to position the nicking site for ssDNA synthesis so as to make it accessible to an AL1 protein-containing complex. This interaction would be analogous to ϕX174 and is consistent with the striking conservation of the loop sequence TAATATTAC in all geminiviruses and its similarity to the *geneA* protein recognition and cleavage site in ϕX174 (Heidekamp *et al.*, 1982). Evidence for this type of model involving cruciform structural elements has also been presented for prokaryotic single-stranded plasmid replication origins (Gruss *et al.*, 1987; Gruss and Ehrlich, 1989). Alternatively, specificity for recognition by the AL1 protein to direct replication could involve recognition of sequences in both the conserved stem-loop and upstream regions. By site-directed mutagenesis, the CAA at the start of the SqLCV-E loop sequence has been replaced with the TT present in the TGMV loop sequence (Lazarowitz and Goodlove, unpublished). As this is the only sequence difference between SqLCV-E and TGMV in the stem-loop (Figure 4), the analysis of the ability of this mutated CR to replicate in the presence of the SqLCV-E and TGMV expression cassettes should directly discriminate between these two models for the interaction between the AL1 protein and 90 nt origin-containing fragment. Fontes and co-workers (1992), using a complementary *in vitro* immunoprecipitation assay, have demonstrated that AL1 is a sequence-specific DNA binding protein; and their studies are consistent with the two models for AL1 recognition to effect viral DNA replication as discussed above.

The Role of BR1 in Viral Host Range

In addition to providing evidence for the importance of viral replication functions in the host range properties of geminiviruses, studies of SqLCV-E and SqLCV-R have also clearly demonstrated the role of viral movement proteins in determining viral host range (Lazarowitz, 1991). The unexpected identification of a naturally occurring mutant genomic component of SqLCV-R, B_R^{td1}, which was no longer infectious for *N. benthamiana*, although it still would infect pumpkin, bean, and squash, makes a persuasive case for the role of the viral BR1 movement protein as a critical determinant of viral host range.

B_R^{td1} has a Host-Dependent Defect in Movement: Although like the highly homologous SqLCV-E, SqLCV-R is infectious for *N. benthamiana*, members of the *Cucurbitaceae* (pumpkin, squash), and *Phaseolus vulgaris* (common green bean); in contrast to SqLCV-E, it replicates less efficiently and has a lower infectivity in the last three host plants than in *N. benthamiana* (Lazarowitz and Lazdins, 1991; Lazarowitz, 1991). Hence, it was surprising when a mutant B_R component, indistinguishable in its restriction map from the wild-type component, was cloned which could no longer infect *N. bethamiana* (Lazarowitz, 1991). Quantitation of viral infectivity in different hosts revealed that B_R^{td1} was clearly not infectious for *N. benthamiana*, and also was partially defective in its ability to infect pumpkin and squash (Figure 6). In contrast to plants agroinoculated with cloned SqLCV-E (A_E+B_E) or wild-type SqLCV-R (A_R+B_R), no disease symptoms were ever observed to develop in *N. benthamiana* agroinoculated with A_R+B_R^{td1}. Squash or pumpkin inoculated with A_R+B_R^{td1} did develop disease symptoms after a delay of several days, and infectivity was ~40. percent that of the wild-type A_R+B_R (Figure 6). In repeated studies, the infectivity of B_R^{td1} in squash or pumpkin has been observed to vary between ~40. percent and ~80. percent that of the wild-type B_R. Interestingly, although there is a delay in the timing of onset of symptoms, the actual appearance of the disease symptoms in pumpkin or squash infected with A_R+B_R^{td1} is indistinguishable from those seen in plants infected with wild-type SqLCV-R.

To determine whether the absence of infectivity in *N. benthamiana* was due to a defect in viral replication or movement, leaf disc studies were done to directly test the ability of B_R^{td1} to replicate in *N. benthamiana* (Lazarowitz, 1991). Leaf discs from *N. benthamiana* were agroinoculated with cloned A_E+B_E, A_R+B_R, or

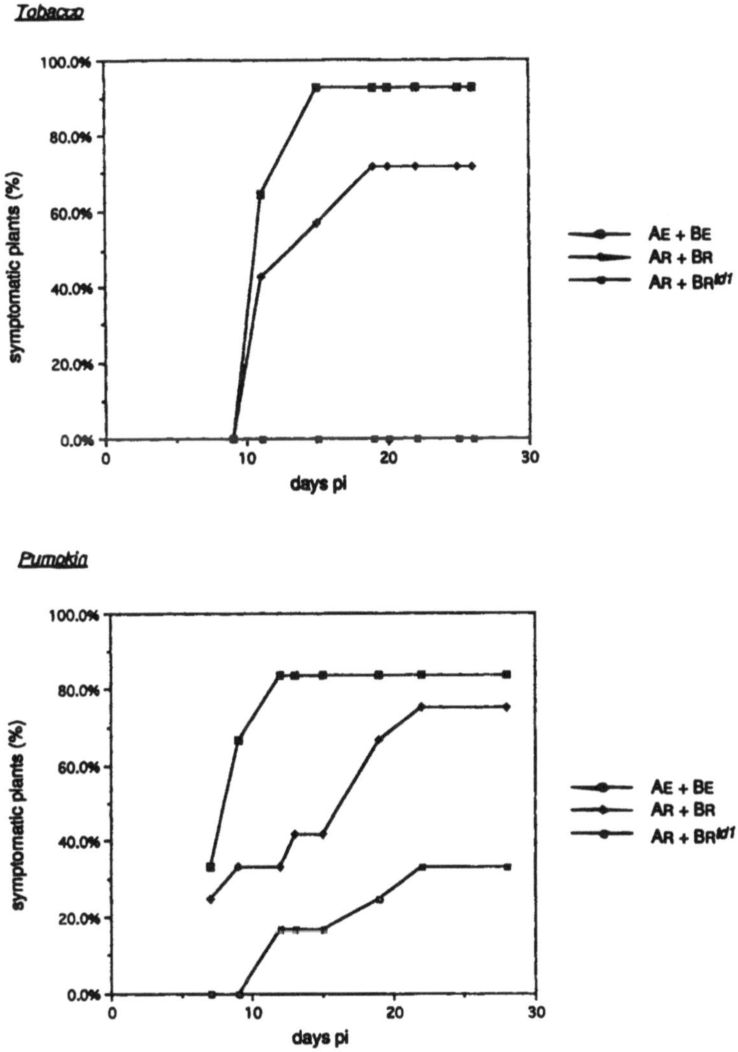

Figure 6. Infectivity of B_R^{td1} in *N. benthamiana* and Pumpkin. Plants were agroinoculated with the cloned DNA components A_E+B_E, A_R+B_R, or A_R+B_R^{td1} as indicated. Inoculations were done on the same day and normalized such that the same amount of *Agrobacterium* was used for each component. All plants were scored for symptoms beginning 5 days post-inoculation (first appearance of symptoms in SqLCV-E inoculated plants) and on each successive day thereafter. Shown are total number of symptomatic plants vs. day post-inoculation.

A_R+B_R^{td1}; and replication of each component was assayed by Southern blotting after 4-6 days. B_R^{td1} was found to replicate to identical levels in the *N. benthamiana* leaf discs as did wild-type B_R and B_E, thus clearly demonstrating that B_R^{td1} has a mutation causing a defect in systemic movement (Lazarowitz, 1991).

B_R^{td1} *is Mutated in the Coding Sequence of the BR1 Movement Protein:* In the absence of any restriction site differences between wild-type B_R and B_R^{td1}, the mutation in B_R^{td1} was mapped by exchanging restriction fragments between the mutant and wild-type genomic components and testing the chimeric components for their infectivity properties on *N. benthamiana*, pumpkin, and squash (Ingham and Lazarowitz, in preparation). In this way, the mutation was mapped to the 450 nt XbaI-BamHI fragment located completely within the NH_2-terminal half of the coding sequence of the movement protein BR1. All chimeric constructs that contained this region from B_R^{td1} were not infectious for *N. benthamiana*, although they did infect pumpkin and squash. Conversely, all chimeric constructs that contained this XbaI-BamHI region from the wild-type B_R component gave infectivity levels of 85 percent - 100 percent on *N. benthamiana*, levels that were indistinguishable from the intact wild-type B_R component. These latter chimeric components were also fully infectious for pumpkin and squash.

Sequence analysis of the XbaI-BamHI fragment from B_R^{td1} has identified three amino acid alterations in the BR1 protein (total length 254 amino acids) of this mutant B component as compared to the wild-type protein (Table 2), with two of these mutations being clustered at positions 9 and 10 at the NH_2-terminus of the protein (Ingham and Lazarowitz, in preparation). Amino acid 9 is changed from a proline to a histidine and amino acid 10 from an asparagine to a tyrosine. Amino acid 98 is changed from an arginine to a cysteine. Within the structural context of the protein, it may be that the alterations at positions 9 and 10 are compensatory ones that could maintain the hydrophilic character of a particular domain at the NH_2-terminus of BR1. The change from an arginine to a cysteine at position 98 would appear to significantly alter the local hydrophilic character at that particular region of the protein. This characterization of B_R^{td1} clearly demonstrates that the viral BR1 movement protein is an important determinant of the host range properties of geminiviruses.

Studies to date have failed to reveal any temperature-sensitive behavior in the mutant BR1 protein encoded by B_R^{td1}. By site-directed

Table 2. Amino Acid Alterations in the BR1 Protein of B_R^{td1}.

amino acid position	wildtype B_R ⟶ mutant B_R^{td1}	
9	P ⟶	H
10	N ⟶	Y
98	R ⟶	C

mutagenesis, each of the three mutations in this protein has been introduced singly and in combination with the other alterations into the wild-type SqLCV-R BR1 protein. These are currently being tested for their infectivity properties on *N. benthamiana*, squash, and pumpkin. The results of these studies will clearly define those alterations in the BR1 protein of B_R^{td1} responsible for its inability to systemically spread in and infect *N. benthamiana*.

Concluding Remarks

The studies of SqLCV reported here have extended our knowledge of the molecular details of geminiviral replication and movement, both of which have been implicated as important determinants of viral host range and pathogenesis (Lazarowitz, 1991; Lazarowitz, in press). With AL1 protein binding and the origin for both ssDNA and dsDNA synthesis being localized to the ~90 nt CR fragment encompassing the canonical stem-loop structure and ~60 nts of 5'-upstream (AL1 proximal) sequence, further site-directed mutagenesis to more precisely define the critical sequence elements within this fragment in addition to the stem-loop structure, and biochemical studies directed at identifying interacting host factors should provide us with a detailed understanding of the mechanism of viral DNA replication. The striking consensus sequence of the stem-loop structure in the geminiviruses infecting dicotyledonous plants (Lazarowitz, 1991; Lazarowitz, in press), as well as the very small size of the viral genome and precedents from other DNA viruses, strongly argue for the involvement of host factors in this process. The identification of such host factors combined with knowledge of the detailed interactions with AL1 protein and the origin to initiate replication could be fundamental in developing strategies for genetically engineering broad-based resistance to these viruses and hold out the potential for the further development of plant expression vectors. Mechanisms by which to engineer resistant plants is of particular interest since engineered coat protein-mediated resistance does not appear to work for the bipartite geminiviruses (Sanderfoot and Lazarowitz, unpublished).

A potentially attractive approach to engineering broad-based resistance to geminivirus infection would be to specifically interfere with viral movement in the plant; and it is in this context that the molecular studies on the movement protein BR1, as well as similar studies of the BL1 movement protein, are important. The bipartite geminiviruses are different from other plant viruses such as TMV in encoding two distinct movement proteins both of which are essential for systemic spread in the plant and in their requirement to move ssDNA, not RNA, cell-to-cell in the plant. Hence, the study of geminiviral movement proteins holds out the potential for providing unique clues concerning the mechanisms by which movement proteins function to facilitate viral spread in the plant and the involvement of these proteins in tissue and host specificity. Our studies of the SqLCV-R mutant B_R^{td1} component clearly demonstrate an important role for the BR1 movement protein in determining viral host range. Knowledge of those amino acid alterations in the B_R^{td1} BR1 protein critical for the inability to systemically move in and infect *N. benthamiana* will provide our first insights into possible epitopes and/or domains in this protein with the potential to interact with host factors. Combined with further site-directed mutagenesis, knowledge of such interactive domains should provide for a rational approach to the introduction of dominant mutations into the protein which would specifically interfere with systemic movement of and, hence, infection by SqLCV and could be more broadly effective for engineering resistance to other geminiviruses.

It is of interest that while the mutation in the BR1 protein of B_R^{td1} drastically affects the host range of SqLCV-R, it does not affect the actual appearance of the disease symptoms in those hosts which are still susceptible to infection by the mutant virus. Recent studies of the expression of the SqLCV-E BR1 and BL1 proteins in transgenic plants are consistent with this finding and lead to the intriguing suggestion that in the bipartite geminiviruses the BR1 protein would appear to be an important determinant of viral host range while the

BL1 protein is important in determining disease symptoms (Pascal *et al.*, in preparation).

Acknowledgements

Different aspects of this research were supported by grants to S. G. Lazarowitz from the National Science Foundation (DMB-8216268) and the National Institutes of Health (R01 AI27449) and funds from the Monsanto Company.

References

Brough, C.L., Hayes, R.J., Morgan, A.J., Coutts, R.H.A., and Buck, K.W. 1988. Effects of mutagenesis *in vitro* on the ability of cloned tomato golden mosaic virus DNA to infect *Nicotiana benthamiana* plants. J. Gen. Virol. 69:503.

Elmer, J.S., Brand, L., Sunter, G., Gardiner, W., Bisaro, D.M., and Rogers, S.G. 1988. Genetic analysis of tomato golden mosaic virus II. The conserved AL1 ORF product is essential for replication. Nucl. Acids Res. 16:7043.

Estessami, P., Callis, R., Ellwood, S., and Stanley, J. 1988. Delimitation of essential genes of cassava latent virus DNA 2. Nucl. Acids Res. 16:4811.

Etessami, P., Saunders, K., Watts, J., and Stanley, J. 1991. Mutational analysis of complementary-sense genes of African cassava mosaic virus. DNA A. J. Gen. Virol. 72:1005.

Fontes, E.P.B., Luckow, V.A., Hanley-Bowdoin, L. 1992. A geminivirus replication protein is a sequence-specific DNA binding protein. The Plant Cell 4:597.

Gardiner, W.E., Sunter, G., Brand, L., Elmer, J.S., Rogers, S.G., and Bisaro, D.M. 1988. Genetic analysis of tomato golden mosaic virus: The coat protein is not required for systemic spread or symptom development. EMBO J. 7:899.

Grimsley, N. and Bisaro, D.M. 1987. Agroinfection. In: *Plant DNA Infectious Agents*. T. Hohn and J. Schell (eds.). Springer-Verlag, Vienna, 87.

Gruss, A. and Ehrlich, S.D. 1989. The family of highly interrelated single-stranded deoxyribonucleic acid plasmids. Microbiol. Rev. 53:231.

Gruss, A., Ross, H., and Novick, R. 1987. Functional analysis of a palindromic sequence required for normal replication of several staphylococcal plasmids. Proc. Natl. Acad. Sci. U.S.A. 84:2165.

Haley, A., Zhan, X., Richardson, K., Head, K., and Morris, B. 1992. Regulation of the activities of African cassava mosaic virus promoters by the AC1, AC2, and AC3 gene products. Virology 188:905.

Hamilton, W.D.O., Bisaro, D.M., Coutts, R.H.A., and Buck, K.W. 1983. Demonstration of the bipartite nature of the genome of a single-stranded DNA plant virus by infection with the cloned DNA components. Nucl. Acids Res. 11:7387.

Hamilton, W.D.O., Stein, V.E., Coutts, R.H.A., and Buck, K.W. 1984. Complete nucleotide sequence of the infectious cloned DNA components of tomato golden mosaic virus: Potential coding regions and regulatory sequences. EMBO J. 3:2197.

Hanley-Bowdoin, L., Elmer, J.S., and Rogers, S.G. 1990. Expression of functional replication protein from tomato golden mosaic virus in transgenic tobacco plants. Proc. Natl. Acad. Sci. U.S.A. 87:1446.
Heidekamp, F., Bass, P.D., and Jansz, H.S. 1982. Nucleotide sequence at the ϕX gene A protein cleavage site in replicative form I DNA's of bacteriophage U3, G4 and a3. J. Virol. 42:91.
Ingham, D.J. and Lazarowitz, S.G. In preparation.
Lazarowitz, S.G. 1987. The molecular characterization of geminiviruses. Plant Mol. Biol. Rep. 4:177.
Lazarowitz, S.G. 1991. Molecular characterization of two bipartite geminiviruses causing squash leaf curl disease: Role of viral replication and movement functions in determining host range. Virology 180:70.
Lazarowitz, S.G. Geminiviruses: Genome structure and gene function. In: *CRC Critical Review in Plant Sciences.* Vol. 11. B. V. Conger (ed.). CRC Press, Inc. Boca Raton, Florida. In press.
Lazarowitz, S.G. and Goodlove, P.E. Unpublished.
Lazarowitz, S.G. and Lazdins, I.B. 1991. Infectivity and complete nucleotide sequence of the cloned genomic components of a bipartite squash leaf curl geminivirus with a broad host range phenotype. Virology 180:58.
Lazarowitz, S.G., Pinder, A.J., Damsteegt, V.D., and Rogers, S.G. 1989. Maize streak virus genes essential for systemic spread and symptom development. EMBO J. 8:1023.
Lazarowitz, S.G., Wu, L.C., Rogers, S.G., and Elmer, J.S. 1992. Sequence-specific interaction with the viral AL1 protein identifies a geminivirus DNA replication origin. The Plant Cell 4:799.
Pascal, E.J., Goodlove, P.E., Wu, L.C., and Lazarowitz, S.G. In preparation.
Revington, G.N., Sunter, G., and Bisaro, D.M. 1989. DNA sequences essential for replication of the B genome component of tomato golden mosaic virus. The Plant Cell 1:985.
Sanderfoot, A.A. and Lazarowitz, S.G. Unpublished.
Stanley, J. and Gay, M.R. 1983. The nucleotide sequence of Cassava Latent Virus DNA. Nature 301:260.
Stenger, D.C., Revington, G.N., Stevenson, M.C., and Bisaro, D.M. 1991. Replicational release of geminivirus genomes from tandemly repeated copies: Evidence for rolling circle replication of a plant viral DNA. Proc. Natl. Acad. Sci. U.S.A. 88:8029.
Sunter, G. and Bisaro, D.M. 1991. Transactivation in a geminivirus: AL2 gene product is needed for coat protein expression. Virology 180:416.
Sunter, G., Hartitz, M.D., Hormudzi, S.G., Brough, C.L., and Bisaro, D.M. 1990. Genetic analysis of tomato golden mosaic virus: ORF AL2 is required for coat protein accumulation while ORF AL3 is necessary for efficient DNA replication. Virology 179:69.
Ugaki, M., Ueda, T., Timmermans, M.C.P., Vieira, J., Elliston, K.O., and Messing, J. 1991. Replication of a geminivirus derived shuttle vector in maize endosperm cells. Nucl. Acids Res. 19:371.

Properties of Genes Influencing Barley Stripe Mosaic Virus Movement Phenotypes

R. G. K. Donald
Department of Biology
University of Pennsylvania
Philadelphia, Pennsylvania 19104

I. T. D. Petty
Department of Microbiology
North Carolina State University
Raleigh, North Carolina 27695

H. Zhou and A. O. Jackson
Department Of Plant Biology
University Of California
Berkeley, California 94720

The genome of barley stripe mosaic virus (BSMV) consists of three RNA's (α, β, and γ) which encode seven polypeptides designated αa, βa, βb, βc, βd, γa, and γb. Molecular genetic analyses reveal that RNA replication in protoplasts requires only the αa and γa encoded proteins, which contain helicase and RNA polymerase motifs respectively. The βb, βc, and βd genes are required for infection of plants; but the capsid protein (βa) and the γb genes are not absolutely essential for systemic spread of the virus. The βb protein, the major nonstructural protein synthesized during the early stages of infection, has a helicase motif and exhibits cooperative RNA binding activity *in vitro*, which is selective in that it has affinity for double-stranded and single-stranded RNA's but not for tRNA or DNA substrates. Mutational analyses indicate that βb has multiple RNA binding sites and that single site changes within the helicase motif abrogate the ability to infect plants. The γb protein, which is a major pathogenicity determinant, has two cysteine-rich motifs (C1 and C2) and has single-stranded RNA binding activity *in vitro*. Single nucleotide changes within the C1 and C2 motifs and within an adjacent basic region of γb elicit several specific changes in the synthesis and cytoplasmic localization of γb that correlate with the symptom phenotypes elicited upon infection of barley.

Introduction

Induction of virus diseases in plants depends on the ability of the virus to spread from initially infected cells throughout the plant. The systemic infection process can be separated into two phases: 1) short distance cell-to-cell movement through plasmodesmata and 2) long distance transport through the vascular system. Although the actual physical entities and their mechanisms of movement have not been determined for any virus, the experimental evidence now available demonstrates that the number and structure of genes involved in movement vary considerably with different virus groups (Hull, 1989; 1991). These studies show that despite the small number of genes encoded by simple RNA and DNA viruses, the elements controlling replication, pathogenesis, and systemic invasion are complex; and many fundamental questions remain to be answered about even the best characterized virus systems. With the hordeivirus, barley stripe mosaic virus (BSMV), we have been able to dissect several elements that are specifically involved in replication, in cell-to-cell movement, and in long distance virus transport and have initiated a series of biochemical characterizations to identify the functions and roles of the genes involved in these processes. In the present overview, we will discuss identification of these factors and speculate on how they may influence replication and movement of BSMV.

A number of contributions have been made in different labs over the past 5 years that have special relevance to mechanisms involved in movement of viruses (for more detailed review, see Hull, 1991; Deom *et al.*, 1992; Maule, 1991). Several distinct groups of plant viruses with diverse movement elements have been proposed based on the number of proteins participating in movement and their amino acid sequence relationships (Atabekov and Taliansky, 1990). According to a recent extensive protein sequence comparison (Koonin *et al.*, 1991) the clearest groupings of putative movement proteins can be obtained by assigning them to two broad families of related proteins. Movement proteins included in Family 1 are encoded by the tobamo-, tobra-, como-, and the DNA caulimo- and geminiviruses. A second distantly related family includes movement proteins encoded by the bromo-, cucumo-, ilar-, and dianthoviruses. The hordei-, furo-, potex-, and carlaviruses form a third group, since three proteins encoded within a triple gene block each appear to be required for systemic movement, yet none of these contains detectable amino acid sequence relatedness with movement proteins belonging to Families 1 and 2 (Koonin *et al.*, 1991).

In TMV infected plants, considerable genetic, cytological, and biochemical evidence now provides a persuasive argument that the 30K protein is involved in cell-to-cell movement through plasmodesmata. Genetic analyses have shown that the 30K protein is essential for cell-to-cell movement (Meshi *et al.*, 1987) and that

TMV 30K mutants can be complemented in transgenic plants expressing the 30K protein (Deom *et al.*, 1987). Cytological analyses have demonstrated that the 30K protein is localized in plasmodesmata (Atkins *et al.*, 1991) and can increase the plasmodesmatal size exclusion limit (Wolf *et al.*, 1989). Thus, these studies suggest that one function of the 30K protein is to facilitate movement of some "virus-specific" entity through plasmodesmata, perhaps in part by modifying secondary plasmodesmata (Ding *et al.*, 1992). Another line of evidence based on a series of *in vitro* biochemical studies have demonstrated that the 30K protein is a single-stranded nucleic acid binding protein (Citovsky *et al.*, 1990). The 30K protein is able to bind single-stranded RNA or DNA nonspecifically and cooperatively along its length; and, in doing so, it stretches the nucleic acid into a strand with a sufficiently small diameter that it theoretically could move through the enlarged plasmodesmatal pores generated as a consequence of virus infection (Citovsky *et al.*, 1992). Similar although less extensive genetic, cytological, and biochemical studies have been conducted with the caulimoviurses (Maule, 1991; Citovsky *et al.*, 1992). These studies *in toto* thus provide a model suggesting that virus movement proteins such as those encoded by TMV and caulimoviruses may have a dual function by forming a nucleoprotein complex and chaperoning the complex through plasmodesmatal pores that have become enlarged as a result of interaction with the movement proteins (Citovsky and Zambryski, 1991). However, we stress that this scenario undoubtedly varies with different viruses since considerable diversity exists among groups, both in the requirement for different numbers of putative movement proteins and for a functional capsid protein requirement for movement.

The 3a gene of the bromoviruses (DeJong and Ahlquist, 1991) appears to be functionally analogous to the TMV 30K protein, although the two classes of proteins share only distant amino acid sequence homology (Koonin *et al.*, 1991). Direct evidence for this hypothesis has recently been obtained by a substitution of the TMV (cowpea strain) 30K protein for the 3a "movement" protein of cowpea chlorotic mottle virus (CCMV). In this instance, despite the fact that the TMV strain is unable to complement CCMV with deletions in the 3a gene when the two viruses are coinoculated to plants, systemic movement of the hybrid bromovirus did occur upon substitution of the tobamovirus 30K protein for the 3a gene (DeJong and Alhquist, 1992). These observations extend other studies demonstrating that functional transfers of the 30K gene can be made between closely related TMV strains (Nejidat *et al.*, 1991) and that TMV can complement movement of defective strains of tobacco rattle virus with lesions in the 29K gene (Ziegler-Graff *et al.*, 1991). This is compatible with findings that binding of the movement proteins to viral RNA's does not require recognition of specific sequences located on individual RNA's (Citovsky and Zambryski, 1991) and provides support for the concept that the cellular proximity of replicating viruses may be a major factor in the ability of one virus to complement movement of another.

Several viruses exemplified by BSMV lack a movement protein with detectable amino acid sequence homology to the Family 1 or 2 movement proteins but require at least three proteins organized in a "triple gene block" to facilitate both localized and long distance movement. These viruses have major differences in their genome organization and clearly belong to different taxonomic groups. BSMV is a tripartite RNA virus with a genome organization (Figure 1) distinct from the tricornaviruses (Jackson *et al.*, 1989; Jackson *et al.*, 1991a; 1991b).

Beet necrotic yellow vein virus (BNYVV) has a bipartite genome organization, with two or three smaller accessory RNA's specific to individual strains which contribute to vector specificity and host determinants (Gilmer *et al.*, 1992; Niesbach-Klösgen *et al.*, 1990), whereas the potexviruses (Morozov *et al.*, 1989) are monopartite RNA viruses. In the case of these viruses (Gilmer *et al.*, 1992; Petty *et al.*, 1990a; Beck *et al.*, 1991; Richards and Tamada, 1992), construction of deletion or frame shift mutations designed to inactivate the proteins have demonstrated that each "triple gene block" protein is required for systemic invasion or symptom production in inoculated plants.

In addition to interaction with a putative movement protein, other virus encoded factors may have major roles in mediating cell-to-cell and long distance virus movement in plants. For example, many viruses require the capsid protein for efficient movement, whereas other viruses exemplified by BSMV (Petty and Jackson, 1990b), the tobraviruses (Hamilton and Baulcombe, 1989), and the tombusviruses (Dalmay *et al.*, 1992) are able to establish systemic infections in the absence of a functional coat protein. In contrast, all four encoded genes of BMV and other bromoviruses are required for efficient systemic infection (DeJong and Ahlquist, 1991); and deletions beyond the first 11 amino acids at the amino-terminus of the BMV coat protein severely compromise long distance movement of the virus

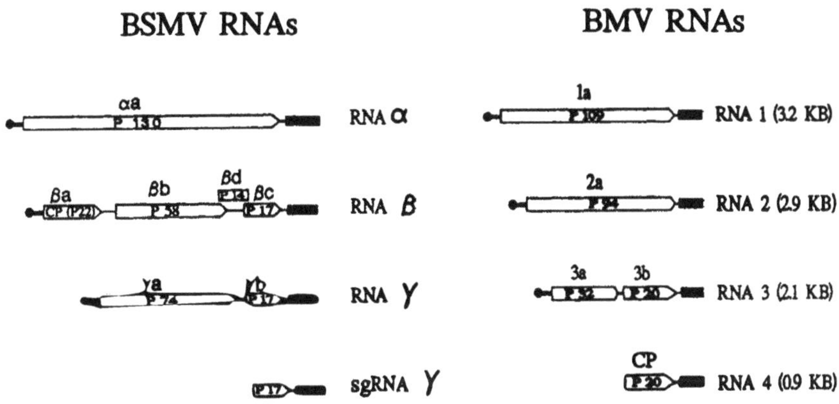

Figure 1. Comparison of the Organization of the Genomic RNA's of Barley Stripe Mosaic Virus (BSMV) and Brome Mosaic Virus (BMV). The filled circles represent the 5'-cap structures preceding the leader sequences, open rectangles represent open reading frames (ORF's), and the solid rectangle represents the 3'-terminal tRNA-like structure. Substantial differences exist between the genetic organization of these two viruses since BSMV encodes seven major proteins whereas BMV encodes four. However, as discussed in the text, the BSMV αa and γa proteins appear to be analogous to the BMV 1a and 2a proteins respectively in sharing sequence domains associated with replicase proteins and their requirement for replication. None of the BMV ORF's have obvious amino acid sequence relatedness to the BSMV RNAβ or the γb ORF's. Reproduced from Jackson *et al.*, 1991a.

in barley plants (Sacher and Ahlquist, 1989). Similar studies with cowpea chlorotic mottle virus (CCMV) mutants have indicated that the coat protein (3b) is required for systemic movement in cowpea leaves but that very limited localized transport can occur in its absence (Allison *et al.*, 1990). In addition, TMV strains containing mutations in the coat protein gene move slowly and erratically in tobacco hosts (Siegel *et al.*, 1962; Dawson, 1992). These studies, thus, suggest that in addition to the so called "movement proteins," efficient systemic movement of some viruses may require either encapsidated virus or additional undefined functions supplied by the capsid protein or RNA sequences encoding the capsid protein.

At least one group of viruses, the comoviruses, may require whole virus particles for cell-to-cell movement. Molecular genetic evidence has incriminated the overlapping 48 and 58K cowpea mosaic virus (CPMV) proteins as well as the capsid proteins in localized movement. With CPMV, a striking array of tubules are associated with the plasmodesmata of infected mesophyll cells and also are found in protoplasts. Microscopy studies suggest that the tubules "grow" into neighboring cells and that the tubules contain substantial amounts of the 48/54K "movement" proteins plus normally appearing CPMV virions (van Lent *et al.*, 1991). Nevertheless, since little direct evidence is available to implicate virus particles in cell-to-cell movement and since, as discussed below, BSMV clearly does not require intact particles for localized or systemic establishment of infections, the CPMV system will not be discussed further.

BSMV (Petty and Jackson, 1990b) and the tobraviruses (Hamilton and Baulcombe, 1989) clearly fall into a different class from those viruses requiring some function of the capsid protein in movement because coat protein deficient mutants are able to quickly establish systemic infections. In BSMV, efficient movement occurs in barley plants when the coat protein is deleted, when frame shift mutations are introduced into the coat protein, or when the capsid protein initiation codon is altered to prevent initiation of translation of the coat protein (Petty and Jackson, 1990b). However, some RNA determinants within the coat protein cistron may be necessary for efficient movement in the experimental dicot host, *Nicotiana benthaminia* (Petty and Jackson, 1990b).

Even though the "triple gene block" proteins βb, βc, and βd (Figure 1) are clearly the primary determinants involved in cell-to-cell movement of BSMV, we (Petty *et al.*, 1990a) have also obtained evidence that the levels of replicase synthesis or the form of the replicase may be secondary determinants that influence long distance spread in both a tobacco host and in the ability of γb null mutants to establish systemic infections in barley. In the case of *N. benthamiana*, the Type strain of BSMV replicates to normal levels in inoculated leaves but fails to move into uninoculated leaves, whereas the ND18 strain elicits an aggressive systemic infection. We have mapped the determinant encoding this phenotype to a short open reading frame (ORF) in the 5'-leader sequence of RNAγ, which our molecular genetic evidence suggests is a translational attenuator of γa replicase protein synthesis (Petty *et al.*, 1990a). Similarly, barley plants infected with γb deletion mutants have a strain specific infection phenotype. Whereas the ND18 strain deletion mutant exhibits erratic mottled systemic symptoms, the Type strain containing identical deletion mutants is unable to establish systemic infections. Removal of a direct nucleotide sequence repeat which encodes a 123 amino acid extension near the amino-terminus of the γa replicase protein permits the Type strain γb mutant to establish infections indistinguishable from those of the ND18 strain deletion mutant (Petty *et al.*, 1990b). These results suggest that both the levels of synthesis and structural differences within the replicase proteins can have a profound influence on the ability of the virus to establish systemic infections.

The results of the studies mentioned above provide a compelling argument that BSMV is a unique tripartite RNA virus which differs markedly from many other viruses in the number and organization of genes involved in cell-to-cell movement and in its requirements for localized and long distance movement throughout the plant. The section below provides an overview of the structure and properties of the BSMV genome and the role of the individual viral determinants contributing to infection of plants.

Organization of the BSMV Genome

Although BSMV is a tripartite RNA virus, it differs in several major respects from the other tripartite viruses, which include the bromoviruses, the cucumoviruses, and the ilarviruses (Jackson *et al.*, 1989; Atabekov and Dolja, 1986). The tricomaviruses have icosahedral particle morphology, with the exception of alfalfa mosaic virus (AlMV) which has a class of bacilliform particles, whereas BSMV virions have helical morphology. In addition, the genome of BSMV has little similarity to those of the tricornaviruses in the number and arrangement of encoded genes. As shown in Figure 1, BSMV encodes seven polypeptides, whereas four are encoded by BMV (Ahlquist *et al.*, 1987). However, some resemblance exists between BSMV RNAα and RNA 1 of BMV since each encodes a similarly-sized replicase component with putative methyltransferase and helicase domains that are related to the alphavirus nsP1 and nsP2 proteins (Ahlquist *et al.*, 1987; Gustafson *et al.*, 1989).

In contrast, BSMV RNAβ, which encodes four proteins, has little relationship in its organization to RNA's of other tripartite RNA viruses (Gustafson and Armour, 1986); and the RNAβ encoded gene products have no discernable similarities to those of BMV RNA 2 or

3. The BSMV capsid protein is encoded by the (βa) ORF adjacent to the 5'-terminus of RNAβ, whereas the coat protein gene of BMV is the 3'-proximal gene of RNA3 and is expressed via synthesis of a subgenomic (sg) RNA. The BSMV capsid protein gene is separated from the βb ORF by a 117 nt intergenic region that contains a promoter for transcription of a subgenomic RNA. The βb, βc, and βd ORF's, which presumably are translated from the subgenomic RNA, are tightly organized to form a "triple gene block" whereby βd overlaps βb and βc. This portion of the BSMV genome has no analogy with those of other tripartite viruses but, as discussed above, is similar in organization to the overlapping gene blocks found in the furoviruses (Richards and Tamada, 1992); potexviruses (Morozov *et al.*, 1989; Forster *et al.*, 1988; Huisman *et al.*, 1988); and potato virus M, a carlavirus (Rupasov *et al.*, 1989). Although BSMV RNAγ is similar to BMV RNA 2 in that both RNA's encode proteins with putative replicase motifs, they differ markedly in organization, since RNAγ is bicistronic (Gustafson *et al.*, 1987) and BMV RNA 2 is monocistronic (Figure 1). Both BMV RNA 3 and BSMV RNAγ express their 3'-terminal gene products via sgRNA's (Atabekov and Dolja, 1986; Jackson *et al.*, 1983). However, the BMV coat protein is translated from sgRNA 4, whereas BSMV sgRNAγ encodes the γb regulatory gene product which appears to be unrelated to the BMV gene products.

Another unusual feature of BSMV is the variation in size and number of RNAγ species present in different strains (McFarland *et al.*, 1983). For example, sequence analysis of ND18 reveals a single RNAγ of about 2.8 KB, but RNAγ of some other strains, such as the Type strain, is about 3.15 KB because of a directly repeated sequence block of more than 350 nt near the 5'-terminus (Gustafson *et al.*, 1987). Additional complexity exists in other strains, such as the Argentina Mild strain, which contains three RNAγ species (McFarland *et al.*, 1983). Two of these correspond in size to those of the Type and ND18 strains, whereas the third appears to be defective and has an internal deletion of 185 nt corresponding to a deletion within the carboxy-terminal portion of the γa protein (Kozlov *et al.*, 1989).

BSMV genomic RNA's share several common features in their noncoding regions. Each of the genomic RNA's has a 7-methylguanosine cap at its 5'-terminus and contains the initial sequence m^7GpppGUA (Gustafson *et al.*, 1989; Gustafson and Armour, 1986; Gustafson *et al.*, 1987). The remaining nucleotides forming the 5'-leaders of BSMV RNA's appear to have little obvious sequence similarity, but they undoubtedly contain important recognition features necessary for initiating and regulating translation and replication.

Immediately downstream of their coding regions, each BSMV RNA contains a highly conserved 238 nt 3'-terminal tRNA-like structure that can be aminoacylated *in vitro* (Atabekov and Dolja, 1986) and *in vivo* (Loesch-Fries and Hall, 1982). This moiety has some structural similarities to the 3'-termini of BMV RNA's, but the nucleotide sequences of the two structures share little relatedness (Kozlov *et al.*, 1984). A second notable feature of the 3'-noncoding region is the presence of a polyadenylate (poly A) sequence of variable length that separates the coding region of each RNA from the tRNA-like structure. The first two adenylate residues of the poly A tract form part of the UAA codons which terminate the 3'-proximal ORF's of each of the RNA's. Tertiary interactions resulting from highly ordered pseudoknot structures within this region of the genome (Pleij *et al.*, 1987) probably function to stabilize configurations necessary for interaction with enzymes mediating replication.

Involvement of BSMV Genes in Replication and Movement

We have constructed recombinant DNA clones from several different strains of BSMV from which infectious RNA transcripts can be derived (Petty *et al.*, 1989). Our experiments utilizing these transcripts reveal that although all three genomic components are required for systemic invasion of plants, only RNA's α and γ are necessary for RNA replication in protoplasts (Petty *et al.*, 1990b). In order to obtain more information concerning the roles of individual gene products in the replication and infection processes, deletion or point mutations were made in each ORF to inactivate each of the encoded proteins. The most important findings of these experiments and their relevance to the function of the individual gene products are discussed below.

αa Protein: The 130 kD (kilodalton) αa protein of BSMV is thought to be an integral component of the viral replicase since it has significant sequence relatedness to analogous proteins involved in replication of RNA (Gustafson *et al.*, 1989). A motif near the amino-terminus of αa (Figure 2) is similar to a methyltransferase domain found in the analogous protein of alphaviruses (Mi *et al.*, 1989) and in several plant RNA viruses belonging to the alphavirus superfamily (Rozanov *et al.*, 1992). Experimental evidence derived from Sindbis virus mutants (Mi and Stollar, 1990) and *in vitro* activity of recombinant

proteins (Mi and Stollar, 1991) suggest that this motif is involved in cap methylation during RNA replication.

The carboxy-terminal portion of the αa protein also contains a motif with sequences related to helicases (Hodgman, 1988) which consists of at least six distinct blocks of amino acids that are conserved among several different groups of RNA viruses (Gorbalenya and Koonin, 1989). Deletion analysis has demonstrated that RNAα is essential for replication of RNA in protoplasts (Petty *et al.*, 1990b). We have also recently been able to detect the presence of the αa protein in BSMV infected barley by immunoblot analyses (Donald *et al.*, 1993). The αa protein first appears in the cytosolic, membrane, and cell wall fractions at about 4 days post inoculation when the symptoms first become evident; and it declines gradually in concentration as infection progresses. These results thus provide additional evidence supporting computer derived predictions that the αa protein is an essential replicase component.

βa (Coat Protein): The 22 kD coat protein of BSMV increases in abundance throughout the infection cycle and becomes one of the most abundant proteins in infected plants (Jackson *et al.*, 1983). The coat protein is encoded by the 5'-proximal ORF on RNAβ from which it appears to be directly translated (Figure 2); and, in this regard, the BSMV coat protein is similar in location to the capsid protein of BNYVV. The capsid protein has an obvious role in formation of virus particles; but in contrast to many viruses such as BMV (Sacher and Ahlquist, 1989), the BSMV capsid protein is dispensable for systemic movement in plants. Barley plants develop systemic infections within 3 to 4 days post inoculation (DPI) with RNAβ mutants containing several different deletions in the coat protein cistron, a stable point mutation that destroys the coat protein initiation codon, or frame shift mutations that truncate the coat protein gene (Petty and Jackson, 1990b). Infections with most of these mutants result in particularly severe symptoms in barley typified by stunting, a pronounced mosaic, and lack of symptom remission which may be followed by a lethal necrosis (Figure 3).

Of particular interest in this regard are previous observations (Pring, 1971; Brakke *et al.*, 1988) that a considerable amount of noncoat protein encapsidated RNA is present in wild-type infections at about the time mosaic symptoms appear. As the initial "shock" phase remisses and a recovery phase characterized by less severe "chronic" symptoms appears, the unencapsidated RNA disappears and presumably becomes encapsidated. Therefore, these correlations

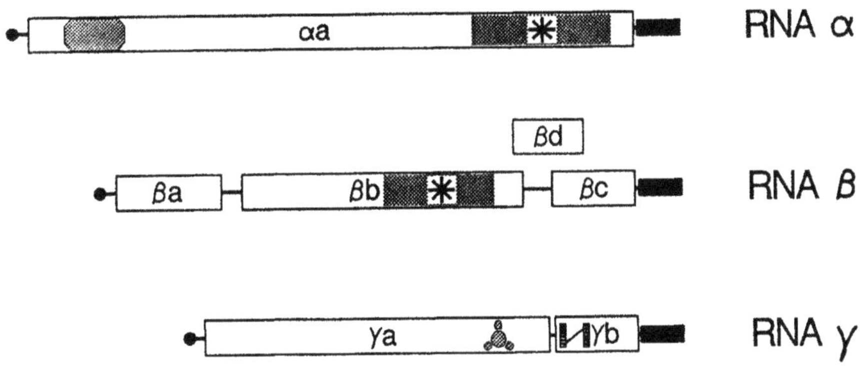

Figure 2. The Structural Organization of the BSMV Genome. The filled circle, open rectangles, and solid rectangles represent the 5'-cap structure, the open reading frames, and the 3'-terminal tRNA-like structure. RNAα encodes a single protein, αa with a putative methyltransferase domain (▢) near the amino-terminus and a nucleotide binding motif (■✱■) near the carboxy-terminus. RNAβ encodes four proteins. These include βa, the coat protein; βb, the major nonstructural protein, which also contains a nucleotide binding motif (■✱■); βc, the 17 kD 3'-proximal gene, which is separated from the βb ORF by 173 nt and terminates with a UAA to initiate the short poly A tract separating the 238 nt tRNA tract from the coding regions; and βd, a 14 kD polypeptide which overlaps the βb and βc ORF's. RNAγ, which varies in size among BSMV strains due to a tandemly duplicated region near the 5'-terminus, encodes two polypeptides. The γa polypeptide contains the GDD domain (⁂) that shares sequence relatedness with other viral proteins thought to be involved in RNA replication. The 17 kD γb protein, which is translated from a subgenomic RNA, contains a cysteine-rich region (⊬) and can affect the expression of genes encoded by RNAβ (Petty *et al.*, 1990b). Reproduced from Jackson *et al.*, 1991b.

suggest that the presence of the coat protein or possibly the encapsidation process may be essential for recovery of the plant from the initial severe shock phase.

Although frameshift and point mutations that destroy the βa ORF clearly demonstrate that the coat protein is dispensable for systemic infection (Figure 4), the βa deletion mutants vary in their ability to establish systemic infections in *N. benthamiana* when coinoculated

Figure 3. **Systemic Symptoms Elicited in Barley by Infection with Wild-Type BSMV and a Coat Protein Deficient Mutant. Left: plants infected with α, β, and γ transcripts derived from the ND 18 strain. Right: plants infected with α and γ transcripts plus transcripts derived from the ΔCP mutant, which contains a site directed mutation converting the βa initiation codon to an AUC codon. Western blot analysis has indicated that the inoculated plants synthesize no detectable coat protein (Petty and Jackson, 1990b). Reproduced from Jackson *et al.*, 1991b.**

with an intact RNAγ or to systemically infect barley in the presence of γb gene deletions (Petty *et al.*, 1990a).

Some of the mutants illustrated in Figure 4 are stable during replication in barley plants, but other mutants undergo deletions *in vivo*. Additional evidence from our studies (Petty *et al.*, 1990a) indicate that although the coat protein *per se* is dispensable, a part of the RNA sequence encompassing the coat protein gene is necessary for infection in dicot hosts as well as in barley inoculated with γb null mutants. Since we have been unable to identify any open reading frames within this region of the genome, these results suggest that the RNA itself contains cryptic features that contribute to the ability to infect *N. benthamiana*, *C. amaranticolor* and perhaps other hosts as well (Petty and Jackson, 1990b).

βb Protein: A disease specific protein of about 60 kD, which appears at early stages of infection and declines as the chronic stages of infection develop, has previously been observed in BSMV-infected barley (Jackson *et al.*, 1983; Brakke *et al.*, 1988). We have shown that this protein is the βb gene product by demonstrating that antibodies raised against the protein react specifically with a recombinant protein formed by fusion of βb to the carboxy-terminus of β-galactosidase (Petty and Jackson, 1990a). Our recent serological analyses (Donald *et al.*, 1993) have shown that the βb protein appears in highest amounts in the cytosolic and the cell wall fractions as symptoms develop, but that it declines precipitously in concentration as infection progresses.

The available evidence suggests that the βb protein has distinct analogies to the larger triple gene block protein encoded by BNYVV and members of the potexvirus group. The βb protein shares considerable amino acid sequence homology with the 42 kD protein of BNYVV as well as the 25 and 26 kD proteins of the potexviruses, potato virus X, and white clover mosaic virus (Morozov *et al.*, 1989; Forster *et al.*, 1988; Huisman *et al.*, 1988). A particularly striking feature is the presence of a nucleotide binding motif (Figure 5) similar to that found on the αa replicase-associated protein.

Deletions within the βb protein (Figure 6) and the single site amino acid mutations within the conserved helicase motif that are illustrated at the bottom of Figure 5 prevent systemic movement by BSMV. Thus, the presence of the helicase motif and the requirement of βb for systemic movement suggest that βb and the related analogues on other viruses with the hallmark "triple gene block" may be involved in some aspect of nucleic acid metabolism. Our (Donald and Jackson, unpublished data) recent results have verified this possibility with βb derivatives expressed in *Escherichia coli*. We have used *in vitro* gel shift analyses which demonstrate that the recombinant βb protein has nucleic acid binding activity for both single- and double-stranded RNA. Significantly, the βb protein does not bind to single- or double-stranded DNA and our RNAse protection data suggest that it has a stronger binding affinity for double-stranded than for single-stranded RNA. These studies thus extend other observations in infected plants where the BSMV βb protein is associated with non-encapsidated viral RNA to form heterodispersely sedimenting nucleoprotein particles (Brakke *et al.*, 1988). Since deletion analyses reveal that the βb protein is essential for systemic infectivity, binds to nucleic acids *in vitro*, and is localized in a cell wall

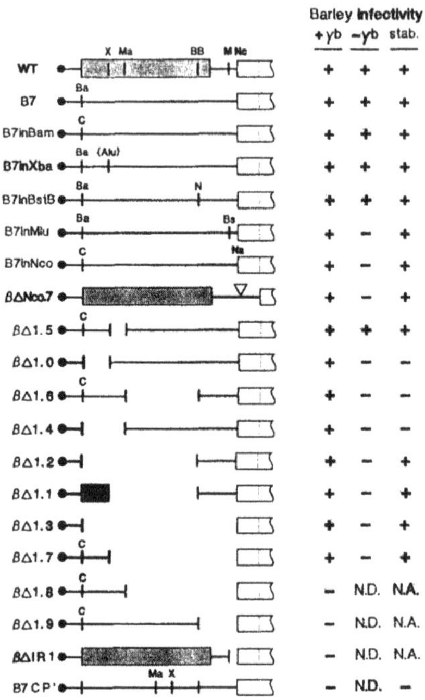

Figure 4. Summary of the Mutations in the Coat Protein Gene and the First Intercistronic Region of BSMV RNAβ and the Ability of the Mutants to Infect Barley. The figure (modified from Petty and Jackson (1990b)) illustrates about 900 nt of the 5'-terminus of RNAβ. The solid circle indicates the 5'-cap, the shaded rectangle indicates the expressed portion of the coat protein gene, and the open rectangle represents the βb gene. The solid line within the βa gene indicates that these regions are not expressed due to mutations that destroy the open reading frame. The designated restriction enzyme sites (Ba, BamHI; BB, BstBI, Bs, BssHII; C, ClaI; M, MluI; Ma, MamI; N, NruI; Nc, NcoI; Ns, NsiI; and X, XbaI) were used for mutation according to Petty and Jackson (1990b). The B7 mutant was constructed by oligonucleotide site-directed mutagenesis, converting the βa initiation codon to AUC. Insertional frameshift mutants were constructed from B7. For example, "B7inBam" represents restriction of B7 with BamHI, filling-in of the ends with the Klenow fragment of DNA polymerase and religation of the blunt ends. The βΔ mutants represent deletions of various sizes. B7CP' represents an inversion of the coat protein sequence between the Cla I (C) and BstBI (BB) restriction sites. A (+) specifies an infectious stable mutant, a (-) an uninfectious or unstable mutant, ND was not done and NA are not applicable.

fraction, βb may be a functional analogue of the TMV 30K protein which functions to chaperone dsRNA rather than ssRNA from cell-to-cell during infection.

In addition to the deleterious effects of the βb mutations on systemic movement, three deletion mutations upstream of βb also appear to affect movement since systemic infections fail to develop in barley plants inoculated with transcripts containing these mutations. These deletions truncate a portion of the coat protein gene and also affect the putative βb subgenomic RNA promoter (Figure 4), which most likely prevents transcription of the βb mRNA. Other deletions within the βa-βb intercistronic region which fuse the 5'-leader sequence directly to the βb ORF are viable, and barley plants infected with these mutants accumulate detectable amounts of the βb protein (Petty and Jackson, 1990b). One of these mutations, which is unusual in that it elicits attenuated symptoms on barley, contains

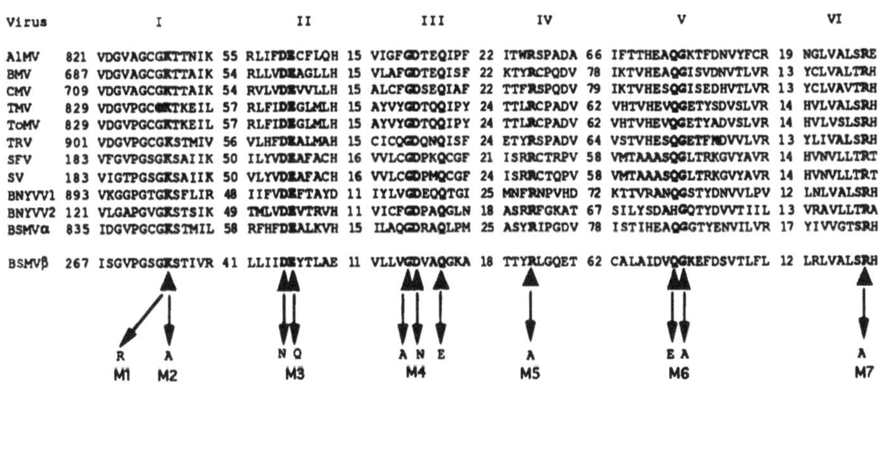

Figure 5. The Putative Nucleotide-Binding Motifs of the BSMV αa and βb Proteins Compared with Related Regions of Other Alphaviruses. The motifs are derived from several sequenced viruses including the plant viruses, alfalfa mosaic virus (AlMV), barley stripe mosaic virus (BSMV), beet necrotic yellow vein virus (BNYYV), brome mosaic virus (BMV), cucumber mosaic virus (CMV), tobacco mosaic virus (TMV), tomato mosaic virus (ToMV), and tobacco rattle virus (TRV), and the animal viruses Sindbis virus (SV) and Semliki forest virus (SFV). Arrows and designations below the sequence indicate the site specific amino acid mutations. Modified from Jackson *et al.*, 1989.

a longer region of 5'-sequence fused to the βb ORF. Barley plants infected with this mutant did not contain detectable amounts of the βb protein in western blots at 6 DPI when plants are normally analyzed, but small amounts were subsequently detected at 8 DPI (Petty and Jackson, 1990b). Thus, our results demonstrate that the βb protein is essential for systemic infection, that single amino acid mutations within the helicase motif can destroy the ability of BSMV to undergo localized movement, and that reduced levels of βb can apparently influence the infection phenotype. In summary, our experiments and studies by others demonstrating that the "triple gene block" proteins of white clover mosaic potexvirus (Beck *et al.*, 1991) and BNYVV (Gilmer *et al.*, 1992) are essential for establishment of systemic infections in plants provide direct experimental evidence for the contribution of the βb homologues to localized movement in plants.

βc Protein: Little is known about the expression or function of the 17 kD protein encoded by the βc gene, and we have been unable to detect the presence of this protein in infected barley by immunoblot analyses (Donald *et al.*, 1993). However, a single mutation within the βc ORF (Figure 6) indicates that the gene product is required for infectivity in barley, *N. benthamiana*, and *C. amaranticolor* (Petty and Jackson, 1990b).

The βc ORF is separated from the βb ORF by a 173 nt sequence, which forms part of the βd ORF; and the first two nucleotides of the poly A tract form the UAA βc termination codon. Despite the fact that βc has overlapping counterparts in furoviruses and potexviruses, it has no obvious sequence homology to analogous polypeptides encoded by members of this group (Morozov *et al.*, 1989). However, βc does have some properties of membrane-bound proteins as do the 15 kD protein of BNYVV and the 8 kD protein of PVX (Morozov *et al.*, 1989).

βd Protein: Several lines of evidence indicate that the βd gene product has an important role in systemic infection of plants. An oligonucleotide-directed mutation of the βd ORF (Figure 6), which alters the 11th and 12th amino acids from Lys-Tyr to Asn-Arg and simultaneously creates a nonviable mutant, suggests that the βd protein is essential for systemic infection (Petty and Jackson, 1990b). In addition to its genome organization, the sequence of the βd gene product has some similarity to the 13 kD proteins of the potexviruses and the 14 kD protein of BNYYV. Such regions of similarity include

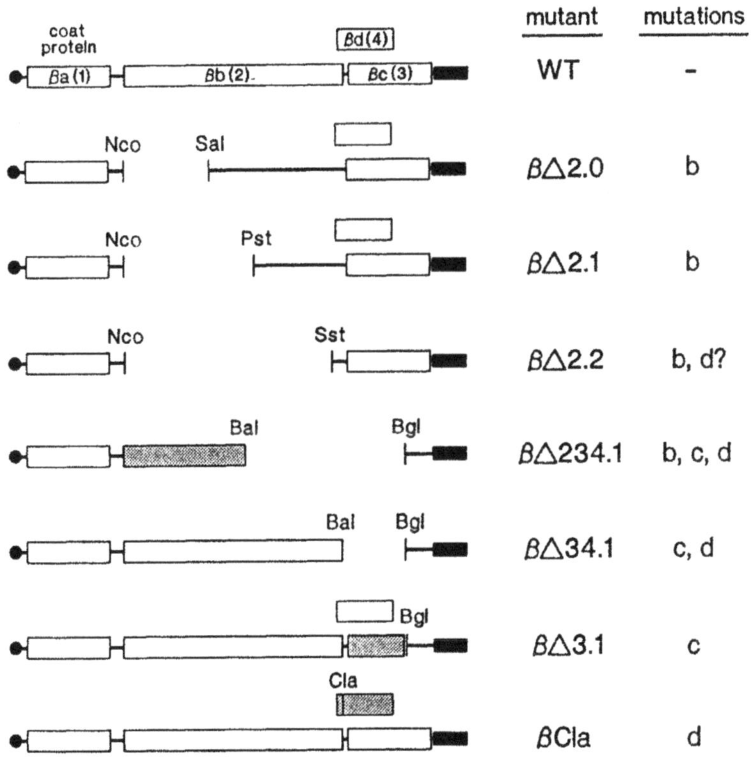

Figure 6. Genetic Organization of BSMV RNAβ and Mutations Within the βb, βc, and βd Genes. The solid circle indicates the 5'-cap structure, open rectangles represent the open reading frames within the ca. 3.3 KB genome, and the solid rectangle specifies the 3'-terminal tRNA-like structure. Mutants were constructed by deleting (Δ) the sequence between the indicated restriction sites (NcoI, SalI, PstI, SstI, BalI, BglII or ClaI) or, in the case of mutant βCla, by altering a Lys-Tyr dipeptide in the βd polypeptide to Asn-Arg. The shaded region indicates that the predicted gene product was truncated or in the case of βCla is nonfunctional. The "mutations" column lists the gene(s) affected by the mutations. The figure is modified from Petty and Jackson (1990b).

amino-terminal and carboxy-terminal stretches of hydrophobic amino acids that are separated by a conserved hydrophilic block of amino acids residing in the central portion of the proteins (Morozov *et al.*, 1987). The hypothesis that these related proteins may be membrane-bound is supported by our recent immunoblot assays. These studies, using antibodies raised against a βd specific peptide, demonstrate that the 14 kD protein can be readily detected in the membrane fraction of infected barley. Similar experiments were previously reported by Niesbach-Klösgen *et al.* (1990), who found the 14 kD BNYVV protein in a cellular fraction enriched for membranes.

γa Protein: Nucleotide sequence analysis (Gustafson *et al.*, 1987) indicates that RNAγ of the Type and ND18 strains differ by more than 360 nt due to a tandem duplication near the 5'-terminus of the genome (Figure 7). In the cloned ND18 and Type strains, this duplication results in γa proteins differing in size by 124 amino acids. Several *in vitro* translation experiments have indicated that the longer RNAγ from the Type and Russian strains directs synthesis of an 87 kD protein corresponding to γa, whereas strains with shorter γRNA's like those of ND18 (Figure 7) direct synthesis of a 74 kD product (Jackson *et al.*, 1983; Dolja *et al.*, 1983b; Gustafson *et al.*, 1981). The deduced sequence of these proteins reveals that the γa protein (Figure 2) contains a motif related to putative RNA polymerases found in many other RNA viruses (Kamer and Argos, 1984; Haseloff *et al.*, 1984). A recent phylogenetic tree using amino acid sequence comparisons of RNA polymerase genes has revealed that these proteins can be placed into three large supergroups (Koonin, 1991). The γa protein shows the greatest similarity to the putative tobacco rattle virus polymerase (Morozov *et al.*, 1989) and can be assigned to supergroup III, which also contains TMV and the tricomaviruses. PVX and BNYVV are also included in supergroup III, but the similarities between their polymerase proteins and the BSMV γa protein are much less pronounced than the relationship of the polymerases of BSMV to those of TMV and the other tripartite RNA viruses (Koonin, 1991). Our demonstration of the requirement for an intact γa gene for replication of BSMV RNA in protoplasts (Petty *et al.*, 1990b) provides direct experimental evidence that γa is required for replication.

The γa protein has been detected in infected plants using immunoblot assays (Donald *et al.*, 1993). Surprisingly, however, the γa and the αa proteins differ in the time course of their expression and in their subcellular distribution. The αa protein is generally distributed in the cytosol, membrane, and cell wall fractions; and its peak concentration coincides with the appearance of symptoms in barley. In contrast, the γa protein is primarily concentrated in the membrane fractions and reaches its highest concentrations 3 to 6 days after the onset of symptoms. These differences in the timing of

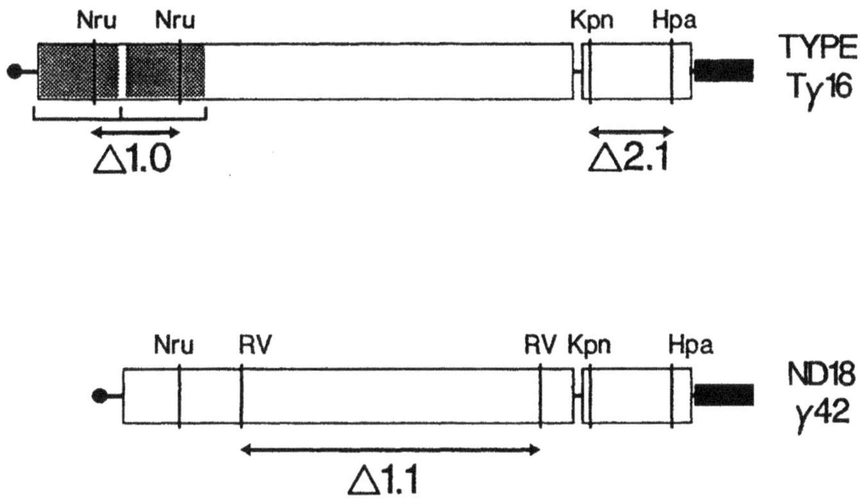

Figure 7. Illustration of the Organization of BSMV RNAγ from the Type and ND18 Strains and Deletion Mutations Introduced into the Typeγ16 and ND18γ42 Clones. The indicated restriction sites (Nru = NruI, Kpn = KpnI, Hpa = HpaI, and RV = EcoRV) were used to construct mutations listed below the clones. The large rectangle indicates the γa open reading frame, and the smaller rectangle specifies the γb open reading frame. The deletions (Δ) made within each clone are shown below each schematic. The shaded 5'-terminal region of Tγ16 illustrates the duplicated amino-terminal region of the γa protein. Reproduced from Jackson *et al.*, 1991b.

expression suggest that differential synthesis of the replicase proteins could be a mechanism employed by BSMV to modulate the synthesis of viral RNA and regulate various aspects of the replicative cycle.

γb Protein: The γb ORF (Figure 2) is separated from the γa ORF by a 42 nt intercistronic region and terminates with a UAA formed from the first two nucleotides of the poly A tract (Gustafson *et al.*, 1987). The sgRNAγ transcripts synthesized during infection are encapsidated by coat protein to form a minor component of virus preparations (Jackson *et al.*, 1983). The site for initiation of transcription of this sgRNA has been mapped to a specific guanosine residue in the short intercistronic region between the γa and γb genes (Gustafson *et al.*,

1987); and hybridization analyses with probes derived from the 3'-end of the genomic RNA indicate that this subgenomic RNA possesses the same 3'-terminal sequence as the genomic RNA (Petty *et al.*, 1990b), rather than having a heterogeneous polyadenylated tail as previously indicated by 3'-terminal labeling and sequencing (Stanley *et al.*, 1984). Translation of the subgenomic RNAγ *in vitro* reveals that it is an efficient mRNA which directs synthesis of a 17 kD product (Jackson *et al.*, 1983; Dolja *et al.*, 1983b). Because the γb protein is present in such low concentrations during infection, it is impossible to differentiate in infected plant extracts by Coomassie blue staining of polyacrylamide gels (Jackson *et al.*, 1983). However, the protein can easily be detected by immunoblot assays in the soluble fraction of infected barley where its concentration remains relatively constant throughout the infection cycle (Donald *et al.*, 1993) .

The γb protein (Figure 2) has some analogies in size and 3'-terminal location to small cysteine-rich proteins encoded by furoviruses, tobraviruses, and carlaviruses (Morozov *et al.*, 1989). However, despite these intriguing features, these proteins share no common statistically significant similarities in overall amino acid sequence; and even within the short stretches surrounding the cysteine residues, only marginally significant similarity appears (Koonin *et al.*, 1991). Instead, the comparisons of Koonin *et al.* (1991) reveal that these proteins are more closely related to proteins derived from other sources. For example, the BNYVV 14 kD protein is similar to the E6 proteins of the papillomaviruses; and the tobravirus cysteine-rich proteins have marginal similarity with the basic amino-terminal domains of high mobility group chromatin proteins. The two cysteine-rich regions near the amino-terminus of γb are separated by basic amino acid residues that are related to a motif present in chloroplast tRNA intron-encoded proteins and a small stretch of amino acids present in reverse transcriptases. The region of similarity in the reverse transcriptases occurs in a tether domain which separates the polymerase and RNAse H domains. All of these proteins possibly have some role in RNA binding and Koonin *et al.* (1991) speculate that the γb protein may participate in RNA folding, so it is possible that altered conformation of BSMV RNA in association with mutated γb could affect symptom phenotype and expression of genes encoded by RNAβ (Petty *et al.*, 1990b).

Infectivity studies utilizing mutants with deletions in the γb ORF (Figure 7) have demonstrated that the γb protein is dispensable for RNA replication in plants (Petty *et al.*, 1990b). Although γb is not necessary for RNA replication *per se*, the patterns of RNA synthesis in protoplasts infected with RNAγb mutants differ from infections with wild-type RNAγ. In particular, the synthesis of RNA's α and β are reduced relative to RNAγ when γb is deleted. A mutant with a deletion in the entire γb gene of the ND18 strain elicits attenuated symptoms in barley plants that are characterized by an erratic mottled mosaic pattern (Petty *et al.*, 1990b). Further analysis reveals that in the presence of this null γb mutant, the concentrations of the capsid and the βb proteins are reduced by two to three orders of magnitude in leaf tissue, although the accumulation of viral RNA is reduced only to about 20 percent of wild-type infections (Petty *et al.*, 1990b). Thus, this result shows that deletion of the γb protein has a differential effect on RNA and protein synthesis and indicates that γb may have some role in regulating synthesis of proteins encoded by RNAβ.

Despite the dispensability of the γb protein for systemic movement in the ND18 strain background, an interesting differential phenotype appears when mutations are made in the Type strain γb gene, since the latter mutants were unable to establish systemic infections in barley (Petty *et al.*, 1990b). Mutagenesis experiments demonstrate that the 372 nt direct repeat present at the 5'-terminus of the Type strain RNAγ (Figure 7) inhibits systemic movement when the γb gene is deleted and that removal of the duplicated 372 nt restores the ability to move systemically. However, a cloned ND18 RNAγ variant with a similar 372 nt direct repeat retains the ability to move systemically in the presence of the γb deletion. Further recombination studies between the Type and ND18 strain γ RNA's have revealed that a second site within the γa gene that falls outside the 372 nt repeat is also an important genetic determinant which can overcome the inhibitory effects of the repeat on systemic movement (Petty *et al.*, 1990b). The salient conclusions from these experiments are that important interactions exist between the γb protein and other BSMV genes, particularly those on RNAβ, and that subtle but important structural determinants within the γa gene can help compensate for detrimental effects of deletions within the γb gene on systemic movement.

Mutational Analyses Affecting the Biochemical and Biological Properties of the βb AND γB Genes

We (Petty *et al.*, 1990b) previously reported that barley plants inoculated with wild-type RNA's α and β in combination with RNAγb deletion mutants developed an attenuated mosaic. Levels of viral RNA present in plants infected with the γb mutant transcripts were reduced only to about 10 to 20 percent of wild-type RNA's; but the level of the RNAβ encoded coat protein (βa) was reduced by about 1000 fold and the 58 kD βb protein could not be detected by western serological blotting, although it was readily detected in plants infected with wild-type virus. These results suggested that expression of the genes encoded by RNAβ is dependent on a trans-acting function of the 17 kD γb protein. Such a role is consistent with the observation that the γb protein has two conserved cysteine-rich motifs located within the amino-terminal half of the protein, which leads to the prediction that γb may exert its effects in part by binding to nucleic acids (Koonin *et al.*, 1991). These combined observations provide the basis for our hypothesis that the βb and γb gene products interact to carry out some functional processes; and they raise the question as to whether these proteins, singly or in combination, may be able to exert their effects by an association with BSMV RNA. To investigate this possibility, we (Donald and Jackson, unpublished data) have conducted RNA binding studies on both the βb and γb proteins and have also compared the biological effects resulting from mutations designed to examine the biochemical properties of the proteins.

Biochemical Properties of βb and its Role in BSMV Movement: Studies from the laboratory of our colleague Patricia Zambryski have demonstrated that the TMV 30K and the cauliflower mosaic virus (CaMV) gene I proteins, which have been shown by genetic tests to be involved in virus movement, have the ability to bind nucleic acids *in vitro*. These findings have resulted in the hypothesis that the "movement" proteins act as chaperones to facilitate the transfer of viral nucleoprotein complexes through plasmosdesmatal connections between cells (Citovsky and Zambryski, 1991). Although the BSMV βb protein has little obvious amino acid sequence relationship to the TMV 30K or CaMV gene I proteins, the requirement for movement and the presence of a helicase motif suggest that βb might be involved in some aspect of nucleic acid metabolism. This notion is reinforced by several properties shared between βb and the largest "triple block" gene of the potex-, carla- and furoviruses (Jackson *et al.*, 1991a; 1991b). These properties include similar organization of the ORF's, the profound effects that deletions within the genes have on systemic movement, and the presence of a deduced RNA helicase motif in each of the proteins. Therefore, in order to test whether the βb protein functions similarly to the TMV and CaMV movement proteins, an *E. coli* T7 expression vector (Rosenberg *et al.*, 1987) was used for expression of high levels of the βb protein. The protein was purified from inclusion bodies and *in vitro* binding assays were used to analyze its ability to bind nucleic acids. Our polyacrylamide gel mobility-shift assays have demonstrated that the βb protein has cooperative RNA binding activity and can protect RNA probes from RNase digestion (Donald and Jackson, unpublished results). Competitive binding experiments with RNA and DNA substrates show that the βb protein has a strong binding preference for RNA, but little nucleotide sequence specificity could be demonstrated by the *in vitro* assays used to measure binding. Although both single-stranded (ss) and double-stranded (ds) RNA's are protected, βb binding has a strong affinity for dsRNA's. In this regard, βb differs from the TMV 30K movement proteins or the CaMV gene I protein, since both have strong binding preference for single-stranded molecules.

To identify regions of the βb protein that contribute to RNA binding, two classes of mutants were constructed in biologically active cDNA clones of RNAβ. The first class of mutations was a series of seven site specific amino acid substitutions targeting conserved residues within the RNA helicase motif (Figure 5). The second mutant class involved two series of deletion mutations to define binding domains within βb (Figure 8).

A series of deletions illustrated in Figure 8 were made in an RNAβ cDNA clone to test the biological effects of the βb deletions on the ability of BSMV to infect barley. These 13 overlapping βb deletions and the helicase motif mutations were then subcloned into a T7 expression vector, and the resultant proteins from *E. coli* were tested for *in vitro* RNA binding activity by gel shift analyses (Citovsky *et al.*, 1990). The results of the bioassay experiments revealed that none of the mutants with conserved amino acid substitutions within the helicase motif could systemically infect barley plants, but the biochemical experiments indicated that the nucleic acid binding activity of the mutated proteins from *E. coli* was not affected. In addition to the lethal results of the helicase mutations, only two of the 13 deletion mutants retained biological activity. One of these mutants (ΔNco7), which lacks 35 amino acids from the amino-terminus, yields a βb protein identical to that characterized by Petty and Jackson (1990a), which requires functional γb *in trans* for infectivity. The second viable mutation (Δ5) has a deletion of 26 amino acids, 24 of which are duplicated in a direct tandem repeat of the Type strain. This mutant elicits an attenuated phenotype in which symptoms are delayed, erratic, and less uniform than wild-type infections.

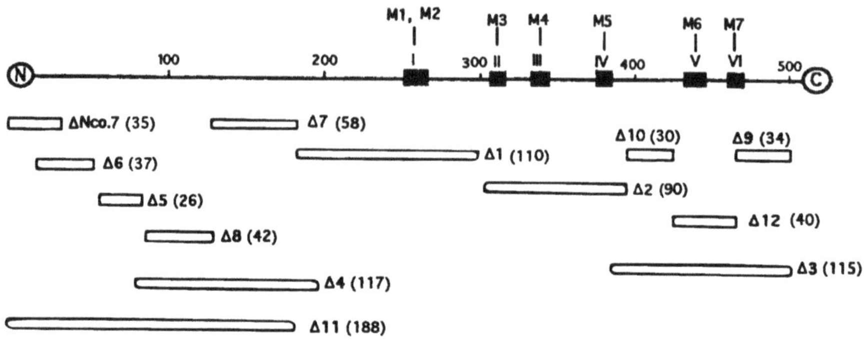

Figure 8. **Deletion Mutations Made to Study RNA Binding by the BSMV βb Protein. The solid line at the top of the figure illustrates the βb protein. The amino (N) and carboxy (C)-terminus are indicated; the numerals identify the amino acid number; and the boxed regions, M1 through M7, indicate the approximate location of amino acid substitutions made within blocks of homology within the helicase motif. The βb open reading frame deletions are designated as open boxes in the lower part of the figure. None of the deletions eliminated RNA binding activity; but as indicated in the text, some of the deletions altered the substrate specificity.**

Despite the loss of biological activity of the mutants, most of the deletions made in the βb ORF had significant ssRNA binding activity that was comparable to the recombinant wild-type βb protein. However, several deletions, which map to two distinct regions of βb (Figure 8) and fell into four distinct classes, did have a profound effect on dsRNA binding activity. Class I (deletions Δ1 and Δ4 and a double mutant Δ4Δ3) had a dramatic effect whereby they simultaneously lost the ability to bind synthetic homopolymeric dsRNA, which normally is a good competitor of binding by wild-type βb protein, and gained the ability to bind single-stranded polyuridine, which normally is not a competitor of wild-type βb binding in competition experiments. A Class II mutant (Δ10) lost the ability to compete for synthetic dsRNA but did not bind single-stranded poly U. Mutants Δ6 and Δ7 formed Class III, which was similar to Class I, but had only a moderate reduction in competitive binding affinity for dsRNA and a lower binding affinity for the single-stranded poly U competitor. Class IV (Δ3 and Δ12) had a reduction in competitive binding for dsRNA relative to the unmutated βb protein, but this reduction was not as severe as in Class II.

A second series of βb deletion mutations, which are illustrated in Figure 9, consisted of amino-terminal deletions fused to the carboxy-terminus of the glutathione transferase gene (Smith and Johnson, 1988). Recombinant proteins derived from these chimeras were expressed in *E. coli* and were used because we could purifiy the fusion proteins to near homogeneity by affinity chromatography. Unfortunately, however, the complete 58 kD product was not stable upon expression in *E. coli* when fused to glutathione transferase and only four of the βb deletion chimeras that were constructed could be recovered in adequate amounts for binding analysis. Although none of the three active proteins (GSTβb-11, GSTβb-12, and GST-βb-13) were able to bind RNA probes cooperatively, they did have single-strand specific binding activity which could be competed for by synthetic polyuridine or unlabeled probe RNA, but not by dsRNA, yeast tRNA, or by either single-stranded or double-stranded DNA templates. The fourth and smallest fusion protein (GST-βb-14) bound only poorly to the templates. These results thus define a C-terminal region with RNA binding activity, but comparisons of the amino acids found within this region or analysis of other amino acid sequences around the region revealed no significant structural similarities with either of the two RNA binding domains identified by Citovsky *et al.* (1992) for the TMV 30K protein.

In summary, several important conclusions can be drawn from the βb experiments. First, the βb protein has cooperative RNA binding to single-stranded RNA probes; and it exhibits competitive binding affinity for ss and dsRNA but not for tRNA, Poly U, or single-stranded or double-stranded DNA. Second, the dsRNA binding activity of the βb protein is influenced by multiple determinants that map to both the N-terminal and C-terminal domains and deletion of either of these domains abolishes synthetic dsRNA binding activity. Third, the carboxy-terminal domain, which apparently is unrelated in sequence or structure to the TMV 30K movement protein, is capable of binding RNA. Glutathione transferase βb fusion protein constructs have confirmed that this domain is independently capable of binding RNA,

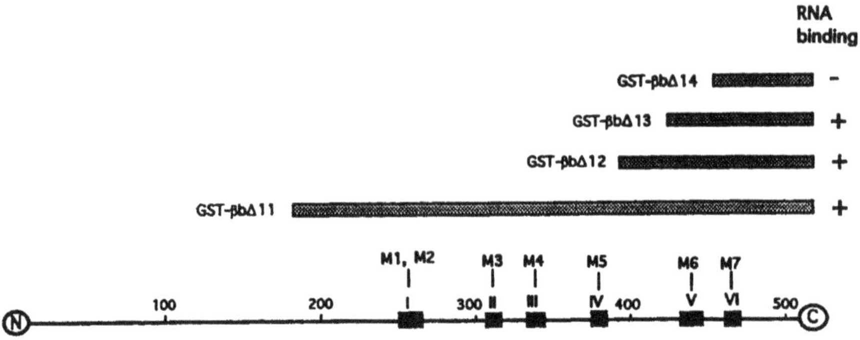

Figure 9. GST Fusion Proteins Made to Study RNA Binding of βb. The glutathione fusion proteins (shaded rectangles) and the extent of their RNA binding are illustrated at the top of the figure. The full-length fusion protein is not shown. The illustration of the βb protein at the bottom of the figure and the identity of the amino acid number and locations of the blocks of homology within the helicase motif are as indicated in Figure 8.

although unlike wild-type βb, it shows no affinity for dsRNA. Our working hypothesis is that amino- and carboxy-terminal domains interact in the native protein to form an RNA binding pocket with competitive binding affinities for ss and dsRNA's. Further studies to investigate this hypothesis in more detail are presently being carried out.

RNA Binding and Regulatory Activity of the γb Protein: In order to assess the biochemical properties of the putative *trans*-activating γb gene product more precisely, we have carried out an additional series of *in vitro* analyses to determine whether the γb protein also has nucleic acid binding properties (Donald and Jackson, unpublished results). For these experiments, as with the βb experiments, we initially attempted to recover the 17 kD γb protein expressed in *E. coli* under the control of the T7 RNA polymerase expression vector pET3a. Unfortunately, attempts to recover biochemically active protein were unsuccessful. We, therefore, utilized γb-glutathione transferase fusions to generate γb mutants suitable for RNA binding experiments.

Gel shift analyses and competition experiments with single-stranded and double-stranded nucleic acids have shown that the γb fusion protein has ssRNA binding activity; but the fusion protein fails to bind synthetic dsRNA (Poly U:Poly A), yeast tRNA, or single- or double-stranded DNA homologous to the BSMV RNA probe. A series of progressive amino-terminal deletions constructed from the GST-γb protein (Figure 10) suggests that the amino acids involved in binding activity are located within the first 44 residues of the amino-terminus of γb.

To more precisely evaluate the binding properties, site-specific mutations have been made in the GST-γb protein to convert each cysteine or histidine residue in the two cysteine-rich motifs to serine or leucine, respectively. In four additional cases, adjacent cysteine residues have both been mutagenized to determine whether these modifications have synergistic effects. The nature and location of each of these mutations is shown in more detail in Figure 10. To determine the biological effects of the sequence modifications, each of the mutants has also been incorporated into an infectious RNAγ clone; and the resulting phenotypes were determined by infecting barley plants (Figure 11). The symptoms elicited by the mutants include a classical wild-type mosaic (W), a bleached mosaic (B) culminating in white leaf symptoms, or a "null" phenotype (Φ) consisting of a mottled, erratic mosaic typical of that observed upon infection with γb deletion mutants (Petty *et al.*, 1990b). Our results thus demonstrate that mutations in the two cysteine clusters generate two distinctive phenotypes. Mutations of cysteine residues 7, 19, and 23 in the C1 cluster and the double mutants having an effect on symptoms all elicited the bleached symptom phenotype (Figure 11). In contrast, mutations within the C2 cluster which had an effect on symptom phenotype invaribaly yielded the null phenotype. These biological results suggest that the mutations may alter functionally distinct properties of the γb protein that depend on the individual cysteine clusters. We have attempted to correlate the biological data with nucleic acid binding ability of the proteins containing each of these mutations by overexpressing and purifying them from *E. coli* and testing their respective nucleic acid binding abilities. The results, however, have shown that none of the mutations within the C1 or C2 domain have major effects on RNA binding. Thus, although the conserved C1 and C2 residues are important determinants of biological activity, they have little influence on the ability of the γb protein to bind nucleic acids as assayed by the *in vitro* gel shift analyses.

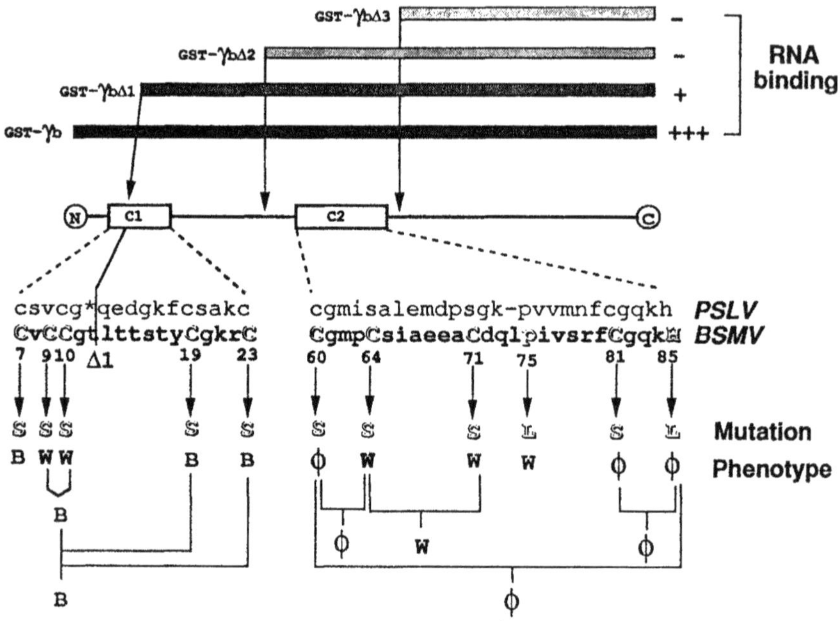

Figure 10. BSMV γb Mutations Used to Study Biological Activity and RNA Binding. GST fusion proteins isolated to show the extent of RNA binding and the relative amounts of estimated binding are indicated at the top of the figure. The γb protein is illustrated in the middle of the figure. The amino acid sequence of the two cysteine clusters (C1 and C2) of BSMV and the related poa semilatent virus (PSLV) γb protein is shown below and the substituted serine (S) and leucine (L) amino acids are indicated. The wild-type (W), the bleached (B), and the null (Φ) phenotypes elicited by the amino acid substitutions are shown below each substitution and are illustrated in Figure 11.

Figure 11. **Phenotypes of Barley Infected with Different BSMV Mutants. The plants were inoculated with wild-type transcripts or with transcripts containing mutations in the γb cysteine rich region and photographed 8 days post inoculation. H = Leaf from an uninoculated plant. W = Wild-type infection resulting from inoculation with ND18 transcripts. B = Bleached symptoms typical of those indicated in the C1 cluster (Figure 10) or in the tether region (Figure 12). The symptom phenotype is delayed and initially is attenuated and sporadic in appearance, but after about a week to 10 days the leaves begin to bleach; and the entire leaf blade may collapse and die, as is the case with the region near the tip of the leaf which is beginning to collapse. (Φ) = The null phenotype typical of infections in which the γb gene has been deleted or γb mutations in the C2 cluster and in tether mutant 29. This phenotype is mottled and erratic, and the uniform mosaic typical of the wild-type symptoms fail to appear.**

In order to investigate the symptom phenotype and RNA binding in more detail, additional amino acid substitutions were made in the basic region separating the C1 and C2 clusters (Figure 12). This motif contains a series of amino acids that are conserved in the related PSLV γb protein (Agranovsky *et al.*, 1992) and have some homology to the "tether" domain of chloroplast intron encoded proteins (Koonin *et al.*, 1991). The mutations designated 26 and 28 when present in a biologically active cDNA yielded the bleached phenotype characterized by the white leaf syndrome found in the C1 cluster (Figure 11). Mutant 29 yielded a null phenotype characteristic of that occurring when the entire γb gene is deleted or critical cysteine residues in the C2 cluster are targeted. In contrast to these effects, mutation 27 failed to affect the phenotype and inoculated plants developed symptoms indistinguishable from wild-type infections.

The RNA binding assays carried out with the GST fusion proteins revealed that mutation of some of the basic amino acids within the tether domain has an effect on the RNA binding activity (Figure 12). Substitutions 26 and 29 have the most severe effects on the ability of the GSTγb protein to form high molecular weight retarded complexes characteristic of the unmutated fusion protein. However, these substitutions apparently do not completely eliminate RNA binding, since some lower molecular weight complexes with the labeled probe were present in some of the assays. The critical amino acids identified in these complexes include the basic residues lysine and arginine within the amino-terminal side and central portion of the tether domain, whereas the basic residues proximal to the C2 cluster have only minimal effect on RNA binding.

These *in vitro* studies provide no direct evidence that RNA binding and the infection phenotypes resulting from mutagenesis can be directly correlated, since mutations that lead to reduced RNA binding have different phenotypes; yielding in one case (Mutation 29) a null phenotype and in another (Mutation 26) a bleached phenotype (Figure 12). Of interest, however, were the results of immunoloblot assays to correlate distribution of the βb and γb proteins in barley infected with the γb mutants. In the case of the γb protein, striking differences in accumulation and subcellular distribution could be correlated with the bleached, null, and wild-type symptom phenotypes. When the mosaic symptoms were elicited by wild-type RNAγ, the γb protein was cytoplasmically localized and constitutively expressed at relatively low amounts from days 4 to 10. The same low level of expression was observed in the 27 and 29 tether domain mutants and in the C2 mutants which yielded the null phenotype. However, the subcellular distribution of mutants 29 and C2 were shifted almost entirely from the cytoplasmic fraction to the low speed (organelle) and cell wall fractions. In addition to these shifts in distribution, striking increases were detected in the abundance of γb proteins containing the tether 26 and 28 mutations and the C1 cluster mutants which elicited the bleached phenotype. From these results, it is obvious that these mutations influence the expression and abundance of the γb protein. Although these changes in concentration and distribution may be the result of altered protein conformation, possibly by exposing hydrophobic domains that may be buried in the wild-type protein, other more biologically significant possibilities also exist. One tantalizing possibility is that the γb protein may regulate its own expression. Another option is that the abnormal partitioning of the mutant γb proteins into the membranes and into components forming the cell wall fraction (plasmodesmata?) contribute to dysfunction of the protein, which possibly may interfere with interactions of γb with one or more of the "triple gene block" proteins to interrupt targeting of RNA movement protein complexes to adjacent cells via plasmodesmatal connections.

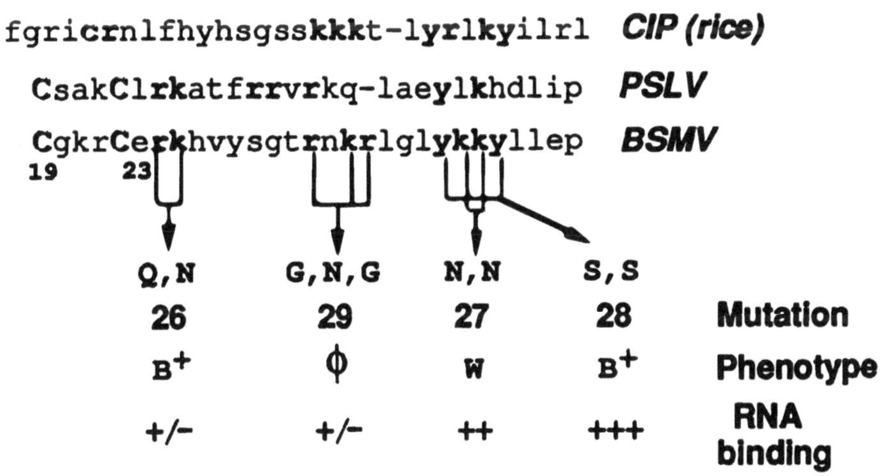

Figure 12. **Illustration of the Basic Tether Region Separating the C1 and C2 Clusters. This region is highly conserved between BSMV and PSLV and corresponds to a conserved region of chloroplastintron proteins, CIP (Koonin *et al*, 1991). The mutagenized amino acid residues include three groups of basic residues and two tyrosines which are found in a similar distribution in the corresponding PSLV and CIP sequence. The symbols indicate the effects of the mutations on symptom pathology and RNA binding. Substituted lysines, arginines, and tyrosines are shown in bold.**

In addition to possible autoregulatory activities which may control its accumulation and cellular distribution, the γb protein has also been observed to have pleiotropic effects on the abundance of the coat protein and the 58 kD βb protein. Reductions in accumulation of the coat protein were observed, and the βb protein could not be detected in the null mutants of the C2 cluster and tether mutant 29 which has a null phenotype. In contrast, the bleached mutants of the C1 cluster and tether mutants 26 and 28 appeared not to drastically alter the expression of the coat protein or βb. These results, thus, extend previous studies (Petty *et al.*, 1990b) suggesting that the γb protein is involved in regulation of BSMV RNAβ gene expression. However, the present results do not clarify whether the regulatory mechanism is direct or indirect and several alternative models exist for such mechanisms of action. Experiments are presently underway to provide more detailed information to elucidate these possible mechanisms.

Acknowledgments

We thank Denise Lapidus for assistance with the figures and for manuscript proofing. This research was supported by USDA competetive grants (FD90-37262-553 and 92-01257) awarded to A. O. Jackson. R. G. K. Donald was the recipient of a McKnight Fellowship Award.

References

Agranovsky, A.A., Karasev, A.V., Novikov, K.V., Lunina, N.A., Loginov, S., and Tyulkina, L.G. 1992. Poa semilatent virus, a hordeivirus having no internal polydisperse poly (A) in the 3'- non-coding region of the RNA genome. J. Gen. Virol. 73:2085-2092.

Ahlquist, P., French, R., and Bujarski, J. 1987. Molecular studies of brome mosaic virus using infectious transcripts from cloned cDNA. Adv. Virus Res. 32:215-242.

Allison, R., Thompson, C., and Ahlquist, P. 1990. Regeneration of a functional RNA virus genome by recombination between deletion mutants and requirement for cowpea chlorotic mottle virus 3a and coat genes for systemic infection. Proc. Natl. Acad. Sci. U.S.A. 87:1820-1824.

Atabekov, J.G. and Dolja, V.V. 1986. Hordeiviruses: Structure and replication. In: *The Plant Viruses. Vol 2. The Rod-shaped Plant Viruses.* M.H.V. van Regenmortel, H. Fraenkel-Conrat (eds.), pp. 397-420. Plenum Press, New York.

Atabekov, J.G. and Taliansky, M.E. 1990. Expression of a plant virus-coded transport function by different virus genomes. Adv. Virus Res. 38:201-248.

Atkins, D., Hull, R., Wells, B., Roberts, K., Moore, P., and Beachy, R.N. 1991. The tobacco mosaic virus 30K movement protein in transgenic tobacco plants is localized in plasmodesmata. J. Gen. Virol. 72:209-211.

Beck, D.L., Guilford, P.J., Voot, D.M., Anderson, M.T., and Forster, R.L.S. 1991. Triple gene block proteins of white clover mosaic potexvirus are required for transport. Virology 183:695-702.

Brakke, M.K., Ball, E.M., and Langenberg, W.G. 1988. A non-capsid protein associated with unencapsidated virus RNA in barley infected with barley stripe mosaic virus. J. Gen. Virology 69:481-491.

Citovsky, V., Knorr, D., Schuster, G., and Zambryski, P. 1990. The P30 movement protein of tobacco mosaic virus is a single-strand nucleic acid binding protein. Cell 60:637-647.

Citovsky, V., Wong, M.L., Shaw, A.L., Prasad, B.V., and Zambryski, P. 1992. Visualization and characterization of tobacco mosaic virus movement protein binding to single-stranded nucleic acids. Plant Cell 4:397-411.

Citovsky, V. and Zambryski, P. 1991. How do plant viruses nucleic acids move through intercellular connections? BioEssays 13:373-379.

Dalmay, T., Rubino, L., Burgyan, J., and Russo, M. 1992. Replication and movement of a coat protein mutant of cymbidium ringspot virus. Mol. Plant-Microb. Interact. 5:379-383.

Dawson, W.O. 1992. Tobamovirus-plant interactions. Virology 186:359-367.34.

DeJong, W. and Ahlquist, P. 1991. Bromovirus host specificity and systemic infection. Sem. in. Virol. 2:97-105.

DeJong, W. and Ahlquist, P. 1992. A hybrid plant RNA virus made by transferring the noncapsid movement protein from a rod-shaped to an icosahedral virus is competent for systemic infection. Proc. Natl. Acad. Sci. U.S.A. 89:6808-6812.

Deom, C.M., Lapidot, M., and Beachy, R.N. 1992. Plant virus movement protein. Cell 69:221-224.

Deom, C.M., Oliver, M.J., and Beachy, R.N. 1987. The 30-kilodalton gene product of tobacco mosaic virus potentiates virus movement. Science 237:389-394.

Ding, B., Haudenshield, J.S., Hull, R.J., Wolf, S., Beachy, R.N., and Lucas, W.J. 1992. Secondary plasmodesmata are specific sites of localization of the tobacco mosaic virus movement protein in transgenic tobacco plants.

Dolja, V.V., Lunina, N.A., Leiser, R.-M., Stanarius, T., Belzhelarskaya, S.N., Koslov, Y.V., and Atabekov, J.G. 1983b. A comparative study on the *in vitro* translation products of individual RNA's from two-, three-, and four-component strains of barley stripe mosaic virus. Virology 127:1-14.

Donald, R.G.K., Zhou, H., and Jackson, A.O. 1993. Serological analysis of barley stripe mosaic virus-encoded proteins in infected barley. Virology 195:659-668.

Forster, R.L.S., Bevan, M.W., Harbison, S-A, and Gardner, R.C. 1988. The complete nucleotide sequence of the potexvirus white clover mosaic virus. Nucleic Acids Res. 16:291-303.

Gilmer, D., Bouzoubaa, S., Hehn, A., Guilley, H., Richards, K., and Jonard, G. 1992. Efficient cell-to-cell movement of beet necrotic yellow vein virus requries 3'-proximal genes located on RNA 2. Virology 189:40-47.

Gorbalenya, A.E. and Koonin, E.V. 1989. Viral proteins containing the purine NTP-binding sequence pattern. Nucleic Acids Res. 17:8413-8440.

Gustafson, G.D. and Armour, S.L. 1986. The complete nucleotide sequence of RNAß from the Type strain of barley stripe mosaic virus. Nucleic Acids Res. 14:3895-3909.

Gustafson, G.D., Armour, S.L., Gamboa, G.C., Burgett, S.G., and Shepherd, J.W. 1989. Nucleotide sequence of barley stripe mosaic virus RNAα: RNAα encodes a single polypeptide with homology to corresponding proteins from other viruses and to the BSMV ßb protein. Virology 170: 370-377.

Gustafson, G.D., Hunter, B.G., Hanau, R., Armour, S.L., and Jackson, A.O. 1987. Nucleotide sequence and genetic organization of barley stripe mosaic virus RNAγ. Virology 158:394-406.

Gustafson, G.D., Larkins, B.A., and Jackson, A.O. 1981. Comparative analysis of polypeptides synthesized *in vivo* and *in vitro* by two strains of barley stripe mosaic virus. Virology 111:579-587.

Hamilton, W.D.O. and Baulcombe, D.C. 1989. Infectious RNA produced by *in vitro* transcription of a full-length tobacco rattle virus RNA-1 cDNA. J. Gen. Virol. 70:963-968.

Haseloff, J., Goelet, P., Zimmern, D., Ahlquist, P., Dasgupta, R., and Kaesberg, P. 1984. Striking similarities in amino acid sequence among nonstructural proteins encoded by RNA viruses that have dissimilar genomic organization. Proc. Natl. Acad. Sci. U.S.A. 81:4358-4362.

Hodgman, T.C. 1988. A new superfamily of replicative proteins. Nature 333:22-23.

Huisman, M.J., Linthorst, H.J.M., Bol, J.F., and Cornelissen, B.J.C. 1988. The complete nucleotide sequence of potato virus X and its homologies at the amino acid level with various plus-stranded RNA viruses. J. Gen. Virol. 69:1789-1798.

Hull, R. 1989. The movement of viruses in plants. Annu. Rev. Phytopathol. 27:213-240.

Hull, R. 1991. The movement of viruses within plants. Semin. Virol. 2:89-95.

Jackson, A.O., Dawson, J.R.O., Covey, S.N., Hull, R., Davies, J.W., McFarland, J.E., and Gustafson, G.D. 1983. Sequence relations and coding properties of a subgenomic RNA isolated from barley stripe mosaic virus. Virology 127:37-44.

Jackson, A.O., Hunter, B.G., and Gustafson, G.D. 1989. Hordeivirus relationships and genome organization. Annu. Rev. Phytopathol. 27:95-121.

Jackson, A.O., Petty, I.T.D., Jones, R.W., Edwards, M.C., and French, R. 1991a. Molecular genetic analysis of barley stripe mosaic virus pathogenicity determinants. Can. J. Plant Pathol. 13:163-177.

Jackson, A.O., Petty, I.T.D., Jones, R.W., Edwards, M.C., and French, R. 1991b. Analysis of barley stripe mosaic virus pathogenicity. Semin. Virol. 2:107-119.

Kamer, G. and Argos, P. 1984. Primary structural comparison of RNA-dependent polymerases from plant, animal, and bacterial viruses. Nucleic Acids Res. 12:7269-7282.

Koonin, E.V. 1991. The phylogeny of RNA-dependent RNA polymerases of positive-strand RNA viruses. J. Gen. Virol. 72:2197-2206.

Koonin, E.J., Boyko, V.P., and Dolja, V.V. 1991. Small cysteine-rich proteins of different groups of plant RNA viruses are related to different families of nucleic acid binding proteins. Virology 181:395-398.

Kozlov, Y.V., Afanasiev, B.N., Rupasov, V.V., Golova, Y.B., Kulaeva, O.I., Dolja, V.V., Atabekov, J.G., and Bayev, A.A. 1989. The complete nucleotide sequence of barley stripe mosaic virus RNA3 and its variability. Molecular Biology 23:1080-1090. (In Russian)

Kozlov, Y.V., Rupasov, V.V., Adyshev, D.M., Belgelarskaya, S.N., Agranovsky, A.A., Mankin, A.S., Morozov, S.Y., Dolja, V.V., and Atabekov, J.G. 1984. Nucleotide sequence of the 3'-terminal tRNA- like structure in barley stripe mosaic virus genome. Nucleic Acids Res. 12:4001-4009.

Loesch-Fries, L.S. and Hall, T.C. 1982. *In vivo* amino-acylation of brome mosaic and barley stripe mosaic virus RNA's. Nature 298:771-773.

Maule, A.J. 1991. Virus movement in infected plants. Crit. Rev. Plant Sci. 9:457-473.

McFarland, J.E., Brakke, M.K., and Jackson, A.O. 1983. Complexity of the Argentina mild strain of barley stripe mosaic virus. Virology 130:397-420.

Meshi, T., Watanabe, Y., Saito, T., Sugimoto, A., Maeda, T., and Okada, Y. 1987. Function of the 30 kD protein of tobacco mosaic virus: Involvement in cell-to-cell movement and dispensability for replication. EMBO J. 6:2557-2563.

Mi, S., Durbin, R., Huang, H.V., Rice, C.M., and Stollar, V. 1989. Association of the Sindbis virus RNA methyltransferase activity with the nonstructural protein nsP1. Virology 170:385-391.

Mi, S. and Stollar, V. 1990. Both amino acid changes in nsP1 of Sindbis virus LM21 contribute to and are required for efficient expression of the mutant phenotype. Virology 178:429-434.

Mi, S. and Stollar, V. 1991. Expression of Sindbis virus nsP1 and methyltransferase activity in *Escherichia coli*. Virology 184:423-427.

Morozov, S.Y., Dolja, V.V., and Atabekov, J.G. 1989. Probable reassortment of genomic elements among elongated RNA-containing plant viruses. J. Mol. Evol. 29:52-62.

Morozov, S.Y., Lukasheva, L.I., Chernov, B.K., Skryabin, K.G., and Atabekov, J.G. 1987. Nucleotide sequence of the open reading frame adjacent to the coat protein cistron in potato virus X genome. FEBS Letters 213:438-442.

Nejidat, A., Cellier, F., Holt, C.A., Gafny, R., Eggenberger, A.L., and Beachy, R.N. 1991. Transfer of the movement protein gene between two tobamoviruses: Influence on local lesion development. Virology 180:318-326.

Niesbach-Klösgen, U., Guilley, H., Jonard, G., and Richards, K. 1990. Immunodetection *in vivo* of beet necrotic yellow vein virus encoded proteins. Virology 178:562-561.

Petty, I.T.D., Edwards, M.C., and Jackson, A.O. 1990a. Systemic movement of an RNA plant virus determined by a point substitution in a 5'-leader sequence. Proc. Natl. Acad. Sci. U.S.A. 87:8894-8897.

Petty, I.T.D., French, R., Jones, R.W., and Jackson, A.O. 1990b. Identification of barley stripe mosaic virus genes involved in viral RNA replication and systemic movement. EMBO J. 9:3453-3457.

Petty, I.T.D., Hunter, B.G., Wei, N., and Jackson, A.O. 1989. Infectious barley stripe mosaic virus RNA transcribed *in vitro* from full-length genomic cDNA clones. Virology 171:342-349.

Petty, I.T.D. and Jackson, A.O. 1990a. Two forms of the major barley stripe mosaic virus nonstructural protein are synthesized *in vivo* from alternative initiation codons. Virology 177:829-832.

Petty, I.T.D. and Jackson, A.O. 1990b. Mutational analysis of barley stripe mosaic virus RNAβ. Virology 179:712-718.

Pleij, C.W.A., Abrahams, J.P., van Belkum, A., Rietveld, R., and Bosch, L. 1987. The spacial folding of the 3'-noncoding region of aminoacylatable plant virus RNA's. In: *Positive Strand RNA Viruses*. M.A. Brinton and R.R. Rueckert (eds.), pp. 299-316. Alan R. Liss, New York.

Pring, D.R. 1971. Viral and host RNA synthesis in BSMV-infected barley. Virology 44:54-66.

Richards, K. and Tamada, T. 1992. Mapping functions on the multipartite genome of beet necrotic yellow vein virus. Annu. Rev. Phytopathol. 30:291-313.

Rosenberg, A.H., Lade, B.N., Chiu, D., Lin, S., Dunn, J.J., and Studier, W.F. 1987. Vectors for selective expression of cloned DNA's by T7 polymerase. Gene 56:125-135.

Rozanov, M.N., Koonin, E.V., and Gorbalenya, A.E. 1992. Conservation of the putative methyltransferase domain: A hallmark of the 'Sindbis-like' supergroup of positive-strand RNA viruses. J. Gen. Virol. 73:2129-2134.

Rupasov, V.V., Morozov, S.Y., Kanyuka, K.N., and Zavriev, S.K. 1989. Partial nucleotide sequence of potato virus M RNA shows similarities to potexviruses in gene arrangement and the encoded amino acid sequences. J. Gen. Virol. 70:1861-1869.

Sacher, R. and Ahlquist, P. 1989. Effects of deletions on the N-terminal basic arm of brome mosaic virus coat protein on RNA packaging and systemic infection. J. Virology 63:4545-4552.

Siegel, A., Zaitlin, M., and Sehgal, O.P. 1962. The isolation of defective tobacco mosaic virus strains. Proc. Natl. Acad. Sci. U.S.A. 48:1845-1851.

Smith, D.B. and Johnson, K.S. 1988. Single-step purification of polypeptides expressed in *E. coli* as fusions with glutathione-S- transferase. Gene 67:31-40.

Stanley, J., Hanau, R., and Jackson, A.O. 1984. Sequence comparison of the 3'-ends of a subgenomic RNA and the genomic RNA's of barley stripe mosaic virus. Virology 139:375-389.

van Lent, J., Storms, M., van der Meer, F., Wellink, J., and Goldbach, R. 1991. Tubular structures involved in movement of cowpea mosaic virus are also formed in infected cowpea protoplasts. J. Gen. Virol. 72:2615-2623.

Wolf, S., Deom, C.M., Beachy, R.N., and Lucas, W.J. 1989. Movement protein of tobacco mosaic virus modifies plasmodesmatal size exclusion limit. Science 246:377-379.

Ziegler-Graff, V., Guilford, P.J., and Baulcombe, D.C. 1991. Tobacco rattle virus RNA-1 29K gene product potentiates viral movement and also affects symptom induction in tobacco. Virology 182:145-155.

Function and Expression of Cucumber Necrosis Virus Genes

D. M. Rochon
Agriculture Canada Research Station
6660 N.W. Marine Drive
Vancouver, B.C. V6T 1X2
Canada

J. Johnston and R. Finnen
Department of Microbiology
University of British Columbia
Vancouver, B.C. V6T 1W5
Canada

M. McLean
Department of Plant Science
University of British Columbia
Vancouver, B.C. V6T 2A2
Canada

Cucumber necrosis virus (CNV), a member of the tombusvirus group, is a spherical virus which encapsidates a monopartite, positive polarity RNA genome. Mutational analyses of the nucleotides surrounding the upstream AUG codon of a bifunctional subgenomic mRNA derived from the CNV 3'-terminus has provided information on the optimal sequence context for expression of plant mRNA's *in vivo*. In addition, this information is being used to determine if expression of the corresponding downstream AUG codon occurs by leaky scanning, a mechanism not yet definitively reported to occur in plants. Mutational studies of the CNV coat protein nucleotide sequence demonstrate that the coat protein is not required for short or long distance movement and is apparently dispensable for mechanical transmission as well. Finally, the possible involvement of the CNV 20 kDa non-structural protein in preventing the generation or accumulation of defective interfering RNA is also discussed.

Introduction

Cucumber necrosis virus (CNV) is one of three tombusviruses whose complete genome has recently been characterized at the molecular level (Rochon *et al.*, 1991; Hearne *et al.*, 1990; Grieco *et al.*, 1989). CNV is an icosahedral virus which encapsidates a single molecule of monopartite positive polarity RNA, approximately 4.7 kb in length.

The CNV genome has the capacity to encode at least five proteins from five distinct open reading frames (Rochon and Tremaine, 1989) and may, in addition, encode a sixth protein from a small open reading frame at the extreme 3'-terminus of the genome (Boyko and Karasev, 1992). We have recently obtained full-length cDNA clones of the CNV genome which can be used as templates for the production of highly infectious synthetic RNA inoculum (Rochon and Johnston, 1991). The availability of these clones has enabled a variety of mutagenesis studies for deciphering the functions of CNV encoded proteins and the means by which the expression of these proteins are regulated. The purpose of this paper will be to summarize what is presently known regarding the molecular biology of CNV and also to describe our most recent unpublished findings. In addition, we will overview what is currently known regarding CNV defective interfering (DI) RNA's and the means by which these RNA's may arise during the course of infection.

Biological Properties of CNV

CNV has been reported only in Canada (McKeen, 1959) where it was determined that this virus was the causative agent of a serious disease of greenhouse grown cucumbers. Natural infections by CNV have not been reported elsewhere and infection has so far only been reported to occur in cucumber. In contrast to its limited natural host range, CNV can be mechanically transmitted to a wide range of host plants. Most infections result in necrotic spots only on the inoculated leaves; however, systemic necrosis occurs consistently in *Nicotiana clevelandii* and *Nicotiana benthamiana*. Systemic necrosis also occurs in cucumber but only sporadically. CNV is soil-transmitted; but unlike other soil-transmitted tombusviruses, transmission of CNV has been demonstrated to be facilitated by zoospores of the soil-inhabiting, root-infecting, chytrid fungus, *Olpidium radicale* (Dias, 1970). We have recently noted that CNV-infected *N. clevelandii* cells contain multivesicular bodies (derived from peroxisomes) as do most other tombusvirus infected plants. These multivesicular bodies are believed to represent sites of viral RNA synthesis (Martelli *et al.*, 1988).

Overview of CNV Genomic Structure and Function

Figure 1 shows the structure of the CNV genome as deduced from its nucleotide sequence. The CNV genome contains five long open reading frames (ORF's 1-5) capable of encoding proteins with predicted molecular weights of 33, 92, 41, 21, and 20 kDa. The 33 kDa protein terminates with an amber codon which, if readthrough, would produce the 92 kDa protein. Whatever its mode of expression, the 92 kDa protein most likely forms at least part of the CNV replicase since a portion of its amino acid sequence contains the "GDD" motif and surrounding hydrophobic amino acid residues found in the polymerases encoded by most RNA viruses examined to date (Kamer and Argos, 1984). In addition, this region and sequences flanking it show significant amino acid sequence similarity with the putative polymerases of a number of plant viruses (the luteovirus, barley yellow dwarf, the carmovirus, carnation mottle, the dianthovirus, red clover necrotic mosaic, the unclassified maize chlorotic mottle virus, and the necrovirus, tobacco necrosis) and the animal virus Hepatitis C (Riviere *et al.*, 1990). A recent comparison of amino acid sequence motifs present in the putative polymerases of a number of RNA viruses of plants, animals, and bacteria has suggested that tombusviruses and the above mentioned viruses along with Hepatitis C; flaviviruses; pestiviruses; and, unexpectedly, certain single-stranded RNA bacteriophages, be grouped in a separate supergroup distinct from the alpha and picornavirus supergroups (Koonin, 1991). A helicase motif has not been found in the CNV 92 kDa protein or in any other CNV encoded protein suggesting that either some portion of the CNV genome encodes a unique helicase or that the helicase function is carried out by a host protein. The role of the 33 kDa protein in CNV replication has not been determined or investigated yet. The 41 kDa protein is the CNV coat protein and is encoded by an internally located ORF. The 21 and 20 kDa proteins are translated from separate nested ORF's at the CNV 3'-terminus. Mutational analyses, described below, have provided insight into the role the CNV coat and 21 kDa proteins play in inter- and intracellular movement. Additionally, mutagenesis studies suggest the involvement of the CNV 20 kDa protein in preventing replication errors since the absence of this protein in CNV infections leads to the rapid *de novo* appearance of defective interfering RNA's (see below).

Expression of CNV Genes

Earlier *in vitro* translation experiments using both wheat germ extracts and rabbit reticulocyte lysates have shown that genomic length CNV RNA serves as the template for the 33 kDa protein and that 2.1-and 0.9 kb subgenomic length RNA's serve as templates for

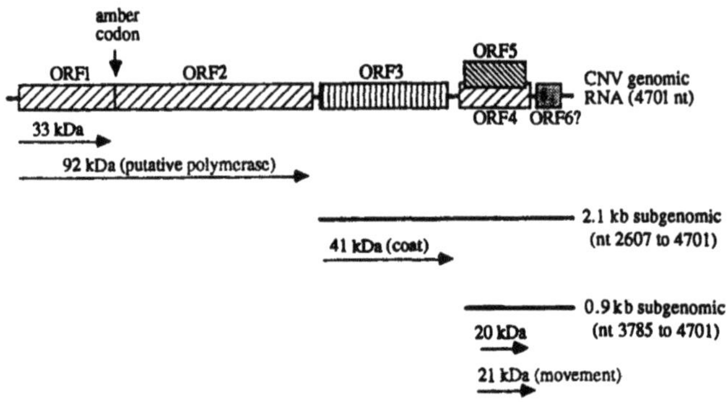

Figure 1. Summary of the Organization, Expression, and Function of CNV Genes and Gene Products. Bars below the CNV genomic structure correspond to the sizes and positions of CNV subgenomic mRNA's. It has not yet been determined if ORF 6 in fact encodes a protein (see text).

the coat protein and the 21/20 kDa proteins, respectively (Johnston and Rochon, 1990). We have not observed synthesis of the 92 kDa putative readthrough protein *in vitro* in either wheat germ extracts or rabbit reticulocyte lysates even in the presence of added exogenous tRNA from a number of sources. The means by which the expression of the 92 kDa protein is regulated is presently unknown. It has recently been suggested that tombuviruses, including CNV, may encode a sixth small protein which would initiate from an AUG codon just downstream from the 21/20 kDa ORF (see Figure 1) (Boyko and Karasev, 1992). This suggestion is based on the finding of significant amino acid sequence similarities deduced from the nucleotide sequences of these regions of four sequenced tombusvirus genomes. We have initiated experiments to test this hypothesis by first changing the putative AUG codon for this small ORF to a non-AUG codon so that the possible biological effect of this change can be assessed.

Function and Expression of the CNV 21 and 20 kDa Proteins: We have recently established that the distinct CNV 21 and 20 kDa proteins are expressed from a bifunctional subgenomic mRNA derived from the CNV 3'-terminus (Rochon and Johnston, 1991). This conclusion is based on both *in vitro* and *in vivo* studies and can be summarized as follows. *In vivo* evidence that both proteins are expressed comes from the results obtained from inoculating *N. clevelandii* plants (a systemic host for CNV) with full length synthetic CNV transcripts which have single point substitutions in the AUG codons for either the 21 or the 20 kDa ORF's. 21 kDa mutant transcripts do not produce symptoms and do not accumulate to detectable levels in inoculated plants as determined by northern blots. This would suggest that the 21 kDa protein is normally expressed during CNV infection. We have recently inoculated *Nicotiana plumbaginifolia* protoplasts with the 21 kDa mutant and have found that this mutant replicates to approximately 60 percent of the level of wild-type transcripts as determined by ELISA using CNV coat protein antibody. The replication of this mutant in protoplasts and its lack of detectable accumulation in plants would suggest that the 21 kDa protein plays a role in the cell-to-cell movement of CNV. Mutant CNV transcripts which lack the AUG codon for the 20 kDa protein replicate well in inoculated *N. clevelandii*; however, plants are persistently infected showing a very mild necrosis unlike the lethal necrosis induced by wild-type synthetic transcripts. The altered symptomatology of 20 kDa mutant infections suggests that the 20 kDa protein is expressed during normal CNV infections. As will be described below we have found that the attenuated symptoms associated with 20 kDa mutant infections is at least partly due to the rapid *de novo* appearance of defective interfering RNA's in infected plants.

Virion derived 0.9 kb subgenomic RNA's and synthetic subgenomic length RNA's corresponding to the 3'-terminal 0.9 kb of CNV RNA both give rise to 21 and 20 kDa sized proteins *in vitro* in wheat germ extracts suggesting that a single mRNA can serve as the template for both the 21 and 20 kDa proteins *in vivo*. In addition, 20 kDa mutant subgenomic length synthetic transcripts produce the 21 kDa protein but not the 20 kDa protein, and 21 kDa mutant transcripts produce the 20 kDa protein but not the 21 kDa protein. This demonstrates that the two proteins produced from wild-type transcripts are indeed the 21 and 20 kDa proteins predicted from the CNV nucleotide sequence. Finally, only a single prominent extension product can be found using an oligonucleotide primer complementary to a region just downstream from the 21 and 20 kDa ORF's and sucrose gradient fractions containing the 0.9 kb subgenomic RNA derived from CNV virions or from infected leaves. Together, these data lead to the conclusion that CNV encodes a bifunctional subgenomic mRNA responsible for the expression of the 21 and 20 kDa proteins.

The finding that two distinct translation initiation events can occur from the same mRNA template leads to the question as to how the 20 kDa protein is expressed in infected plants: its AUG codon is not 5'-proximal and, therefore, should not be readily accessible to eukaryotic ribosomes. According to Kozak, in certain circumstances, scanning ribosomes may occassionally be able to "leak" past the first AUG codon on a template and instead initiate at the second AUG (Kozak, 1989). These circumstances may arise if the first AUG is in a sub-optimal context for efficient recognition by ribosomes. Alternatively, ribosomes may be able to initiate at an internal AUG codon directly, independent of scanning, such as has been recently described for initiation of poliovirus polyprotein synthesis (Pelletier and Sonenberg, 1988). To assess whether ribosomes access the 20 kDa initiator codon via leaky scanning, we first needed to determine the optimal sequence context for initiation at the 21 kDa AUG. Once determined, we could then alter this sequence context and determine the effect these changes have on expression of the 20 kDa AUG. Figure 2 shows the sequences surrounding the 21 and 20 kDa AUG codons in CNV RNA. The consensus sequence for optimal expression of mRNA's in animals as determined by Kozak (1989) is CACCAUGG and that for plants as determined by Lutcke *et al.* (1987) (see below) is AACAAUGGC. According to Kozak, the most important determinant for translational efficiency in animal mRNA's is the -3 position which should be a purine, and preferably A. However, according to Lutcke, the -3 position does not appear to be important in plants as determined by *in vitro* studies in wheat germ extracts. The consensus sequence for plants proposed by Lutcke is based on computer assisted comparisons of plant mRNA sequences. Based on this data, Lutcke suggested that because the frequency of occurrence of a G and a C in the +4 and +5 positions in plant mRNA's resembles the frequency for an A in the -3 position in animal mRNA's, the +4 and +5 positions might represent important determinants of codon selection in plants. The 5'-terminal region of the CNV 0.9 kb subgenomic mRNA encoding the 21 and 20 kDa proteins was, therefore, examined to determine what constitutes a good context for initiation of translation of plant mRNA's *in vivo*. To do this, a construct (pCGUS-1) was developed (Figure 2) which contains sequences corresponding exactly to the 5'-untranslated region of the 0.9 kb CNV subgenomic mRNA up to and including the 21 kDa AUG codon. The CaMV promoter was placed upstream of this region, and the reporter gene β-glucuronidase in pAGUS-1 (kindly provided by R. Gesteland, University of Utah) was placed downstream and in frame with the CNV 21 kDa AUG codon. Oligonucleotide-directed mutagenesis was then used to introduce single point substitutions into three selected positions surrounding the CNV 21 kDa AUG resulting in the -3 position being either an A or a U, the +4 position being either a G or a U, and +5 position being either a C or an A. A total of 8 pCGUS constructs (see Figure 2) were then used for further studies.

The GUS activities of each protoplast transfected construct are summarized in Figure 2. These preliminary data suggest that the presence of either an A in the -3 position or a G in the +4 position relative to the AUG codon are important to obtain high levels of expression. For constructs containing pyrimidines in both these positions, a significant decrease in expression was observed. In general, these data are in agreement with both the Kozak and Lutcke consensus sequences and confirm the regulatory role of nucleotides flanking the AUG codon in plant mRNA's. The results do, however, indicate that purine to pyrimidine changes in the -3 and +4 positions have a similar impact on translational activity, which differs somewhat from that observed for animal mRNA's.

As mentioned previously, another objective of the above study is to determine if leaky scanning is the means by which expression of the 20 kDa protein is obtained in CNV infections. To do this, a larger fragment corresponding to the 5'-non-coding region of the CNV 0.9 kb subgenomic RNA extending past the 21 kDa initiation codon and including the 20 kDa codon was inserted in pAGUS-1 as above with the 20 kDa AUG codon in frame with the GUS gene. Nucleotide substitutions were again introduced flanking the 21 kDa start codon such that the resulting transcripts contain either a G or a U in the +4 position and a C or an A in the +5 position (see Figure 2). The GUS activities obtained from transcripts containing a G in the +4 position were consistently lower relative to the level obtained from transcripts containing a U in this position. Since the GUS activities of these constructs are inversely related to the efficiency of recognition of the 21 kDa context as determined above, these data are consistent with the concept that leaky scanning is the means by which the 20 kDa AUG codon is accessed for translation initiation.

Role of the CNV Coat Protein in Virus Movement

The three-dimensional structure of the capsids and the constitutive coat protein subunits of tomato bushy stunt virus, the type member of the tombusvirus group have been determined at high resolution by X-ray crystallography (Harrison, 1983). The capsids, comprised of 180 identical subunits, form a T=3 icosahedron. Each coat protein subunit folds into three distinct domains: the random (R), shell (S), and protruding (P) domains. The R domain is flexibly tethered to the rest of the subunit by the arm (a), and the S and P domains are connected by a short 5 amino acid hinge (h). The P domain projects outwards from the shell domain in pairs, and it has been suggested that dimer interaction between such domains may function to enhance particle stability (Harrison, 1983). Tombus-, carmo-, and probably dianthoviruses as well, differ from other spherical viruses in containing a P domain which may provide unique structural or biological properties to these groups of plant viruses.

In order to define the biological functions and/or structural basis (s) of the CNV coat protein protruding domain, a CNV mutant which lacks most of the coding sequence for this domain [designated PD(-); see Figure 3] has been produced and its infectivity analyzed and compared to wild-type transcripts on the systemic CNV host, *N. clevelandii.* Both wild-type transcripts and PD(-) inoculated plants develop necrotic local lesions 5-7 days post inoculation; however, the lesions on PD(-) transcript-inoculated leaves are always smaller than those on wild-type infected plants indicating that efficiency of replication, accumulation, or movement in PD(-) may be somewhat impaired by the lack of the P domain. Systemic infection occurred with the PD(-) mutant, as it does in wild-type infections; however, this process was somewhat delayed in PD(-) mutant infections [8-12 days in wild-type, 16-24 days in PD(-)]. Analysis of viral RNA species in inoculated and systemically infected leaf extracts has revealed that PD(-) length RNA (*ca.* 4.4 kb) is rarely detectable. Instead, an RNA species of about 3.8 kb predominates. Since 3.8 kb is approximately the size of CNV template which lacks the coding sequence for the entire coat protein and since a variety of immunological procedures utilizing CNV antibody (raised to virions) failed to detect CNV coat protein in PD(-) infected plants it seemed likely that the prominent 3.8 kb RNA represented further coat protein coding sequence deletion mutants of PD(-) which occurred *in planta* due to the absence of selection for protruding domainless coat protein. To determine if the 3.8 kb RNA, in fact, lacks the entire CNV coat protein sequence, RT-PCR (reverse transcriptase initiated polymerase chain reaction), using primers which flank the coat protein coding region, was used to amplify any remaining coat protein sequences using total RNA extracted from systemically infected leaves of PD(-) sap inoculated plants. The most prominent RT-PCR product was approximately 270 nucleotides, which is ca. 1200 nucleotides smaller than the RT-PCR product obtained from wild-type infected plants. This difference in

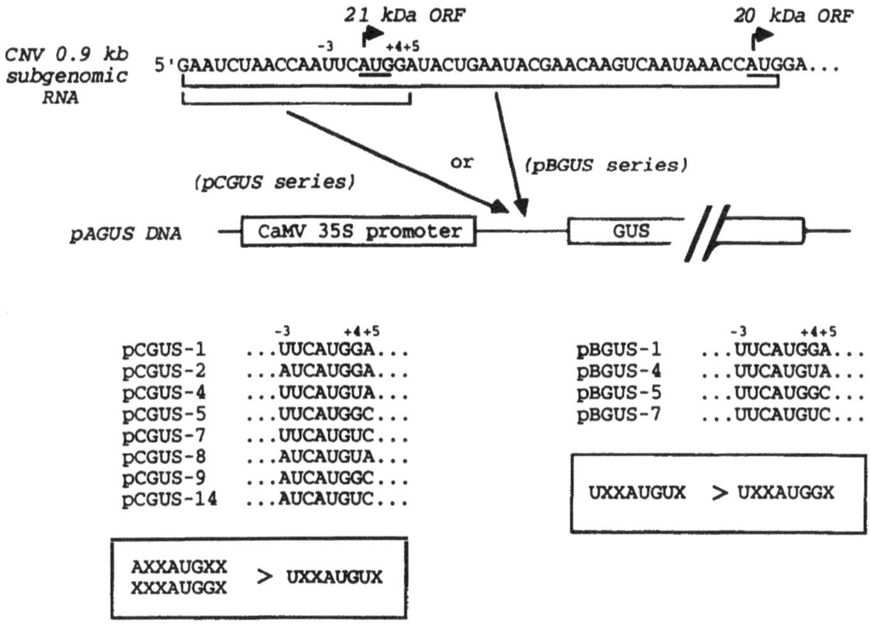

Figure 2. **Summary of GUS Constructs Used for Determining the Optimal Sequence Surrounding the CNV 21 kDa Codon and for Determining if the 20 kDa Codon is Accessed by a Leaky Scanning Mechanism. The sequence at the 5'-terminus of the 0.9 kb bifunctional subgenomic mRNA encoding the CNV 21 and 20 kDa proteins is shown at the top of the diagram. The AUG codons which define the start point for translation of each protein are underlined. The nucleotides (-3, +4, and +5) surrounding the 21 kDa AUG codon which were mutated in this study are indicated. Mutagenized regions of either of the two bracketed regions of the CNV sequence were placed between the CaMV promoter and the GUS gene in the plasmid pAGUS such that the AUG codons of either the 21 kDa protein (the pCGUS series) or the 20 kDa protein (the pBGUS series) were in-frame with the GUS gene. The sequences of eight such pCGUS constructs and four such pBGUS constructs are shown on the left and right of the diagram, respectively. The boxed regions below the constructs summarize the relative GUS activities for each set of data. The -3, +4, and +5 nucleotide positions with respect to the AUG codons are indicated.**

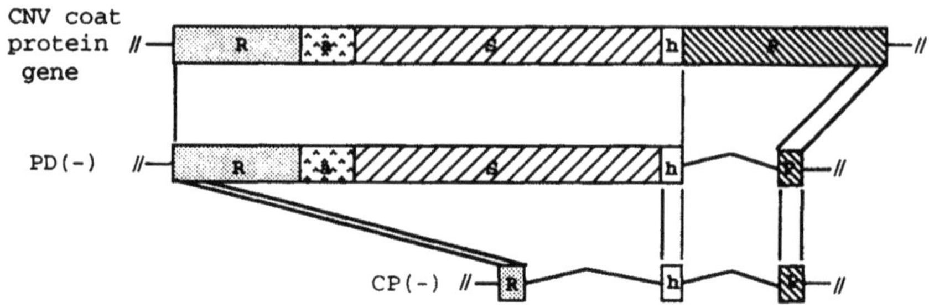

Figure 3. Structure of the Region Encoding the CNV Coat Protein and of Corresponding Regions of the Coat Protein Mutants PD(-) and CP(-). The regions of the coat protein sequence encoding the random (R), arm (a), shell (S), hinge (h), and protruding (P) domains are indicated. Deletions in PD(-) and CP(-) relative to the original CNV coat protein gene are shown using bent lines (see text for details).

size corresponds to a complete coat protein deletion. Eight independent clones from two separate PCR experiments using two different leaf RNA extracts were sequenced. The sequences of each clone were found to be highly similar and revealed that very little coat protein coding sequence in PD(-) infected tissue, in fact, remained; these sequences being limited to 27 nucleotides at the coat protein gene 5'-terminus corresponding to the extreme amino terminus of the R domain, 14 nucleotides of an internal location corresponding to the 5 amino acid hinge which connects the shell with the protruding domain, and 33 nucleotides of the coat protein gene 3'-terminus corresponding to portions of the protruding domain which were undeleted in the original PD(-) clone. To determine if CNV can indeed replicate and spread in the absence of these coat protein sequences, a clone of the RT-PCR product described above was used to replace the analogous region in a wild-type CNV cDNA clone. Transcripts of this new clone [designated CP(-)] were found to replicate and spread systemically in *N. clevelandii* in a manner indistinguishable from that of wild-type transcripts. These results demonstrate that the CNV coat protein is not required for efficient cell-to-cell or systemic movement. In addition, they suggest that CNV particles which lack the coat protein protruding domain are not functional [since further deletion derivatives of coat protein apparently have a selective replicative advantage over those containing coat protein sequences in PD(-)], perhaps because they are not stable in the absence of P domain dimer contacts (see above). It is worth pointing out that when sap from systemically infected leaves of wild-type and PD(-)-infected plants is used as inoculum, signficiant differences in symptoms between the different inocula are no longer apparent; all local lesions are of the same appearance and both inocula cause systemic symptoms to appear at the same time. Similar results have recently been obtained in transcript inoculated plants using CP(-) transcript as inoculum. This suggests that the original smaller lesion size and delay in systemic symptom expression observed in PD(-) transcript inoculated plants may be due to a lower rate of replication and that a higher rate can be resumed once excess non-functional coat protein coding sequence has been deleted. The compensatory function of the coat protein in full length transcripts may explain, at least in part, why coat protein is maintained in normal CNV infections. In addition, it seems likely that virus particles should be important in interplant movement and despite some unsuccessful attempts to infect cucumber roots with zoospores of *O. radicale*, the fungal vector of CNV, we still wish to analyze the involvement of CNV coat protein sequences, especially protruding domain sequences, in fungal transmission. Finally, we wish to point out that sap from PD(-) infectious material can be consistently and successfully sap transmitted to give a large number of local lesions on the host *Chenopodium quinoa*; however, at a much reduced efficiency compared with wild-type inoculum, even after the sap has been left over 60 minutes. No lesions were observed under similar conditions using healthy sap amended with high concentrations of otherwise highly infectious CNV RNA. This shows that during infection CNV RNA is partially stabilized in plants in the absence of coat protein. A candidate for a stabilizing (nucleic acid binding) protein is the CNV 21 kDa protein which we have already suggested (see above) may be involved in intracellular movement.

CNV Defective Interfering RNA's

Laboratory stocks of CNV produce a mild, systemic necrosis on *N. clevelandii;* and these plants are persistently infected, whereas wild-type synthetic transcript inoculum produces a very severe necrosis and the plants dessicate and die usually in less than 2 weeks (Rochon and Johnston, 1991). Northern blot analysis of natural CNV and synthetic transcript infected plants has revealed the presence of high levels of an approximately 400 nucleotide long RNA in CNV infected plants but not in synthetic transcript inoculated plants. In addition, a 400 nucleotide long RNA can also be found in CNV virion RNA preparation (Rochon and Johnston, 1991). Low molecular weight, symptom attenuating defective interfering (DI) RNA's have been well characterized in tombusvirus infections (Morris and Knorr, 1990; Hillman, *et al.*, 1987; Rubino *et al.*, 1990); and so it seems reasonable that the 400 nucleotide long RNA is a DI RNA responsible for the observed symptom attenuation associated with infections by CNV laboratory stocks. To determine if this is, in fact, the case, the 400 nucleotide long DI RNA was cloned from CNV virion RNA preparations using RT-PCR and 5'- and 3'-primers homologous and complementary to the CNV 5'- and 3'-terminus, respectively. The 5'-primer incorporated the bacteriophage T7 RNA polymerase promoter sequence immediately adjacent to CNV 5'-sequence to enable subsequent transcription of the cloned sequence. The 3'-primer incorporated a SmaI site for restriction enzyme cleavage prior to run-off transcription. An approximately 400 nucleotide long PCR product was obtained which was subsequently cloned into the plasmid pUC19. Run-off transcripts of one such clone (designated DI42) were found to replicate to high levels when co-inoculated with wild-type DI-free synthetic transcripts and also resulted in a dramatic attenuation of symptoms typical of that observed in plants inoculated with CNV laboratory cultures. The sequence of this RNA and of three other cloned *ca.* 400 nucleotide long RNA's were determined, and it was found that these molecules are linear deletion mutants of CNV genomic RNA with a structure (see Figure 4) very similar to that originally found for DI's associated with the cherry strain of tomato bushy stunt (TBSV-Ch;) (Morris and Knorr, 1990; Hillman *et al.*, 1987) and more recently with cymbidium ringspot tombusvirus (CyRSV) (Rubino *et al.*, 1990). Interestingly, a smaller DI-like RNA of 305 nucleotides (DI48) was also found in RNA extracts of CNV laboratory stocks. The sequence of this DI shows that it is highly similar to other CNV DI RNA's except that it lacks a large portion of

CNV 3'-sequence (with the exception of the 3'-terminal 18 nucleotides) present in all tombusvirus DI RNA's sequenced to date. This DI was, however, unable to replicate in the presence of genomic helper RNA in inoculated plants. This demonstrates the necessity of these sequences for DI RNA accumulation and provides insight into *cis*-acting sequences involved in CNV RNA replication.

The presence of DI RNA's in laboratory cultures of TBSV-Ch, CyRSV, and CNV and now with several other non-tombusvirus plant viruses and most animal viruses as well, has prompted questions regarding their origin. DI's are known to appear in animal cell culture as a consequence of several serial passages of virus inoculum at high multiplicity of infection. In some of these cases, the original virus inoculum may have already contained low levels of DI particles; and undiluted passage may have favored replication because more cells were co-infected with both DI and helper virus. *De novo* generation of DI RNA's from genomic RNA templates has been suggested to occur in two major steps: (1) synthesis of a defective RNA via an aberrant transcriptional event follwed by (2) selective amplification through replication (Schlesinger, 1988). A generalized mechanism for the synthesis of DI RNA's from template molecules involving a "leaping replicase" complex has been described (Holland, 1991). That model

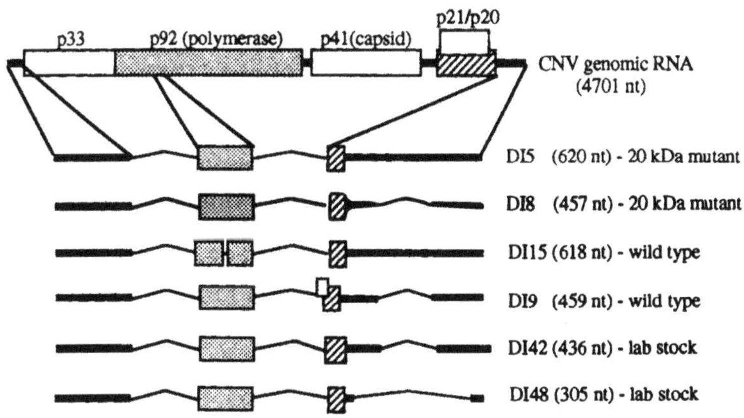

Figure 4. Structure of Naturally Occuring and *de novo* Generated DI RNA's Associated with Infection by CNV Laboratory Stock, Wild-Type, and 20 kDa Mutant Infections. The sizes and approximate locations of DI RNA sequence relative to CNV genomic RNA is shown. The bent lines between the blocks of DI RNA sequence correspond to deletions in DI RNA's relative to CNV genomic RNA. The origin of each type of DI is indicated to the right.

proposes that the transcribing polymerase along with its nascent RNA pauses during RNA synthesis and then resumes nascent RNA synthesis at a different site either on the same RNA molecule or on a different RNA molecule. Some experimental insight into the possible mechansim of *de novo* DI generation has recently been obtained in this laboratory (Rochon, 1991). Two CNV mutants which do not express the 20 kDa non-structural protein (one mutant contains a point substitution in the 20 kDa start codon and the other an introduced stop codon 15 nucleotides downstream and in-frame with the 20 kDa start codon) were found to rapidly generate high levels of DI RNA's [with structures similar to those associated with laboratory cultures or generated *de novo* after several serial high multiplicity of infection (m.o.i.) passages; see Figure 4] a few days after a single high m.o.i. passage or in transcript inoculated plants 21-35 days post inoculation. This is in contrast to the 8-15 high m.o.i. passages required for DI formation by wild-type transcripts. Thus, the CNV 20 kDa protein likely normally plays a role in CNV RNA replication. Perhaps the 20 kDa protein is an accessory protein whose role is to suppress the replicase from switching or falling off templates during RNA replication. It is possible that other plant or animal viruses encode a protein with a similar function; and, if so, conditions which interfere with the synthesis of this may enable DI formation.

In future experiments, we would like to distinguish whether 20 kDa mutant infections lead to rapid DI formation via an increased frequency of errors in replication or, instead, by rapid accumulation of aberrant replication products which arise only infrequently. We have, in fact, noted a low level of DI-sized RNA species in early passages of wild-type infections; however, PCR analysis and subsequent cloning and sequencing have not confirmed these species as being authentic DI RNA's. We would also like to determine if, in fact, the 20 kDa protein is associated with the CNV replicase. Finally, we are examining if certain environmental conditions, such as exposure to high temperature, encourage DI RNA's to arise *de novo* during CNV infection.

Acknowledgements

We thank Helene Sanfacon for critically reading the manuscript. The work described in this paper was partially supported by NSERC Grant OGP0043840.

References

Boyko, V.P. and Karasev, A.V. 1992. Tombusvirus genome may encode the sixth small protein near its 3'-terminus. Virus Genes 6:143-148.

Dias, H.F. 1970. Transmission of cucumber necrosis virus by *Olpidium cucurbitacaerum* Barr & Dias. Virology 40:828-834.

Grieco, F., Burgyan, J., and Russo, M. 1989. The nucleotide sequence of cymbidium ringspot virus RNA. Nucleic Acids Res. 17:6383.

Harrison, S.C. 1983. Virus structure: High resolution perspectives. Adv. in Virus Res. 28:175-240.

Hearne, P.Q., Knorr, D.A., Hillman, B.I., and Morris, T.J. 1990. The complete genome structure and synthesis of infectious RNA from clones of tomato bushy stunt virus. Virology 177:141-151.

Hillman, B.I., Carrington, J.C., and Morris, T.J. 1987. A defective interfering RNA that contains a mosaic of a plant virus genome. Cell 51:427-433.

Holland, J.J. 1991. Defective virual genomes. In: *Fundamental Virology.* Vol. 1. B.N. Fields and D.M. Knipe (eds.), p.151-165. Raven, New York.

Johnston, J.C. and Rochon, D.M. 1990. Translation of cucumber necrosis virus *in vitro*. J. Gen. Virol. 71:2233-2241.

Kamer, G. and Argos, P. 1984. Primary structural comparison of RNA-dependent polymerases from plant, animal, and bacterial viruses. Nucleic Acids Res. 12:7269-7283.

Koonin, E.V. 1991. The phylogeny of RNA-dependent RNA polymerases of positive strand RNA viruses. J. of Gen. Virol. 72:2197-2206.

Kozak, M. 1989. The scanning model for translation: An update. J. Cell. Biol. 108:229-241.

Lutcke, H.A., Chow, K.C., Mickel, F.S., Moss, K.A., Dern, H.F., and Scheele, F.A. 1987. Selection of AUG codons differ in plants and animals. EMBO J. 6:43-48.

Martelli, G.P., Gallitelli, D., and Russo, M. 1988. Tombusviruses. In: *The Plant Viruses.* R. Koenig (ed.), pp. 13-72. Plenum Publishing Corp., New York.

McKeen, C.D. 1959. Cucumber necrosis virus. Canadian J. Bot. 37:913-925.

Morris, T.J. and Knorr, D.A. 1990.Defective interfering viruses associated with plant virus infections. In: *New Aspects of Positive Strand RNA Viruses.* M.A. Brinton and F.X. Heinz (eds.), pp. 123-127. Am. Soc. Microbial., Washington.

Pelletier, J. and Sonenberg, N. 1988. Internal initiation of translation of eukaryotic mRNA directed by a sequence derived from poliovirus RNA. Nature (London) 334:320-325.

Riviere, C.J., Pot, J., Tremaine, J.H., and Rochon, D.M. 1990. Coat protein of melon necrotic spot carmovirus is more similar to those of tombusviruses than those of carmoviruses. J. Gen. Virol. 71:1887-1896.

Rochon, D.M. 1991. Rapid *de novo* generation of defective interfering RNA by cucumber necrosis virus mutants that do not express the 20 kDa non-structural protein. PNAS, U.S.A. 88:11153-11157.

Rochon, D.M. and Johnston, J.C. 1991. Infectious transcripts from cloned cucumber necrosis virus RNA: Evidence for a bifunctional subgenomic mRNA. Virology 181:656-665.

Rochon, D.M., Johnston, J.C., and Riviere, C.J. 1991. Molecular analysis of the cucumber necrosis virus genome. Can. J. Plant Path. 13:142-145.

Rochon, D.M. and Tremaine, J.H. 1989. Complete sequence of the cucumber necrosis virus genome. Virology 169:251-259.

Rubino, L., Burgyan, J., Grieco, F., and Russo, M. 1990. Sequence analysis of cymbidium ringspot satellite and defective interfering RNA's. J. Gen. Virol. 71:1655-1660.

Schlesinger, S. 1988. The generation and amplification of defective interfering RNA's. In: *RNA Genetics, Retroviruses, Viroids, and RNA Recombination.* Vol. 2. E. Domingo, J.J. Holland, and P. Ahlquist (eds.), pp. 167-185. CRC, Boca Raton, Florida.

The Role of DNA B in Geminivirus Spread and Symptom Induction

J. Stanley, A. von Arnim, T. Frischmuth, S. Roberts, and K. Saunders
Department of Virus Research
John Innes Institute
Norwich NR4 7UH
United Kingdom

The majority of geminiviruses that infect dicotyledonous plants have two genomic components (DNA's A and B). The two essential genes located on DNA B are required for virus spread throughout the plant and play a significant role in the induction of symptoms. The DNA B gene products (BV1 and BC1) of African cassava mosaic virus (ACMV), a typical bipartite geminivirus, have been detected in infected *Nicotiana benthamiana* extracts. BV1 behaves as a predominantly soluble protein of 29 kDa while BC1 occurs as multiple species of 37, 39, and 42 kDa, suggesting that it is post-translationally modified. In common with several plant virus spread proteins, BC1 is associated with cell wall fractions. In inoculated leaves of *N. benthamiana*, the proliferation of a BV1 mutant, but not a BC1 mutant, is greatly impaired, indicating that BV1 is important for virus amplification in these tissues. Because neither BV1 nor BC1 contributes directly to viral DNA replication in *Nicotiana tabacum* protoplasts, our results suggest that BV1 amplifies virus in inoculated leaves by facilitating cell-to-cell spread; and BC1 is additionally required for systemic spread of infection. Regions of DNA B that modulate symptom induction have been mapped to predominantly intergenic sequences by constructing recombinants between two phenotypically distinct strains of the related geminivirus tomato golden mosaic virus (TGMV). Co-inoculation experiments have demonstrated that ACMV DNA B genes facilitate the spread of other geminiviruses. However, TGMV BL1 (the homologue of ACMV BC1) and a BL1/BC1 chimera disrupt the systemic spread of ACMV, suggesting that a novel resistance strategy might be developed based on the inhibition of virus spread.

Introduction

Geminiviruses are widely distributed throughout tropical and subtropical areas where many cause significant yield decreases in a variety of economically important crops (Harrison, 1985; Brown and Bird, 1992). In some instances, increased understanding of the etiology of the diseases and the epidemiology of the insects responsible for their transmission has contributed to the production of disease-free plants and virus-resistant lines. Nonetheless, the diseases caused by these pathogens frequently represent a major barrier to the production of crops that have limited inherent resistance. Consequently, continued efforts are being made to identify sources of natural resistance for conventional breeding purposes. Recently, two strategies have been explored for the development of geminivirus-resistant plants based on our knowledge of the molecular biology of the pathogens and using contemporary plant transformation techniques. In the first of these, we exploited the characteristics of a naturally-occurring defective genomic component of African cassava mosaic virus (ACMV) that has been shown to have an adverse effect on virus infection (Stanley and Townsend, 1985). It was demonstrated that an extrachromosomal copy of the defective DNA could be mobilized from an integrated dimeric template in infected transgenic *N. benthamiana*. Amplification of the defective DNA in *trans* by the infecting virus disrupted viral DNA replication, resulting in reduced levels of virus and causing amelioration of symptoms (Stanley *et al.*, 1990; Frischmuth and Stanley, 1991). In the second approach, the proliferation of tomato golden mosaic virus (TGMV) was disrupted by the expression of an antisense construct, encompassing the virus-encoded gene responsible for replication, in transgenic *N. tabacum* (Day *et al.*, 1991).

Both of these strategies use viral DNA replication as the target for disrupting the virus infection cycle. However, interfering with the progress of infection by blocking virus proliferation at the level of spread offers an alternative approach that might be exploited. Little information is available concerning the movement of geminiviruses from their nuclear location, the form that the viruses adopt whilst spreading, and the virus and host factors involved in the process. Undoubtedly, the ability of a virus to spread within a plant will play a significant role in the determination of host range and tissue tropism which, in turn, will affect the character and severity of induced symptoms. Therefore, the molecular dissection of virus-encoded determinants of spread and virus/host interactions that take place during this process should not only help in understanding the basis of pathogenesis but also might suggest alternative strategies with which to alleviate or prevent the adverse effects of the disease.

Geminiviruses provide versatile models for the investigation of the determinants of pathogenesis (Stanley, 1991). Because they have DNA genomes, they are easily manipulated; and the cloned genomic components can be readily re-introduced into plants by mechanical means or by agroinoculation for the purpose of screening mutants. In addition, many closely related but phenotypically distinct group members have been described that exhibit unique but overlapping host ranges and induce distinct symptoms in common hosts. We have exploited these features to define and characterize genes that play a central role in geminivirus spread and to examine their contribution to symptom induction.

Detection and Possible Functions of DNA B Gene Products

The genomic organization of ACMV, a typical bipartite member of the group, is shown in Figure 1. Complementary- (C) and virus-sense (V) genes of both genomic components are expressed bidirectionally from an intergenic region (Townsend *et al.*, 1985). The common region, an approximately 200 nucleotide sequence that is highly conserved between the components, contains *cis* elements for the control of transcription and replication and is located within the intergenic region. DNA A encodes the coat protein (AV1) (Townsend *et al.*, 1985) as well as the only viral gene product required for viral DNA replication (AC1) (Etessami *et al.*, 1991). Although DNA A can replicate autonomously and synthesize virus particles (Townsend *et al.*, 1986; Klinkenberg and Stanley, 1990), DNA B is required for virus spread and symptom induction in plants (Stanley, 1983). Mutagenesis studies have demonstrated that both DNA B genes (BV1 and BC1) are essential for systemic infection of *N. benthamiana*; and their coding regions have been defined by expressing functional gene products from DNA A-based coat protein replacement vectors (Etessami *et al.*, 1988), an approach that we have since used to deliver other spread genes into plants (discussed below).

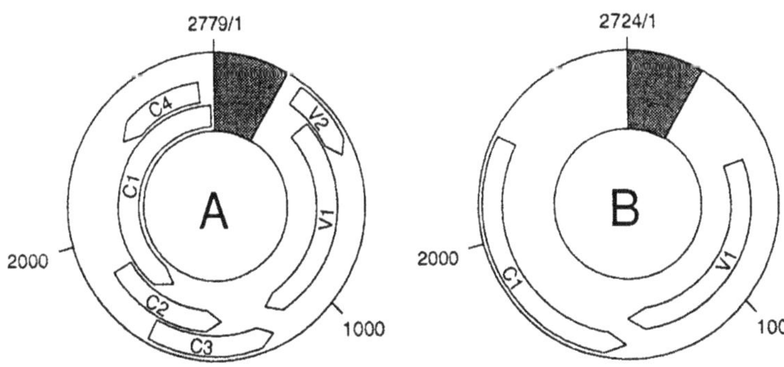

Figure 1. Organization of the Genome of ACMV. The position and orientation of genes are shown in relation to the common region (shaded).

Using polyclonal antisera raised against synthetic oligopeptides predicted from the gene sequences we have identified both DNA B gene products in extracts of ACMV-infected *N. benthamiana* (von Arnim *et al.*, 1993). A protein of 29 kDa was detected using BV1 antiserum (Figure 2), consistent with the size predicted from the gene sequence (Stanley and Gay, 1983). It was not present in extracts of TGMV-infected plants, demonstrating that it is not a virus-induced host-encoded protein. Fractionation of cell extracts by differential centrifugation indicated that a considerable proportion of BV1 was located in the soluble fraction. Residual BV1 in washed pellet fractions could be solubilized by sequential treatment with Triton X-100 and by boiling in SDS/urea. BV1 shows homologies with the coat protein (Kikuno *et al.*, 1984) and both proteins are highly basic. Therefore, BV1 might have biochemical properties in common with the coat protein, specifically the ability to bind to viral DNA. Co-operative binding to single-stranded nucleic acids has been demonstrated for the spread proteins of tobacco mosaic virus (TMV), cauliflower mosaic virus (CaMV), and red clover necrotic mosaic virus (Citovsky *et al.*, 1990; Citovsky *et al.*, 1991; Osman *et al.*, 1992) and might be a common feature of this type of protein.

In contrast to BV1, BC1 antiserum detected proteins of 37, 39, and 42 kDa in addition to other less abundant higher molecular weight species (Figure 2), all of which are larger than the anticipated size of 34 kDa (Stanley and Gay, 1983). Once again, the antiserum did not detect these proteins in extracts of TGMV-infected plants, indicating that they are specific to ACMV. *In vitro* translation of BC1 in a reticulocyte lysate produced a single protein of 35-36 kDa, suggesting that it is post-translationally modified in plants. Although we have not yet determined the nature of the modification, our preliminary data indicate that BC1 is not glycosylated and does not remain associated with nucleic acids under the extraction conditions. BC1 was located exclusively with pellet fractions of cell extracts, resembling the behavior of the spread proteins of TMV, alfalfa mosaic virus, and CaMV under similar fractionation conditions (Godefroy-Colburn *et al.*, 1986; Deom *et al*, 1987; Albrecht *et al.*, 1988). These spread proteins have been localized to the cell wall where they are believed to modify the size-exclusion limit of plasmodesmata to allow virus spread (Stussi-Garaud *et al.*, 1987; Tomenius *et al.*, 1987; Linstead *et al.*, 1988). It is possible that BC1 performs a similar function in the spread of ACMV. Interestingly, weak homologies have recently been described between BC1 and the spread proteins of other plant viruses including those of TMV and CaMV (Koonin *et al.*, 1991) that might reflect functional similarity.

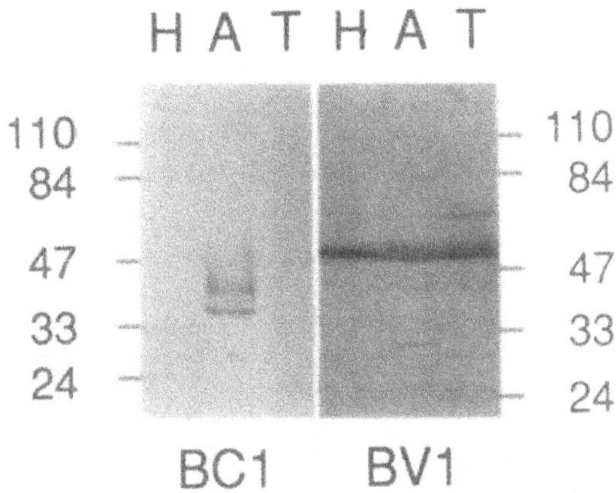

Figure 2. Western Blots of Proteins Extracted from Healthy Plants (H) and from Plants Infected with ACMV (A) and TGMV (T), Probed with Antisera Raised Against Synthetic Oligopeptides Predicted from the Sequences of BC1 and BV1. The positions of marker proteins (kDa) are indicated. Modified from von Arnim *et al.* (1993).

Analysis of the behavior of DNA B mutants in plants and cell culture has provided additional information concerning the function of the gene products. We have shown that a BC1 mutant accumulated to higher levels than a BV1 mutant when co-agroinoculated with DNA A onto *N. benthamiana* leaf discs (Frischmuth and Stanley, 1991). Similarly, when mechanically co-inoculated with DNA A onto *N. benthamiana* plants, the accumulation of the BC1 mutant in inoculated leaves again exceeded that of the BV1 mutant and was comparable to the level of complementing BV1 and BC1 mutants (von Arnim *et al.*, 1993). However, viral DNA accumulation has not been detected in inoculated leaves in the absence of DNA B (von Arnim and Stanley, 1992a). The results indicate that both DNA B genes contribute to virus amplification in these tissues and that BV1 plays a major role in this process. Furthermore, the replication of DNA A and its ability to replicate DNA B *in trans* in *N. tabacum* protoplasts remained unaffected when BV1 and BC1 coding sequences were disrupted (Figure 3), indicating that

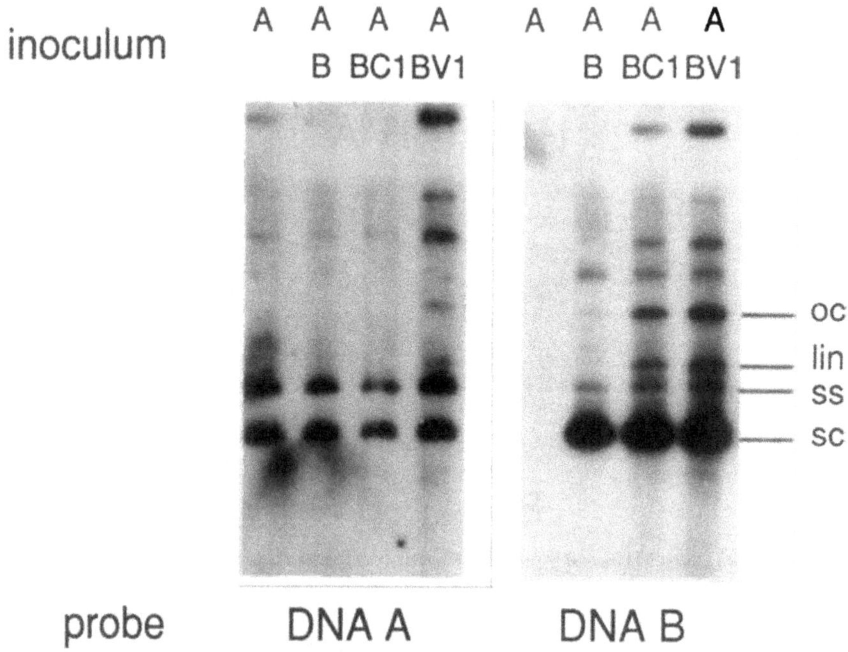

Figure 3. Southern Blot Analysis of Viral DNA's Extracted from *N. tabacum* Protoplasts Inoculated with Combinations of ACMV DNA A, DNA B and DNA B Mutants Containing Disrupted BC1 and BV1 Genes (inoculum). The positions of single-stranded (ss) and double-stranded supercoiled (sc), linear (lin) and open-circular (oc) viral DNA forms are indicated. Modified from von Arnim *et al.* (1993).

neither gene product participates directly in viral DNA replication. This suggests that virus amplification in inoculated leaves is a result of BV1-mediated cell-to-cell spread. The presence of a proportion of BV1 in pelletable fractions of cell homogenates (von Arnim *et al.*, 1993) might reflect its functional association with cell walls during this process. Although the BC1 mutant accumulated to high levels, it was unable to move out of the inoculated leaf, suggesting that BC1 participates in long-distance vascular spread of the virus. BC1 might have an independent role in this process or, as suggested from DNA B gene complementation studies (described below), it might function co-operatively with BV1. At the present time, it is unclear why bipartite geminiviruses require the action of two gene products to elicit systemic spread of infection, particularly as both proteins appear to have features in common with single spread proteins of other plant viruses.

Contribution of DNA B to Symptom Induction

The genome organization of the monopartite geminiviruses beet curly top virus (Stanley *et al.*, 1986) and tomato yellow leaf curl virus (Navot *et al.*, 1991) closely resembles that of DNA A of the bipartite viruses. In the absence of a component equivalent to DNA B these viruses can induce distinctive leaf curl symptoms in a host such as *N. benthamiana*. While this illustrates the fact that DNA B genes are not prerequisite for geminivirus symptom induction, it is likely that these viruses have developed novel or better adapted genes for infection of this host (Stanley and Latham, 1992; Stanley *et al.*, 1992). Not only do the DNA B genes play an essential role in symptom induction by mediating bipartite geminivirus spread but they also serve to modulate symptom type. This has been demonstrated by constructing pseudorecombinants between the common and yellow vein strains of TGMV (csTGMV and yvTGMV). In addition to causing plant stunting and leaf curl symptoms, these strains induce distinct patterns of chlorosis in *N. benthamiana*, *Nicotiana glutinosa* and *Nicotiana clevelandii*. In *Datura stramonium*, the strain difference is even more pronounced with csTGMV inducing severe stunting, leaf curl and extensive chlorosis and yvTGMV inducing only small chlorotic lesions and no leaf curl (Figure 4, top).

By reassortment of the genomic components, we have demonstrated that the symptom differences segregate with DNA B (von Arnim and Stanley, 1992b). Furthermore, using hybrid DNA B components we have shown that differences between the BR1 coding sequences and most of the BL1 coding sequences (BR1 and BL1 are the homologues of ACMV BV1 and BC1, respectively) are not responsible for the unique phenotypes (Figure 4, bottom). While we cannot yet rule out a contribution by amino- or carboxy-terminal sequences of BL1, we consider it most likely that the strain differences result from the timing or efficiency of expression rather than the protein composition *per se*. Consistent with this idea is the observation that the majority of strain sequence differences are located within the left side of the common region within the leader sequence of a complementary-sense transcript that might participate in BL1 expression (Sunter and Bisaro, 1989). In a similar manner, the modulation of gene expression has also been implicated in the determination of symptom severity as well as host range for strains of the monocot-infecting geminivirus maize streak virus (Boulton *et al.*, 1991). An alternative explanation is that strain sequence differences affect the efficiency of DNA A-mediated *trans*-replication

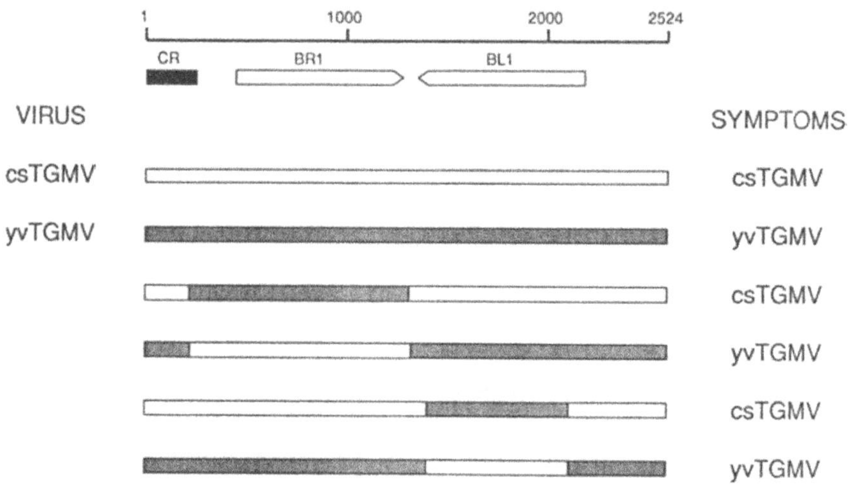

Figure 4. **Top: *D. stramonium* Leaves Systemically Infected with csTGMV (top) and yvTGMV (right), and the Pseudorecombinants yvTGMV DNA A and csTGMV DNA B (bottom) and csTGMV DNA A and yvTGMV DNA B (left). Bottom: Recombinants constructed *in vitro* from csTGMV and yvTGMV DNA B components. The symptom phenotypes produced by co-inoculation with TGMV DNA A (either strain) are indicated.**

of DNA B, which in turn affects the levels of both DNA B gene products. Indeed, the severity of symptoms induced in *N. benthamiana* has been correlated with fluctuations in the level of DNA B of both ACMV and TGMV (Stanley *et al.*, 1990; Revington *et al.*, 1989). Although the overall levels of viral DNA in *N. benthamiana* infected with each of the TGMV strains were comparable (von Arnim and Stanley, 1992b), the observed pattern of chlorosis might reflect subtle differences in the temporal or tissue-specific control of replication. The construction of additional recombinants will allow us to target specific sequences for mutagenesis in order to define elements responsible for strain phenotypic differences.

Complementation of DNA B Genes

Bipartite geminiviruses produce viable pseudorecombinants only when genomic components are exchanged between closely related strains, an example of which has been described above. The reason for the inability of less closely related viruses to exchange components in this way is probably due to incompatibility of gene products during viral DNA replication and/or spread. Indeed, co-inoculation of ACMV and TGMV components cnto *N. benthamiana* leaf discs has indicated that DNA A of one virus is unable to *trans*-replicate DNA B of the other (Roberts and Stanley, unpublished). TGMV AL1 (the ACMV AC1 homologue) has been shown to recognize and bind to part of the common region (Fontes *et al.*, 1992), containing intergenic sequences that are poorly conserved between viruses, suggesting one possible source of specificity. This has been confirmed recently by demonstrating that AL1 of TGMV and squash leaf curl virus will drive replication only when a specific part of the cognate common region is present (Lazarowitz *et al.*, 1992). In view of the overlapping host ranges of many bipartite geminiviruses and the conservation of the DNA B gene products between these viruses, we were interested to know if such specificity extended to virus spread functions.

We adopted a convenient method for screening DNA B gene products for their ability to complement the systemic spread of other viruses by co-agroinoculating combinations of cloned genomic components onto *N. benthamiana*, initially attempted using ACMV and TGMV. We were unable to detect ACMV DNA A in systemically infected leaves when co-inoculated with TGMV DNA A and DNA B, suggesting that TGMV DNA B gene products are unable to support the systemic spread of ACMV (Figure 5). However, in the reciprocal

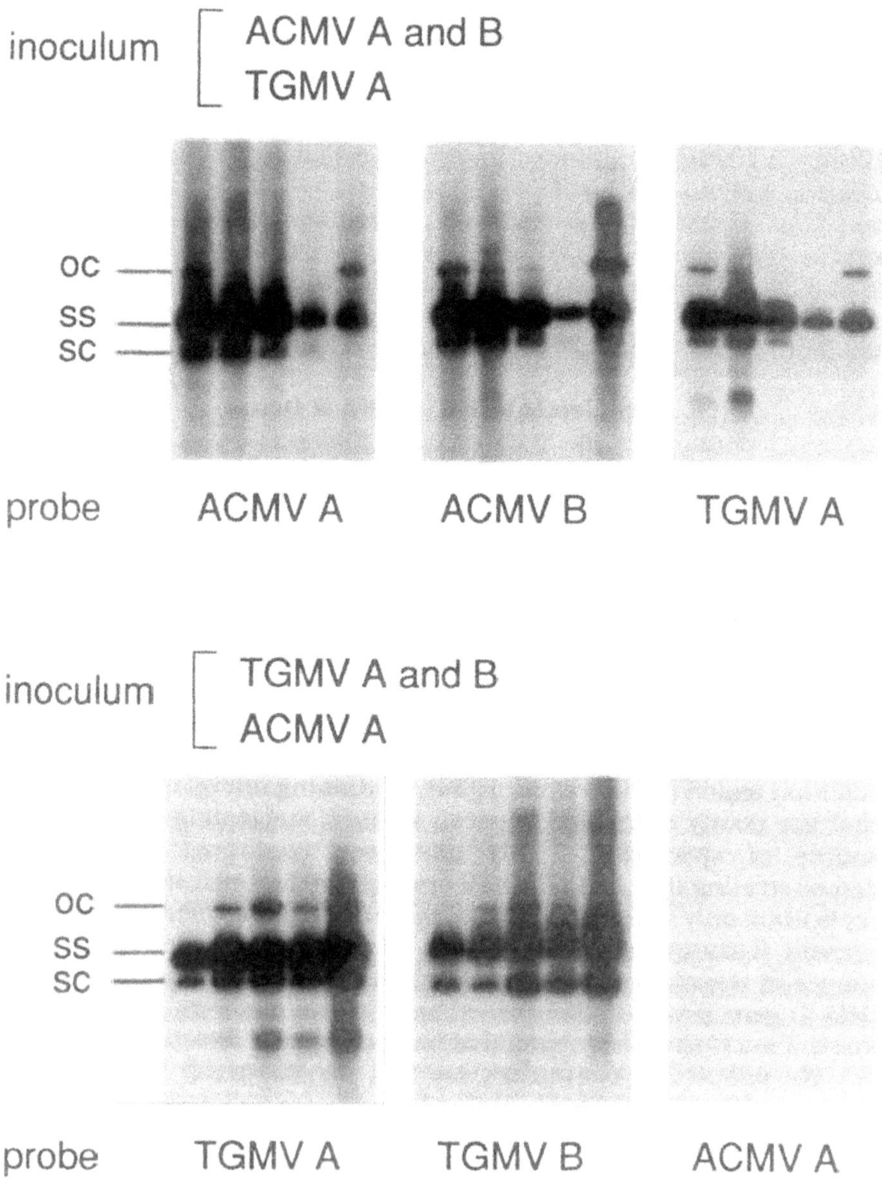

Figure 5. **Southern Blot Analysis of Viral DNA's Extracted from Systemically Infected *N. benthamiana* Leaves (5 examples of each) Co-inoculated with Cloned ACMV and TGMV Components (inoculum). The component-specific probes used to detect the viral DNA are indicated below the blots.**

experiment, TGMV DNA A was routinely found in systemically infected leaves when co-inoculated with ACMV DNA A and DNA B; and the level of the TGMV component was comparable to that found in a normal TGMV infection. Therefore, ACMV DNA B gene products are able to facilitate the spread of TGMV, indicative of a greater flexibility in the interaction between the gene products and the viral DNA in this instance.

This investigation was extended to include the related geminiviruses Indian cassava mosaic virus (ICMV) and abutilon mosaic virus (AbMV). The results summarized in Table 1 indicate that ACMV was able to spread DNA A of AbMV and ICMV in addition to TGMV and, in a reciprocal experiment, ICMV could spread ACMV DNA A. Like TGMV, AbMV was unable to spread ACMV DNA A although it facilitated the spread of TGMV DNA A; and, in the reciprocal experiment, TGMV supported the spread of AbMV DNA A. Other virus component combinations remain to be tested. On the strength of this analysis it would appear that the DNA B gene products of the African and Indian viruses ACMV and ICMV are more flexible in their ability to functionally interact with the genomic components of other viruses than those of the New World viruses TGMV and AbMV. Their different behavior might be due to host range adaptation (both ACMV and ICMV were originally isolated from cassava); or it might reflect divergent evolution of geographically isolated viruses, in which case it would be interesting to extend this investigation to include viruses from other locations. Throughout this series of co-inoculation experiments, induced symptoms were invariably typical of the helper virus, serving to emphasise the importance of DNA B as a phenotypic determinant.

In addition to *N. benthamiana*, ACMV can systemically infect and cause severe symptoms in *N. clevelandii*, *N. tabacum* and *D. stramonium*. Of these hosts, ICMV is unable to systemically infect *N. tabacum* and *D. stramonium*. Factors affecting host range determination were investigated by co-agroinoculation of ACMV and ICMV components into discriminating hosts. Our results using *N. clevelandii* indicate that ACMV can spread DNA A of ICMV and, in the reciprocal experiment, ICMV can spread DNA A of ACMV, consistent with the data derived from *N. benthamiana*, described above. However, ACMV was unable to support ICMV DNA A infection in the discriminating hosts *N. tabacum* and *D. stramonium*. Using a leaf disc assay, ICMV replicated as efficiently as ACMV in *N. benthamiana*. ACMV replicated less efficiently in *N. tabacum* than in *N. benthamiana*, although sufficiently well for the induction of systemic

Table 1. Complementation of Genomic Components for Systemic Spread in *N. benthamiana*.

inoculum clones[a]		symptom type	progeny components[a,b]	
helper	dependent		helper	dependent
AA + AB		ACMV	AA + AB	
AA + AB	TA	ACMV	AA + AB	TA
AA + AB	AbA	ACMV	AA + AB	AbA
AA + AB	IA	ACMV	AA + AB	IA
IA + IB		ICMV	IA + IB	
IA + IB	AA	ICMV	IA + IB	AA
TA + TB		TGMV	TA + TB	
TA + TB	AA	TGMV	TA + TB	
TA + TB	AbA	TGMV	TA + TB	AbA
AbA + AbB		AbMV	AbA + AbB	
AbA + AbB	AA	AbMV	AbA + AbB	
AbA + AbB	TA	AbMV	AbA + AbB	TA

[a] Genomic components are from ACMV (AA and AB), ICMV (IA and IB), TGMV (TA and TB) and AbMV (AbA and AbB).

[b] Viral DNA's in systemically infected leaves were identified by Southern blot analysis using probes specific to each virus component.

infection of this host. In contrast, in a parallel experiment we were unable to detect ICMV replication in *N. tabacum* leaf discs. The results suggest that ICMV host range is limited by the inability of the virus to replicate efficiently in non-hosts. Obviously, under these conditions we have been unable to determine the contribution of virus spread to host range in this instance.

Our results have indicated that TGMV is unable to facilitate the spread of ACMV DNA A. To investigate if individual DNA B genes of TGMV can functionally complement their ACMV counterparts, we adopted the approach of expressing the gene products from an ACMV DNA A coat protein replacement vector (von Arnim and Stanley, 1992a). Because the coat protein is not required for systemic infection of *N. benthamiana* (Stanley and Townsend, 1986), non-essential sequences downstream of the promoter can be replaced with alternative coding sequences that are expressed as transcriptional fusions. Using constructs DNA A:BC1+ and DNA A:BV1+ (Table 2), this approach has been used successfully for the functional expression of ACMV DNA B genes (Etessami *et al.*, 1988). When extended to TGMV DNA B genes, we found that *de novo* synthesized viral DNA could not be detected in the inoculated leaves following co-inoculation of the ACMV DNA A-based constructs DNA A:BC1+ with DNA A:BR1+ and DNA A:BL1+ with DNA A:BV1+ (Table 2) suggesting that, individually, TGMV BL1 and BR1 are unable to complement their ACMV counterparts BC1 and BV1. In contrast, *de novo* synthesized viral DNA was detected when DNA A:BL1+ and DNA A:BR1+ were co-inoculated and symptoms developed on the inoculated leaves, indicating that the gene products were being expressed and that they act in a co-operative manner. However, virus was unable to spread systemically from the inoculated leaves either because of a defect in expression of the TGMV genes or, more likely, as a result of incompatibility between the gene products and ACMV DNA. The latter explanation is consistent with the inability of TGMV to spread ACMV DNA A in the co-inoculation experiments described above.

The possibility that insertion of the TGMV DNA B genes into ACMV DNA A had a detrimental effect on infectivity of the latter was investigated by co-inoculation of the chimeric constructs with ACMV DNA B. Since a full complement of ACMV genes necessary for replication and spread are present under these conditions, it was anticipated that all of the chimeric constructs would be infectious, as previously demonstrated for DNA A:BC1+ and DNA A:BV1+ when co-inoculated with DNA B (Etessami *et al.*, 1988). Certainly, constructs containing the BR1 gene in either orientation (DNA A:BR1+ and DNA A:BR1-) and the BL1 gene in the antisense orientation (DNA A:BL1-) produced severe systemic infection and remained stable in *N. benthamiana* (Table 2). Somewhat surprisingly, however, the infectivity of construct DNA A:BL1+, containing the BL1 gene in the sense orientation, was significantly reduced and systemic symptoms were frequently mild. Analysis of progeny in the inoculated leaves showed that the virus could replicate efficiently and was maintained in its original form. In addition, chlorotic lesions developed in the inoculated leaves with similar timing to those associated with wild-type infections. On the occasions when plants supported a systemic infection, analysis of progeny from in the systemically infected leaves showed that rearrangements had occurred that disrupted the BL1

Table 2. Infectivity of ACMV DNA A Coat Protein Replacement Constructs Expressing DNA B Genes of ACMV (BC1 and BV1) and TGMV (BL1 and BR1) in *N. benthamiana*

inoculum clones[a]		local symptoms	systemic infectivity (%)	systemic symptoms	progeny
DNA A:BC1+	DNA A:BV1+	yes	91	severe	parental
DNA A:BC1+	DNA A:BR1+	no	0	-	-
DNA A:BL1+	DNA A:BV1+	no	0	-	-
DNA A:BL1+	DNA A:BR1+	yes	0	-	parental
DNA A:BR1-	DNA B	yes	91	severe	parental
DNA A:BR1+	DNA B	yes	93	severe	parental
DNA A:BL1-	DNA B	yes	98	severe	parental
DNA A:BL1+	DNA B	yes	16	severe	mutation
				mild	mutation
DNA A:BL1+Δ	DNA B	yes	95	severe	parental and mutation
DNA A:BC1/BL1+	DNA B	yes	100	severe	parental
DNA A:BL1/BC1+	DNA B	yes	7	mild	mutation

[a] + and - refer to sense and antisense orientation of the gene, respectively, relative to the coat protein promoter of ACMV DNA A. Modified from von Arnim and Stanley (1992a).

coding sequences. The results suggested that BL1 expression was responsible for the disruption of virus proliferation and that the block occurred at the level of virus systemic spread. The latter suggestion is consistent with the proposed function of its ACMV counterpart, BC1, in systemic spread. The contribution of BL1 to the phenotype was confirmed by introducing a frame-shift mutation into the BL1 coding sequence (DNA A:BL1+Δ) which served to restore systemic infectivity to the wild-type level (Table 2). Subsequent analysis of progeny in systemically infected tissues verified the presence of parental-type chimera in addition to recombinants that had been produced by disruption of the BL1 coding sequence.

In an attempt to define a domain within BL1 that is responsible for the suppression of ACMV systemic spread, chimeras were constructed between TGMV BL1 and ACMV BC1 coding sequences. When co-inoculated with DNA B, a construct containing a chimera produced from the amino-terminus of BC1 and the carboxy-terminus of BL1 (DNA A:BC1/BL1+) was as infectious as wild-type ACMV, produced severe systemic symptoms and remained stable in systemically infected tissues (Table 2). In contrast, the reciprocal construct containing a chimera produced from the amino-terminus of BL1 and the carboxy-terminus of BC1 (DNA A:BL1/BC1+) mimicked the phenotype of DNA A:BL1+. The progeny of this particular construct replicated and remained stable in inoculated leaves but was impaired in its ability to systemically infect *N. benthamiana* and induced only mild symptoms. Once again, progeny virus in systemically infected tissues contained rearrangements that served to disrupt the BL1/BC1 coding sequences. Therefore, we can conclude that sequences important for the inhibitory effect are located within the amino-terminal half of BL1. The chimera BL1/BC1 is essentially a modified version of BC1 which behaves as a dominant negative mutant of the protein.

Conclusions

Classical experiments involving co-inoculation of helper and dependent viruses will usually only reveal positive complementation between gene products. However, because the use of a coat protein replacement vector ensured that expression of the complementing gene occurred in every infected cell, we have been able to demonstrate the inhibitory effect of a gene product from one virus (TGMV BL1) on the systemic spread of another virus (ACMV). The obvious implication of such a result is that derivatives of spread proteins might be adapted for the control of virus infection. Our initial efforts to define more precisely the domain involved in the inhibitory effect have implicated the amino-terminal sequences of BL1. This investigation might be extended by screening additional BL1/BC1 chimeras and BC1 mutants for inhibitory phenotypes. It will be necessary to correlate the results of such a study with biochemical analyses that define functional domains of the protein that participate in, for example, binding to viral DNA, co-operative multimerisation, and recognition of host factors. While the underlying basis of inhibition is not yet understood, the fact that both BL1 and the chimera BL1/BC1 effectively interfere with virus proliferation in the presence of the functional gene product BC1 suggests that it is not due simply to non-productive binding to, and saturation of, specific recognition sites involved in virus spread. It is more likely that multimeric protein arrays, possibly produced as a result of co-operative binding to single-stranded nucleic acids as seen for other plant virus spread proteins (Citovsky *et al.*, 1990; Citovsky *et al.*, 1991; Osman, *et al.*, 1992), are disrupted by the inclusion of a limited number of copies of the inhibiting protein. If this proves to be the case, it suggests that the use of dominant negative mutants of spread proteins might find general application in the control of plant virus diseases.

We are currently investigating the ability of geminivirus spread proteins and their derivatives to disrupt the systemic spread of infection in transformed plants. Our results on the distribution of ACMV BC1 within cell fractions suggests that it is associated with the cell wall where it might modify the structure of plasmodesmata to allow virus spread, as suggested for other spread proteins (Stussi-Garaud *et al.*, 1987; Tomenius *et al.*, 1987; Linstead *et al.*, 1988). If this is the case, constitutive expression of the gene might be detrimental to plant development; and as a consequence, transformants will be selected for low or non-expression. The problem might be avoided by the use of a promoter that is active specifically in infected cells. Our recent results have demonstrated that expression from the coat protein promoter of ACMV is switched on by the product of gene AC2 (Saunders, Lucy, and Stanley, unpublished), consistent with observations on the control of coat protein gene expression by TGMV AL2 (Sunter and Bisaro, 1991). Consequently, transgenes under the control of the coat protein promoter will not be expressed in non-infected cells and, therefore, will not adversely affect the development of transformed plants. Also, genes might be expressed from an extrachromosomally replicating vector that is mobilized from an integrated template and amplified in *trans* by the infecting virus, as has already been demonstrated for defective ACMV DNA (Stanley *et al.*, 1990). Hence, amplification of transgene expression will occur only in infected cells by a combination of vector replication and promoter induction. Our preliminary results using such a vector have been encouraging. We have observed the induction and amplification of coat protein expression from a mobilized vector following infection of transgenic *N. benthamiana* leaf discs with an ACMV coat protein deletion mutant (Frischmuth and Stanley, unpublished). It remains to be seen if temporal control of expression from the coat protein promoter in infected cells will be suitable for the function of a dominant negative mutant as an inhibitor of virus spread. The feasibility of this approach is currently under investigation.

References

Albrecht, H., Geldreich, A., de Murcia, J.M., Kirchherr, D., Mesnard, J., and Lebeurier, G. 1988. Cauliflower mosaic virus gene I product detected in a cell wall-enriched fraction. Virology 163:503-508.

Boulton, M.I., King, D.I., Donson, J., and Davies, J.W. 1991. Point substitutions in a promoter-like region and the V1 gene affect the host range and symptoms of maize streak virus. Virology 183:114-121.

Brown, J.K. and Bird, J. 1992. Whitefly-transmitted geminiviruses and associated disorders in the Americas and the Caribbean Basin. Plant Disease 76:220-225.

Citovsky, V., Knorr, D., Schuster, G., and Zambryski, P. 1990. The P30 movement protein of tobacco mosaic virus is a single-stranded nucleic acid binding protein. Cell 60:637-647.

Citovsky, V., Knorr, D., and Zambryski, P. 1991. Gene I, a potential cell-to-cell movement locus of cauliflower mosaic virus, encodes an RNA binding protein. Proc. Natl. Acad. Sci. U.S.A. 88:2476-2480.

Day, A.G., Bejarano, E.R., Buck, K.W., Burrell, M., and Lichtenstein, C.P. 1991. Expression of an antisense viral gene in transgenic tobacco confers resistance to the DNA virus tomato golden mosaic virus. Proc. Natl. Acad. Sci. U.S.A. 88:6721-6725.

Deom, C.M., Oliver, M.J., and Beachy, R.N. 1987. The 30 kDa gene product of tobacco mosaic virus potentiates virus movement. Science 237:389-394.

Etessami, P., Callis, R., Ellwood, S., and Stanley, J. 1988. Delimitation of essential genes of cassava latent virus DNA 2. Nucleic Acids Res. 16:4811-4829.

Etessami, P., Saunders, K., Watts, J., and Stanley, J. 1991. Mutational analysis of complementary-sense genes of African cassava mosaic virus DNA A. J. Gen. Virol. 72:1005-1012.

Fontes, E.P.B., Luckow, V.A., and Hanley-Bowdoin, L. 1992. A geminivirus replication protein is a sequence-specific DNA binding protein. The Plant Cell 4:597-608.

Frischmuth, T. and Stanley, J. 1991. African cassava mosaic virus DI DNA interferes with the replication of both genomic components. Virology 183:539-544.

Godefroy-Colburn, T., Gagey, M-J., Berna, A., and Stussi-Garaud, C. 1986. A non-structural protein of alfalfa mosaic virus in the walls of infected tobacco cells. J. Gen. Virol. 67:2233-2239.

Harrison, B.D. 1985. Advances in geminivirus research. Ann. Rev. Phytopathol. 23:55-82.

Kikuno, R., Toh, H., Hayashida, H., and Miyata, T. 1984. Sequence similarity between putative gene products of geminiviral DNA's. Nature 308:562.

Klinkenberg, F.A. and Stanley, J. 1990. Encapsidation and spread of African cassava mosaic virus DNA A in the absence of DNA B when agroinoculated to *Nicotiana benthamiana*. J. Gen. Virol. 71:1409-1412.

Koonin, E.V., Mushegian, A.R., Ryabov, E.V., and Dolja, V.V. 1991. Diverse groups of plant RNA and DNA viruses share related movement proteins that may possess chaperone-like activity. J. Gen. Virol. 72:2895-2903.

Lazarowitz, S.G., Wu, L.C., Rogers, S.G., and Elmer, J.S. 1992. Sequence-specific interaction with the viral AL1 protein identifies a geminivirus DNA replication origin. The Plant Cell 4:799-809.

Linstead, P.J., Hills, G.J., Plaskitt, K.A., Wilson, I.G., Harker, C.L., and Maule, A.J. 1988. The subcellular location of the gene I product of cauliflower mosaic virus is consistent with a function associated with virus spread. J. Gen. Virol. 69:1809-1818.

Navot, N., Pichersky, E., Zeidan, M., Zamir, D., and Czosnek, H. 1991. Tomato yellow leaf curl virus: A whitefly-transmitted geminivirus with a single genomic component. Virology 185:151-161.

Osman, T.A.M., Hayes, R.J., and Buck, K.W. 1992. Cooperative binding of the red clover necrotic mosaic virus movement protein to single-stranded nucleic acids. J. Gen. Virol. 73:223-227.

Revington, G.N., Sunter, G., and Bisaro, D.M. 1989. DNA sequences essential for replication of the B genome component of tomato golden mosaic virus. The Plant Cell 1:985-992.

Stanley, J. 1983. Infectivity of the cloned geminivirus genome requires sequences from both DNA's. Nature 305:643-645.

Stanley, J. 1991. The molecular determinants of geminivirus pathogenesis. Seminars in Virology 2:139-149.

Stanley, J., Frischmuth, T., and Ellwood, S. 1990. Defective viral DNA ameliorates symptoms of geminivirus infection in transgenic plants. Proc. Natl. Acad. Sci. U.S.A. 87:6291-6295.

Stanley, J. and Gay, M.R. 1983. Nucleotide sequence of cassava latent virus DNA. Nature 301:260-262.

Stanley, J. and Latham, J.R. 1992. A symptom variant of beet curly top geminivirus produced by mutation of open reading frame C4. Virology 190:506-509.

Stanley, J., Latham, J.R., Pinner, M.S., Bedford, I., and Markham, P.G. 1992. Mutational analysis of the monopartite geminivirus beet curly top virus. Virology 191:396-405.

Stanley, J., Markham, P.G., Callis, R.J., and Pinner, M.S. 1986. The nucleotide sequence of an infectious clone of the geminivirus beet curly top virus. EMBO J. 5:1761-1767.

Stanley, J. and Townsend, R. 1985. Characterization of DNA forms associated with cassava latent virus infection. Nucleic Acids Res. 13:2189-2206.

Stanley, J. and Townsend, R. 1986. Infectious mutants of cassava latent virus generated *in vitro* from intact recombinant DNA clones containing single copies of the genome. Nucleic Acids Res. 14:5981-5998.

Stussi-Garaud, C., Garaud, J., Berna, A., and Godefroy-Colburn, T. 1987. *In situ* location of an alfalfa mosaic virus non-structural protein in plant cell walls: Correlation with virus transport. J. Gen. Virol. 68:1779-1784.

Sunter, G. and Bisaro, D.M. 1989. Transcription map of the B genome component of tomato golden mosaic virus and comparison with A component transcripts. Virology 173:647-655.

Sunter, G. and Bisaro, D.M. 1991. Transactivation in a geminivirus: AL2 gene product is needed for coat protein expression. Virology 180:416-419.

Tomenius, K., Clapham, D., and Meshi, T. 1987. Localization by immunogold cytochemistry of the virus coded 30 K protein in plasmodesmata of leaves infected with tobacco mosaic virus. Virology 160:363-371.

Townsend, R., Stanley, J., Curson, S.J., and Short, M.N. 1985. Major polyadenylated transcripts of cassava latent virus and location of the gene encoding coat protein. EMBO J. 4:33-37.

Townsend, R., Watts, J., and Stanley, J. 1986. Synthesis of viral DNA forms in *Nicotiana plumbaginifolia* protoplasts inoculated with cassava latent virus (CLV); evidence for the independent replication of one component of the CLV genome. Nucleic Acids Res. 14:1253-1265.

von Arnim, A., Frischmuth, T., and Stanley, J. 1993. Detection and possible functions of African cassava mosaic virus DNA B gene products. Virology 192:264-272.

von Arnim, A. and Stanley, J. 1992a. Inhibition of African cassava mosaic virus systemic infection by a movement protein from the related geminivirus tomato golden mosaic virus. Virology 187:555-564.

von Arnim, A. and Stanley, J. 1992b. Determinants of tomato golden mosaic virus symptom development located on DNA B. Virology 186:286-293.

Figwort Mosaic Virus, a Caulimovirus, Regulates the Expression of its Genes via Transactivation of a Polycistronic mRNA

S. Gowda, H. B. Scholthof, and R. J. Shepherd
Department of Plant Pathology
University of Kentucky
Lexington, Kentucky 40546-0091

The 7700 bp genome of figwort mosaic virus (FMV) is transcribed as two RNA's. The major transcript, which spans the entire circular DNA genome, has five of its genes in a closely spaced tandem array (genes I through V). The second transcript is monocistronic and spans only gene VI. Experiments involving transient expression in plant protoplasts, following electroporation of viral DNA's containing a reporter gene, show that gene VI is a regulatory gene that transactivates expression of genes on the full-length transcript. A *cis* element of the long nontranslated leader region (600 bp) is required for the transactivation response. Gene VI may promote coupled translation of the five closely spaced genes of the full-length transcript. In addition, *cis*-acting sequences of the smaller transcript have a negative effect on the expression of this mRNA. This negative effect is relieved by the presence of gene VI suggesting autoregulation of this gene of FMV.

Introduction

Figwort mosaic virus (FMV), a member of the caulimovirus group (Shepherd *et al.*, 1987), has a double-stranded DNA genome of 7743 base pairs (Richins *et al.*, 1987). Its genome is arranged into two sets of genes distinguised by two corresponding RNA transcripts (Scholthof *et al.*, 1992). The major transcript is polycistronic in nature and encompasses the entire viral genome. This RNA is transcribed from a promoter located in the distal end of gene VI and extending into the large intergenic region. The second transcript is monocistronic and spans only the gene VI portion of the genome (Figure 1). Genes I through V are arranged in a head-to-tail array on the polycistronic transcript with only few nucleotides between the coding regions of adjoining genes (example: genes I/II, IV/V), while in some instances (example: genes II/III, III/IV) the start codon overlaps the stop codon of the preceding gene as though they were translationally coupled, a common feature observed with prokaryotic genes (Gronenborn, 1987).

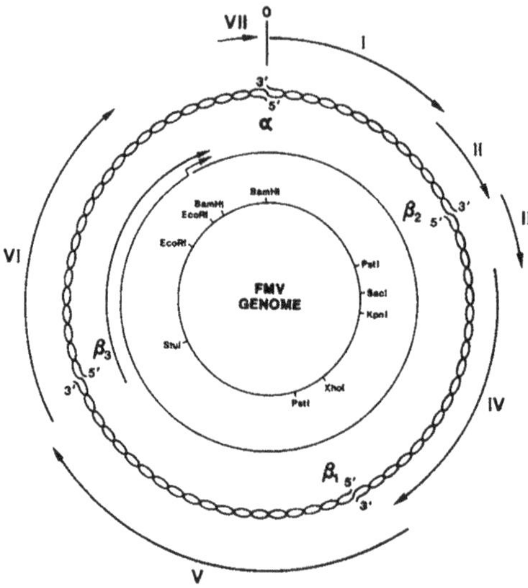

Figure 1. Physical Map of Figwort Mosaic Virus. The circular double-stranded DNA of 7743 base pairs is indicated by the interwoven lines. Virion DNA has four interruptions at specific sites, one (α) in the transcribed minus strand and three (β1, β2, and β3) in the complementary plus strand. Some of the representative restriction sites are indicated. The peripheral heavy arrows and Roman numerals indicate the major open reading regions which occur on the full-length RNA transcript. The two RNA transcripts are represented by the thin lines of the inner circle.

No other subgenomic RNA's other than the gene VI transcript have been found for the caulimoviruses with the exception of soybean chlorotic mottle virus (Soy CMV). A sequence inside gene III of Soy CMV has been reported to promote transcription of a subgenomic RNA (Hasegawa *et al.*, 1989). None of the conserved genes of the major transcript of caulimoviruses occupy a 5'-proximal position. Therefore, the conventional rule for the translation of eukaryotic messenger RNA's may not apply to these polycistronic RNA's. Although translational coupling may explain the expression from the internal genes I through V, the involvement of other factors such as gene VI transactivation and the *cis* elements in the 5'-leader and the smaller transcript indicate that an unusual but as yet undefined mechanism may be involved in the translation of these genes. We will briefly review some of these regulatory events with the caulimoviruses with special emphasis on gene expression of FMV.

Results and Discussion

Preliminary primer extension studies have indicated that the transcription start site for the major transcript is at position 6940 (unpublished observation). The full-length transcript has a 5'-leader (spanning most of the large intergenic region) of 540 nucleotides upstream from the start of the first significant open reading region (gene VII). The smaller transcript is approximately 1700 nucleotides long and is transcribed from the promoter located in the smaller intergenic region positioned between genes V and VI (Figure 1). The approach used to study the transient gene expression in protoplasts consisted of cloning a reporter gene (chloramphenicol acetyltransferase, CAT) at various positions within the FMV genome and electroporation of restructured DNA into protoplasts from suspension cultures of *Nicotiana edwardsonii*, a systemic host for FMV. After 24 hours, the cells were analyzed for CAT gene expression.

Expression of Genes on the Polycistronic Transcript is Dependent on the Transactivation by Gene VI Protein: The leader region of the major transcript of FMV contains small open reading regions varying in lengths from 3 to 23 codons between the 5'-end of the transcript and the start codon of gene VII. In addition, this region of the RNA probably folds into an extensive secondary structure as indicated by computer analysis. The leader of the 35S (full-length) transcript of cauliflower mosaic virus (CaMV) has similar features (Futterer *et al.*, 1988). These open reading regions and secondary structure may be responsible for the repression of translation of downstream cistrons of the full-length transcript of FMV and CaMV (Baughman and Howell, 1988; Hohn *et al.*, 1989; Futterer *et al.*, 1989). The repressing effect of the leader region in FMV on the expression of a CAT gene fused downstream of gene VII (pFMV32 CAT) has been evaluated compared to the high level of reporter gene expression when the latter is placed immediately downstream of the major RNA promoter (pFMV20 CAT) (Gowda *et al.*, 1989). The repressing effect was evident even with short segments of the intergenic region (Gowda *et al.*, 1989). However, addition of the cloned full-length genome of FMV with its intact gene VI to the electroporation mixture along with pFMV32 CAT relieved the repressive effect of the 5'-leader region and elevated the CAT activity by 6-9 fold (Gowda *et al.*, 1989; Gowda *et al.*, 1990) (Figure 2). The absence of any activation of CAT gene expression in pFMV32 CAT upon co-electroporation with either 3'- or

5'-truncated versions of gene VI suggest that the protein of gene VI, rather than its transcript, participates in transactivation (Gowda *et al.*, 1989; Gowda *et al.*, 1990). Operation of an apparently similar phenomenon of transactivation of the genes on the polycistronic RNA of CaMV (Bonneville *et al.*, 1989; Futterer and Hohn, 1991; Futterer and Hohn, 1992) suggest that post-transcriptional transactivation is a common mode of regulation of gene expression for the caulimoviruses.

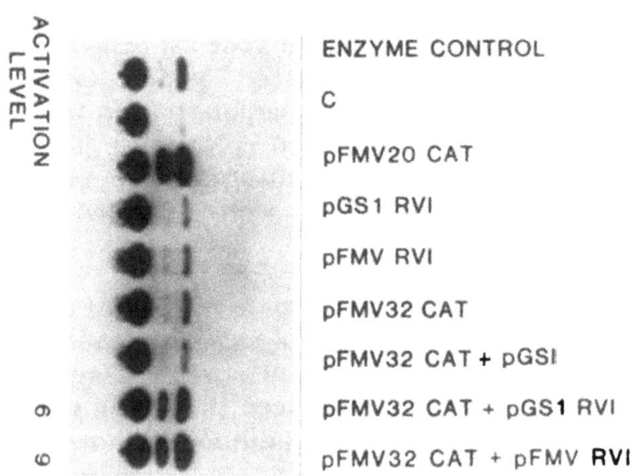

Figure 2. Transactivation by FMV Gene VI Expressing Plasmids, pGS1 RVI and pFMV RVI. Supercoiled DNA's of the reporter gene plasmids and transactivating plasmids were mixed and electroporated into protoplasts from suspension cell cultures of *N. edwardsonii*. After 24 hours, $2x10^5$ cells were used for each CAT assay. The positive CAT expression plasmid, pFMV20, containing the reporter gene placed immediately downstream of the major RNA promoter (devoid of 5'-leader region and gene VII of FMV RNA) was used as a positive control to monitor the electroporation efficiency. For details of electroporation and quantitation of the CAT assays see Gowda *et al.*, 1989.

Gene VII is Required in cis for the Expression of Downstream Genes on the FMV Polycistronic Transcript: The plasmid pFMV32 CAT with the promoter for the major transcript, the large intergenic region (5'-leader) and ORF VII positioned upstream of the reporter gene expressed CAT activity only when co-electroporated with gene VI containing plasmids of FMV (Gowda *et al.*, 1989; Gowda *et al.*, 1990; Gowda *et al.*, 1991). Similar plasmids with varying lengths of the leader devoid of gene VII failed to express CAT even when co-electroporated with gene VI. This strongly suggests that gene VII sequences are required for the expression of downstream genes of the major transcript of FMV (Figure 3). Further, there was no enhanced expression when plasmids with truncated leader regions were supplied with gene VII in *trans*, even in the presence of gene VI, thus showing the need for *cis* acting sequences of gene VII for the transactivation response (sample data not provided).

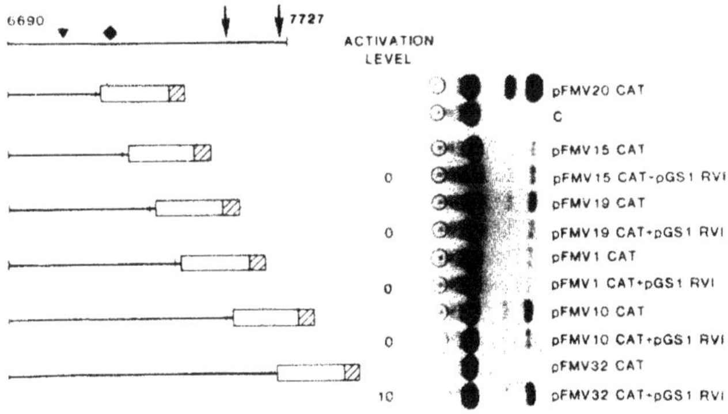

Figure 3. Gene VI Transactivation of Plasmids with Varying Lenghts of the 5'-Leader of the Major Transcript of FMV. The thick solid line at the top indicates the genetic map of the leader region showing the position of the 'TATA' box (▾) and poly A signal (◆). Gene VII (↓↓) starts at nucleotide 7499 and ends at 7700 (Gowda *et al.*, 1989). The relative lengths of the leader sequence attached to the 5'-region of the CAT gene is shown by thin solid lines. The open box represents the CAT gene and the hatched area indicates the rubisco gene 3'-terminator. All clones start at nucleotide 6690 of the FMV genome, nearly 200 bases upstream of the 'TATA' box. In pFMV20 CAT the position of the CAT gene is at nucleotide 7003; in pFMV15 CAT it is at nucleotide 7105; in pFMV19 CAT it is at nucleotide 7223; in pFMV1 CAT it is at nucleotide 7316; in pFMV10 CAT it is at nucleotide 7504; and in pFMV32 CAT it is at nucleotide 7667 (Gowda *et al.*, 1989).

Additional evidence for the requirement of gene VII sequences in *cis* for the transactivation response comes from the experiments evaluating expression of CAT as a downstream gene of the major RNA transcript. The position of gene VII in various truncated leader constructs in relation to the transactivation response was tested (Gowda *et al.*, 1991). Gene VII when placed between CAT and the promoter as in pFMV20-VII CAT (Figure 5) greatly reduced expression of the CAT gene (94.1 percent reduction compared to pFMV20 CAT). Repression of CAT expression in this case may be the direct result of the gene VII open reading region. However, when pFMV20-VII CAT was mixed and co-electroporated with a gene VI containing plasmid, an increase in the CAT expression was observed. In addition, when gene VII was inserted between the leader and the CAT gene in plasmids such as pFMV4 CAT and pFMV52 CAT, the resulting plasmids (pFMV4-VII CAT and pFMV52-VII CAT) become responsive to gene VI transactivation (Figure 4). Various constructs containing

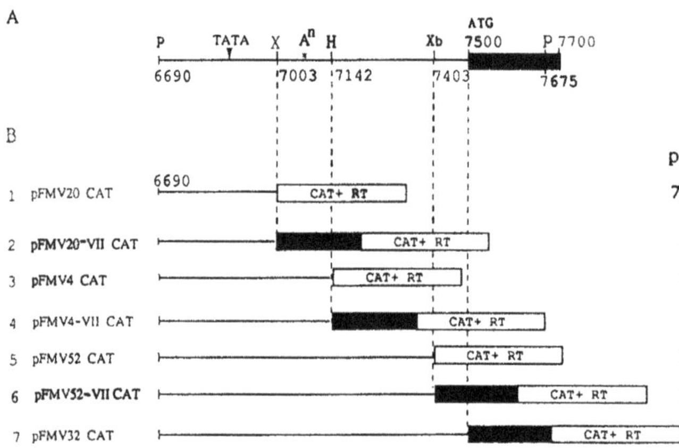

	CAT activity pGS1	CAT activity pGS1 RVI	activation level
1 pFMV20 CAT	74.1	83.3	1.1
2 pFMV20-VII CAT	4.3	47.8	11.1
3 pFMV4 CAT	6.1	6.8	1.1
4 pFMV4-VII CAT	2.2	34.4	15.6
5 pFMV52 CAT	2.3	3.4	1.4
6 pFMV52-VII CAT	2.9	40.2	13.8
7 pFMV32 CAT	3.2	64.5	20.1

Figure 4. Gene VI Transactivation of Truncated Leader Plasmids of FMV With and Without Gene VII in *cis* (Gowda *et al.*, 1991). A: A schematic representation of a portion of FMV genome (nucleotide position 6690-7700 is shown). This region represents the sequence containing the 3'-end of gene VI (gene VI stop codon is at nucleotide position 6902), the complete leader region and gene VII (thick solid box). Positions of the 'TATA' box, poly A signal and the start and stop of gene VII are also indicated. B: The positions of the CAT gene with the rubisco terminator (CAT + RT, open box) in pFMV20 CAT, pFMV4 CAT, and pFMV52 CAT are at nucleotide 7003, 7142, and 7403, respectively, on the leader. In pFMV20-VII CAT, pFMV4-VII CAT, and pFMV32 CAT, gene VII is positioned between the truncated leader and CAT gene at positions indicated by the broken vertical lines. 20 μg of plasmid DNA was electroporated into the suspension cell protoplasts of *N. edwardsonii*. CAT activities are presented in the right panel. In some cases, 50 μg of the transactivating plasmid (pGS1 RVI) was mixed with the expression plasmid and co-electroporated as indicated (Gowda *et al.*, 1991).

gene VII sequences and other short stretches of the leader region were made and evaluated in an effort to define the nature of the *cis* element involved in the gene VI transactivation response. Gene VII truncation from its 3'-end was tested in an effort to define the minimal gene VII sequences needed for the transactivation response. In addition, frame shift mutations in gene VII, mutations of its AUG codon and in-frame deletions were used to further define the *cis* requirement for gene VI transactivation. These experiments revealed that the open reading region itself, corresponding to gene VII of FMV, was not required for the response. Instead only a 45 base pair region within gene VII region was needed for gene VI enhanced expression of downstream genes of the major transcript.

All Major Genes Are Expressed from the Polycistronic Transcript and Respond to Transactivation by Gene VI: In order to understand whether or not genes I through V are expressed from the polycistronic major transcript of FMV, plasmids which resembled the native full-length viral genome with the CAT gene in several more distant downstream positions (Figure 5) (Scholthof *et al.*, 1992) were generated from a redundant end clone of a naturally occurring deletion mutant of FMV (Scholthof *et al.*, 1991). These plasmids were tested in protoplast transient expression assays. Expression from CAT in a variety of downstream positions was observed including some in which CAT was inserted as far downstream as gene V; for example, in pH64 and pH44-3 (Figure 5). Moreover, the data from these plasmids suggest that the transactivation response is dependent on the promoter for the major transcript. Removal of the promoter; e.g., in pH44ΔE, abolishes CAT activity in these experiments (Figure 5). No evidence for the activity of downstream transcription units, with the exception of the gene VI unit, has been observed. Moreover, the only evidence for the splicing of the major transcript is that which generates a deletion mutant of FMV (Scholthof *et al.*, 1991). In fact, splicing appears to be incompatible with the function of the major transcript as a replicative intermediate. It seems more likely that FMV genes I through V are expressed from translation of the major transcript as a polycistronic messenger RNA (Scholthof *et al.*, 1992). In addition, *in vitro* experiments with reticulocyte lysates and artifical transcripts of FMV have also shown that internal cistrons were translated readily (Ranu *et al.*, 1992). The reporter gene was translated efficiently in all positions except when placed immediately downstream of the 5'-leader (data not shown). In these experiments,

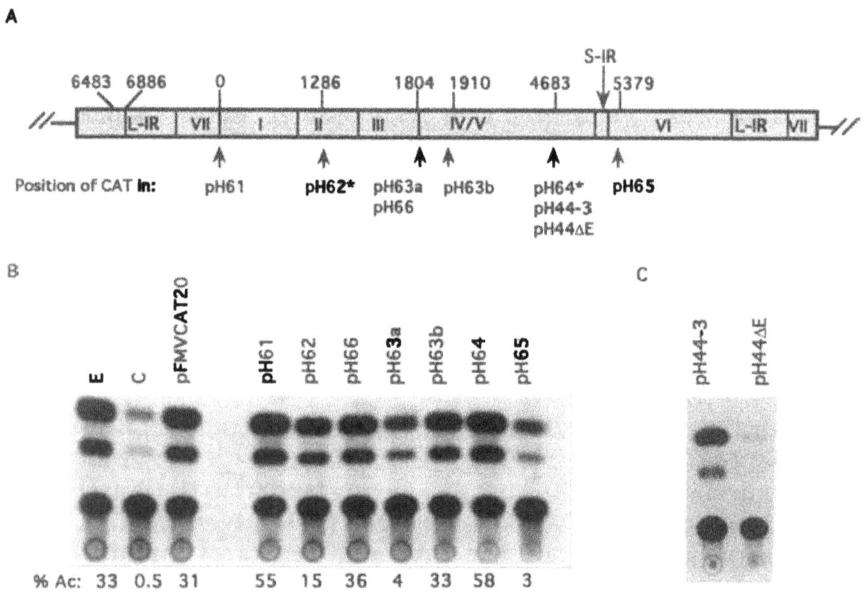

Figure 5. Gene Expression in Partially Redundant Clones of the FMV Genome with CAT Gene at Several Downstream Positions. A: Schematic representation of the partially redundant clone (1.2 mer) of FMV (shaded box) used in the protoplast transient expression experiments. Roman numerals indicate viral genes; the numbers represent the nucleotide positions (denoted by arrows) at which the CAT gene is placed in various plasmids. The asterisks indicate when the CAT gene is present as a separate cistron with its own start codon and not in phase with the gene in which it is inserted. L-IR and S-IR represent the large and the small intergenic region respectively. The *Eco* RI bounded segment between nucleotide positions 6483-6886 has been deleted in pH44ΔE. B: CAT assays of protoplasts electroporated with 50 μg of DNA of plasmids from panel A. Lanes E, C and pFMV CAT20 labeled samples represent the CAT enzyme (0.1 unit), a mock-electroporated sample of protoplasts and protoplasts electroporated with a positive expression plasmid, respectively. The numbers at the bottom indicate the percent acetylation of [^{14}C]-chloramphenicol used in the CAT assay. C: CAT assays of protoplasts electroporated with pH44-3 and pH44ΔE. (Scholthof *et al.*, 1992).

translation of three native viral genes (I, IV, and VI) from artifically generated polycistronic transcripts of FMV confirmed the previous protoplast experiments with viral DNA constructs (Ranu *et al.*, 1992). These observations suggest that the full-length RNA of FMV serves as a polycistronic RNA as others have proposed (Gronenborn, 1987; Sieg and Gronenborn, 1982; Dixon and Hohn, 1984).

In other experiments, the involvement of gene VI in the expression of genes on the partially redundant clones containing CAT at different downstream positions (Figure 6) was evaluated after deletion of a portion of the 5'-end of gene VI. We had previously observed that deletion of this region destroyed the activity of gene VI in the transactivation response (Gowda *et al.*, 1989). Electroporation of these plasmids (pH60 series) resulted in very poor expression of the CAT gene (Figure 6) compared to that of plasmids with an intact gene VI. However, the low level of CAT expression could be dramatically increased by supplying the gene VI function in *trans* by co-electroporation with pGS1 RVI (Figure 6). The latter is a plasmid containing the gene VI coding region of FMV with 5'-/3'-expression signals consisting of the CaMV 35S promoter and rubisco terminator (Gowda *et al.*, 1990). These observations confirmed the initial role of gene VI in the expression of the major genes of the polycistronic transcript of FMV.

cis-Acting Elements Present on Both Viral Transcripts Regulate Viral Gene Expression: As mentioned above, the expression of the downstream genes on the polycistronic transcript in FMV involves transactivation by the product of gene VI (Scholthof *et al.*, 1992; Gowda *et al.*, 1989) and a *cis* element in gene VII (Gowda *et al.*, 1989; Gowda *et al.*, 1991). Gene VI has been observed to transactivate the expression of all genes of the full-length transcript (Scholthof *et al.*, 1992; Gowda *et al.*, 1989; Bonneville *et al.*, 1989; Futterer and Hohn, 1991). Other experiments indicate that the expression of gene VI is inhibited by *cis*-acting sequences within gene VI itself (Scholthof *et al.*, 1992) (Figure 7). This inhibition could be relieved at least partially by supplying the gene VI function as a separate plasmid during electroporation (Scholthof *et al.*, 1992). The promoter for the major transcript of FMV appears to down regulate the expression of gene VI (Scholthof *et al.*, 1992).

The mechanism underlying the gene VI mediated transactivation process has not yet been defined. Several observations suggest it may be translational in nature, perhaps involving some mechanism which

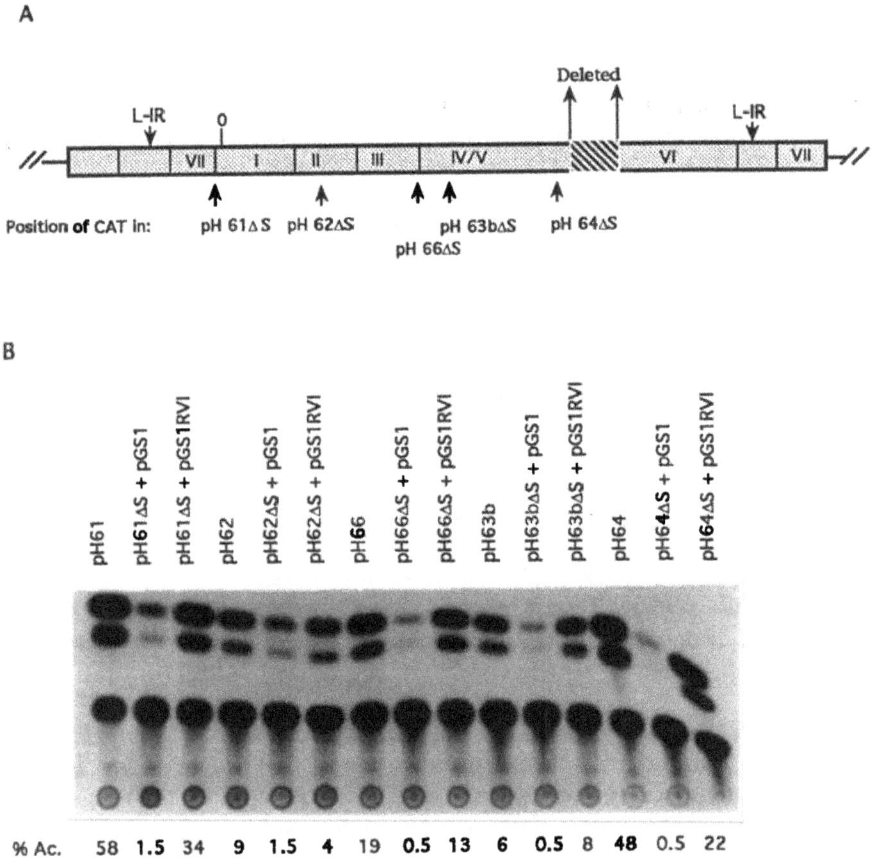

Figure 6. Transactivation of CAT Gene Expression in Various Downstream Positions of the FMV Genome. A: Schematic representation of the partially redundant genome of the plasmids used in the transient expression assays in protoplasts of *N. edwardsonii*. The explanations of the symbols used are the same as those in Figure 5. The *Stu* I-*Sna* BI fragment (nucleotide numbers 4795-5379) containing the promoter and the 5'-end of gene VI was removed from parental clones pH61 through pH66 to generate deletion mutants pH61ΔS through pH66ΔS. B: Results of the CAT assays of suspension cell protoplasts electroporated with the 50 μg of the relevant plasmid DNA's indicated above the respective lanes (Scholthof *et al.*, 1992).

promotes reinitiation by ribosomes at the junction of successive genes. However, this would not account for all the data, particularly that for transactivation of gene VI itself which is not closely spaced to other major genes of the FMV genome.

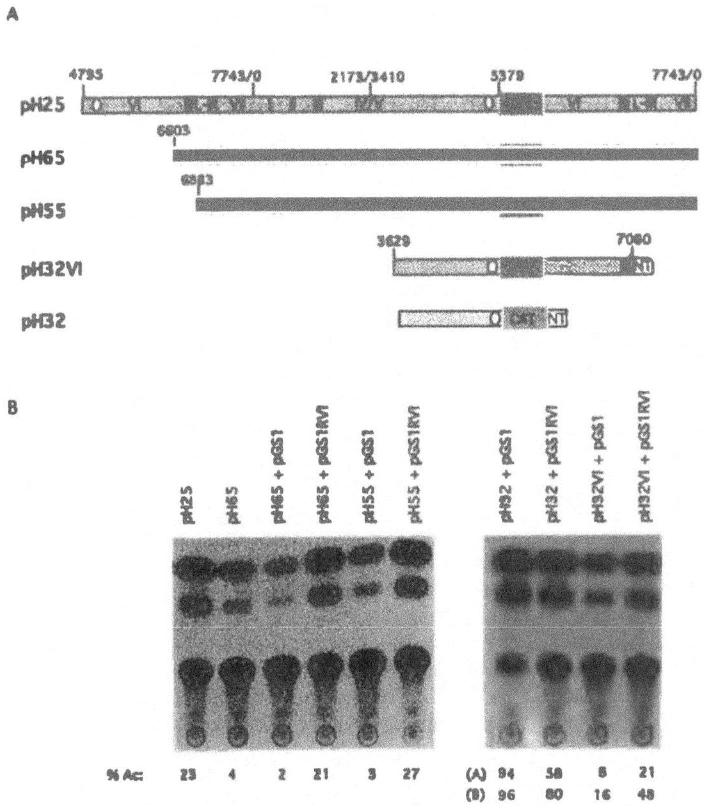

Figure 7. Regulation of Expression of Gene VI of FMV Through Transactivation and Participation of a *cis* Element in a Downstream Portion of the Genome. A: Schematic representation of the plasmids used in the protoplast transient expression experiments. The linear representation of the partial redundant clone of the deletion mutant (Scholthof *et al.*, 1991) containing the CAT gene in frame with gene VI at the *Stu* I site (nucleotide position 5379 on the FMV genome) and other mutant plasmids used in the study. Position 0 is same as position 7743 on the circular genome. The Roman numerals indicate the positions of the major genes. The open circle represents the position of the gene VI promoter. The black circle indicates the position of the promoter for the major transcript. The shaded box represents the FMV polyadenylation signal. The box with NT denotes the nopaline synthase gene polyadenylatin signal. B: CAT assay of *N. eduardsonii* suspension cell protoplasts electroporated with equimolar amounts of DNA of plasmids as indicated above each line. 50 µg of pGS1 or the same plasmid with gene VI of FMV (pGS1 RVI), was added before co-electroporation (Scholthof *et al.*, 1992).

Acknowledgements

The authors wish to thank F. C. Wu for the help with transient expression studies and Jennifer Kiernan for maintaining the suspension cell cultures. This work was supported by U.S. Department of Agriculture Grants 88-37263-3847 and 90-37262-5307. We are also grateful to Kentucky State THRI project 5-41086 for facilities and support.

References

Baughman, G. and Howell, S.A. 1988. Cauliflower mosaic virus 35S RNA leader region inhibits translation of downstream genes. Virology 167:125.

Bonneville, J.-M., Sanfacon, H., Futterer, J., and Hohn, T. 1989. Posttranscriptional *trans*-activation in cauliflower mosaic virus. Cell 59:1135.

Dixon, L.K. and Hohn, T. 1984. Initiation of translation of the cauliflower mosaic virus genome from a polycistronic mRNA: Evidence from deletion mutagenesis. EMBO J. 3:2731.

Futterer, J., Gordon, K., Bonneville, J.-M., Sanfacon, H., Pisan, B., Penswick, J., and Hohn, T. 1988. The leader sequence of cauliflower mosaic virus large RNA can be folded into a large stem-loop structure. Nucleic Acids Res. 16:8377.

Futterer, J., Gordon, K., Sanfacon, H., Bonneville, J.-M., and Hohn, T. 1989. Differential inhibition of downstream gene expression by the cauliflower mosaic virus 35S RNA leader. Virus Genes 3:45.

Futterer, J. and Hohn, T. 1991. Translation of polycistronic mRNA in the presence of the cauliflower mosaic virus transactivator protein. EMBO J. 10:3887.

Futterer, J. and Hohn, T. 1992. Role of an upstream open reading frame in the translation of polycistronic mRNA's in plant cells. Nucleic Acids Res. 20:3851.

Gowda, S., Scholthof, H.B., Wu, F.C., and Shepherd, R.J. 1991. Requirement of gene VII in *cis* for the expression of downstream genes on the major transcript of figwort mosaic virus. Virology 185:867.

Gowda, S., Wu, F.C., Scholthof, H.B., and Shepherd, R.J. 1989. Gene VI of figwort mosaic virus (caulimovirus group) functions in posttranscriptional expression of genes on the full-length RNA transcript. Post Natl. Acad. Sci. U.S.A. 86:9203.

Gowda, S., Wu, F.C., Scholthof, H.B., and Shepherd, R.J. 1990. Gene VI of figwort mosaic virus activates the expression of internal cistrons of the full-length polycistronic RNA transcript. In: *Viral Genes and Plant Pathogenesis*. T.P. Pirone and J.G. Shaw (eds.), pp. 79. Springer-Verlag, New York.

Gronenborn, B. 1987. The molecular biology of cauliflower mosaic virus and its application as plant gene vector. In: *Plant DNA Infectious Agents*. T. Hohn and J. Schell (eds.), pp. 1. Springer-Verlag, New York.

Hasegawa, A., Verver, J., Shimada, A., Saito, M., Goldbach, R., Van Kammen, A., Miki, K., Kameya-Iwaki, M., and Hibi, M. 1989. The complete sequence of soybean chlorotic mottle virus DNA and the identification of the novel promoter. Nucleic Acids Res. 17:9963.

Hohn, T., Bonneville, J.-M., Futterer, J., Gordon, K., Pisan, B., Sanfacon, H., Schultze, M., and Jiricny, J. 1989. The use of 35S RNA as either messenger or replicative intermediate might control the cauliflower mosaic virus replication. In: *Molecular Biology of Plant Pathogen Interactions. UCLA Symposium on Molecular and Cellular Biology.* Vol. 101. B. Staskawicz, P. Ahlquist, and O. Yoder. (eds.), pp. 153. Liss, New York.

Ranu, R.S., Gowda, S., Scholthof, H.B., Wu, F.C., and Shepherd, R.J. 1992. *In vitro* translation of the full-length RNA transcript of figwort mosaic virus (caulimovirus), a eukaryotic polycistronic mRNA. In: *Translational Control* (Abstract) 204. Cold Spring Harbor Laboratory, Cold Spring Harbor, New York.

Richins, R.D., Scholthof, H.B., and Shepherd, R.J. 1987. Sequence of figwort mosaic virus DNA (caulimovirus group). Nucleic Acids Res. 15:8451.

Scholthof, H.B., Gowda, S., Wu, F.C., and Shepherd, R.J. 1992. The full-length transcript of a caulimovirus is a polycistronic mRNA whose genes are transactivated by the product of gene VI. J. Virology 66:3131.

Scholthof, H.B., Wu, F.C., Gowda, S., and Shepherd, R.J. 1992. Regulation of caulimovirus gene expression and the involvement of *cis*-acting elements on both viral transcripts. Virology 190:403.

Scholthof, H.B., Wu, F.C., Richins, R.D., and Shepherd, R.J. 1991. A naturally occuring deletion mutant of figwort mosaic virus (caulimovirus group) is generated by RNA splicing. Virology 184:290.

Shepherd, R.J., Richins, R.D., Duffus, J.E., and Handley, M.K. 1987. Figwort mosaic virus: Properties of the virus and its adaptation to a new host. Phytopathology 77:1688.

Sieg, K. and Gronenborn, B. 1982. Evidence for a polycistronic messenger RNA encoded by CaMV. In: *Structure and Function of Plant Genomes*, pp. 184. NATO/FEBS Adv. Course.

The Role of Structural Domains in the Regulation of Viroid Pathogenicity

R. W. Hammond and R. A. Owens
Molecular Plant Pathology Laboratory
Plant Sciences Institute
Agricultural Research Service, USDA
Beltsville, Maryland 20705

T. Sano
Faculty of Agriculture
Hirosaki University
Hirosaki 036, Japan

T. O. Diener
Center for Agricultural Biotechnology and Department of Botany
University of Maryland
College Park, Maryland 20740

Viroid pathogenicity is a complex biological property whose expression is influenced by both the viroid and host genomes. The molecular response of plants to infection by viroids is manifested by a variety of disease symptoms - the type and severity of which are influenced by the type and strain of viroid and the host onto which it is inoculated as well as the physiological state of the plant and the environment in which it is grown. One direct approach that we and others have taken towards an understanding of viroid pathogenicity is a dissection of the viroid genome in order to assign functions to particular regions of the genome. As a part of an investigation of the role of individual structural domains in viroid pathogenesis and replication, we have introduced a series of point mutations into cloned, infectious cDNA copies of viroid RNA's as well as constructed intra- and interspecific chimeras between certain viroids. Quantitative comparisons of symptom development and progeny accumulations reveal that the T_L (left terminal) domain contains a determinant required for severe veinal necrosis in tomato, the T_L and P (pathogenicity) domains are both important for epinasty and stunting in tomato, and the V (variable) and T_R (right terminal) domains also play an important role in pathogenicity by contributing to the regulation of viroid replication/accumulation.

Introduction

Viroids are the smallest known agents of infectious disease. They are responsible for a number of economically significant diseases in vegetatively propagated crops, and much effort has been devoted to their molecular characterization in hopes of understanding how these unusual pathogens interact with their hosts to cause disease (Diener, 1987). Viroids are characterized as being autonomously replicating, circular, single-stranded RNA's which range in size from 246 to 375 nucleotides. They lack both the protein coat characteristic of viruses and any detectable messenger RNA activity. In a susceptible plant host, they can replicate to levels of up to 10,000 copies per cell (Schumacher *et al.*, 1983). In many cases, viroid replication is accompanied by the production of symptoms in the host; in other cases, viroid replication can lead to the accumulation of considerable viroid concentrations without the production of noticeable disease symptoms. The complexity of the interaction of the viroid with its host has been made apparent from results of numerous studies which have been carried out in several laboratories in order to understand and elucidate the mechanisms of viroid pathogenesis. In our laboratory, mutational analysis of potato spindle tuber (PSTVd) and tomato apical stunt viroids (TASVd) has allowed us to genetically isolate such normally interrelated processes as replication, cell-to-cell movement, and disease induction. In addition, the construction of novel interspecific chimeric viroid molecules between TASVd and citrus exocortis viroid (CEVd) has allowed us to determine the contribution of individual structural domains to viroid pathogenicity.

Molecular Structure of Viroids

In order to understand the rationale behind the design of the experiments described below, it is important to first have an appreciation for the complex nature of viroid molecular structure. Viroids were the first circular RNA's discovered in nature (Sänger *et al.*, 1976). In addition to being covalently closed circles, the RNA genomes of known viroids are characterized by an unusually extensive secondary structure and a highly cooperative thermal denaturation profile (Riesner, 1987). The secondary structure of viroids has been determined by theoretical calculations and experimental results to be a quasi double-stranded rod of short helices and loops, commonly referred to as the native structure (Figure 1a). In addition, evidence of tertiary interactions, such as non-Watson Crick base-pairing, have been discovered in viroids (Branch *et al.*, 1985).

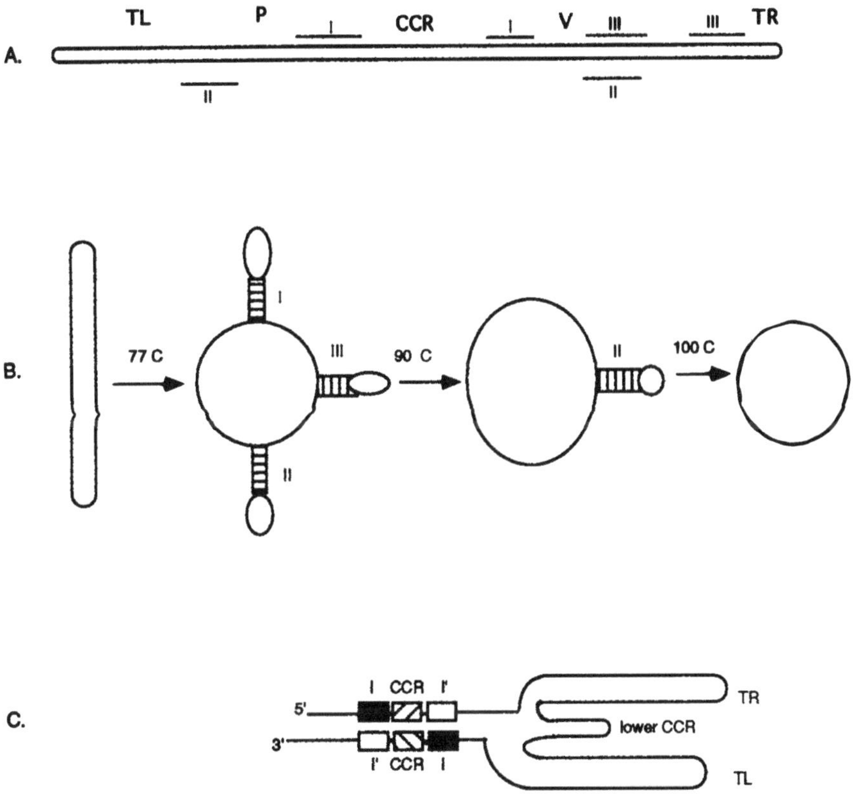

Figure 1. **Proposed Structural Conformations of Viroids and Their Replicative Intermediates. A: Native conformation of viroids illustrates a rod-like secondary structure present in solution at 25 C, 1 M ionic strength. The five structural domains proposed by Keese and Symons (1985) are indicated above the native structure: T_L, left terminal loop; P, pathogenicity domain; CCR, central conserved region; V, variable region; T_R, right terminal loop. The positions of nucleotides involved in formation of secondary hairpins I - III are indicated. B: Denaturation mechanism of PSTVd at 1 M ionic strength. The molecule undergoes a transition from left to right with increasing temperature. The formation and disruption of secondary hairpins is illustrated. The dissociation of hairpin II at 100 C leads to circular RNA. This figure was based upon Figure 3 of Riesner (1990) and illustrates the proposed mechanism of denaturation of Henco *et al.* (1979). C: Structural model of the (+) strand viroid replicative intermediate containing the proposed processing site involving hairpin I and the CCR (Riesner, 1990; Steger *et al.*, 1986). The thermodynamically stable tri-helical segment containing 28 base pairs is marked by slashed boxes.**

Viroids also have the ability to form alternative structures and these structural transitions of viroids have been studied experimentally. As many as three secondary hairpins may form upon disruption of the native structure of PSTVd and related viroids when denatured (Figure 1b) (Henco *et al.*, 1979). Conservation of the ability to form these alternative structural interactions appears to be essential for viroid infectivity and implies that they serve important functions *in vivo* (Riesner, 1990). Mutational analyses involving secondary hairpins I and II support this conclusion (Hammond and Owens, 1987; Loss *et al.*, 1991; Owens *et al.*, 1991).

Viroids replicate via a rolling circle type of mechanism and oligomeric forms of both polarities, presumably the replication intermediates, have been detected in infected plants. Infectivity data and experimental analysis of the structure of the oligomeric intermediates suggest that sequences within the central conserved region (CCR) and hairpin I can potentially rearrange to form an extremely stable trihelical structure containing 28 base pairs (Figure 1c) (Tabler and Sänger, 1984; Steger *et al.*, 1986; Diener, 1986). The biological significance of this structure and the subsequent formation of metastable replication intermediates has been suggested as playing a role in the processing of the oligomeric forms to the monomeric, mature viroid (Riesner, 1990; Diener, 1986).

The secondary structural models described above for viroid molecules do not take into account the biochemical interactions that the RNA must certainly have with host components, be they proteins (e.g., the polymerase responsible for their replication) or other nucleic acids. Evidence for the interactions of viroids with host proteins has been presented by the isolation of viroid-protein complexes from infected plants and the *in vitro* reconstruction of these complexes (Klaff *et al.*, 1989). Although RNA binding activity has been shown for these proteins, the proteins have not been purified; and the nature and sequence/structural specificity of the binding has not been identified. Sequence specific nucleic acid binding proteins most probably play a central role in maintenance of viroid structure, especially the metastable replication intermediates.

Viroid structure, therefore, can be looked upon as being dynamic and not static; and yet studies of structure/function relationships which have focused primarily upon assignment of functional roles to components (domains) of the native structure provide evidence that this conformation plays a significant role in the disease process in some unknown way. The basis for dividing the viroid genome into domains arose from comparative sequence analysis of the genomes of several viroids isolated from different hosts and causal agents for different diseases. This comparison suggests that viroid evolution has involved repeated RNA rearrangements and exchanges of modules of genetic information (Keese and Symons, 1985). As shown in Figure 1a, the rod-like native structure of viroids may be composed of five structural/functional domains: a highly conserved central region (CCR) whose alternative conformations may regulate the replication cycle, as described earlier; flanking pathogenicity (P) and variable (V) domains; and the two terminal domains (T_L and T_R). The mutagenesis studies described in some detail below refer to these domains as points of reference.

Pathogenesis

When a tomato host (*Lycopersicum esculentum* Mill. cv. Rutgers) is infected by one of the PSTVd group of viroids (including CEVd and TASVd), symptoms may include stunting, epinasty, and veinal necrosis. These types of symptoms have also been noted with many plant viral diseases; therefore, the effects of viroid infection appear to be a general secondary influence on host metabolism. However, there do appear to be some specific effects of viroid infection on host metabolism - for example, in the accumulation of specific host proteins (Galindo *et al.*, 1984) and the activation of a host-encoded 68K protein kinase (Hiddinga *et al.*, 1988).

Since viroids do not apparently encode any proteins which may influence symptoms, as in the case of plant viral diseases; e.g., tobacco mosaic virus (Dawson *et al.*, 1988), cauliflower mosaic virus (CaMV) (Schoelz and Shepherd, 1988), and cucumber mosaic virus (CMV) (Shintaku *et al.*, 1992), one or more of the structural features of viroids must be involved in determining its pathogenicity. Comparative sequence analysis of naturally-occurring PSTVd and CEVd isolates inducing a range of symptoms in the same host suggests that the P domain is the primary determinant of symptom expression because most of the nucleotide sequence variation occurs in this region (Schnölzer *et al.*, 1985; Visvader and Symons, 1985). Schnölzer *et al.* (1985) proposed that virulence is determined by the interaction of nucleotides within a portion of the P domain of PSTVd with unidentified host factors. The calculated thermodynamic stability of this domain, and, therefore, its accessibility to host factors, seemed to decrease in severe as compared to mild isolates. The possible role of structural elements outside the P domain in regulating viroid symptom expression was unclear. Visvader and

Symons, on the other hand, reported that symptom severity is not correlated with the thermodynamic stability of the P domain in CEVd (Visvader and Symons, 1985). If the P domain does interact with host components to influence disease symptoms, what would the nature of these host components be? Recently, Sänger and co-workers (Haas *et al.*, 1988) have noted remarkable sequence similarities between a tomato 7S nuclear RNA and human 7SL RNA, a small RNA that is an integral part of the signal recognition particle mediating protein translocation. Sequences near the 5'-terminus of tomato 7S RNA are complementary to the lower strand of the CCR and P domains in PSTVd. Evidence for the interaction of viroids with signal recognition particle RNA's is being sought through sequencing studies of 7S RNA's isolated from several plant species (i.e., symptomatic and asymptomatic hosts as well as nonhosts) (Haas *et al.*, 1988).

These earlier observations of structure/function relationships provided a conceptual framework for later studies utilizing site-directed mutagenesis and chimeric viroid molecules to address certain fundamental aspects of viroid/host interactions.

Analysis of Viroid/Host Interactions

Mutagenesis: Site-directed and random mutagenesis of cloned, infectious viroid cDNA's have been used in a direct approach towards determining the effect of single and multiple nucleotide substitutions upon viroid replication and symptom expression (Loss *et al.*, 1991; Owens *et al.*, 1991; Owens *et al.*, 1986; Hammond and Owens, 1987; Hammond, 1992; Owens, 1990). In many cases, the results from infectivity trials with these mutants were unexpected. For example, in the case of PSTVd, although many mutations rendered the viroid molecule noninfectious, many others were infectious and stable, where still others rapidly reverted to the wild-type sequence. A number of novel PSTVd variants have been isolated, and a comparison of the structural and biological effects of selected mutations provides circumstantial evidence for the role of individual structural domains in replication and pathogenicity and for the importance of alternative structural interactions in viroid replication.

PSTVd: The effect of point mutations on the replication and pathogenicity of PSTVd has been examined in a number of studies (Loss *et al.*, 1991; Owens *et al.* 1991; Haas *et al.*, 1988; Owens *et al.*, 1986; Hammond and Owens, 1987; Hammond, 1992). Initial experiments were designed to create small insertions and deletions in the P domain of PSTVd, a portion of the sequence known to vary among naturally occurring strains. These changes completely abolished infectivity (Owens *et al.*, 1986). Subsequent experiments, using either oligonucleotide-directed or chemical mutagenesis, were performed to introduce a number of single and multiple nucleotide substitutions into all five domains of the PSTVd (Owens *et al.*, 1991; Hammond and Owens, 1987). Studies by Hammond and Owens (1987) and Owens *et al.* (1991) have shown that nucleotide substitutions in four of the five structural domains of PSTVd can abolish infectivity. Because several of these mutations were expected to have only minimal effects upon the rod-like structure of PSTVd, the relationship between structure and infectivity appears to be quite complex.

Recent mutational analyses of the relationship between structure and function have focused upon pathogenicity (Owens *et al.*, 1991; Hammond, 1992) and the biological significance of certain alternative structural interactions (Loss *et al.*, 1991; Owens *et al.*, 1991). In one study, a series of nucleotide substitutions were introduced into the P domain of the intermediate strain of PSTVd in order to examine their effect upon viroid infectivity and pathogenicity with the presence of all four mutations resulting in the sequence of a previously reported severe strain of PSTVd (Figure 2a) (Hammond, 1992). Eight of the resulting mutant cDNA's were characterized for infectivity and symptom expression in tomato, and the secondary structure of their corresponding RNA's was examined by computer analysis and chemical mapping (Figure 2b). No correlation could be found between thermodynamic stability of the P domain and symptom severity. The combined results question a previously proposed correlation between thermodynamic instability and PSTVd pathogenicity (Schnölzer, 1985) and suggest that alternate conformations and/or sequence-specific interactions with host factors play a role in symptom development.

A large collection of PSTVd mutants distributed throughout the PSTVd genome was created by random chemical mutagenesis (Owens *et al.*, 1991). It was hoped from this study that one or more *in vitro* structural interactions could be inferred from viroid function *in vivo*. Lethal as well as nonlethal mutations were generated throughout the genome. Infectivity analyses and subsequent characterization of progeny revealed that although some mutations reverted to wild-type, others were stable. Several mutations in the P and T_L domains were stably maintained. One stable mutation which may promote significant

A

B

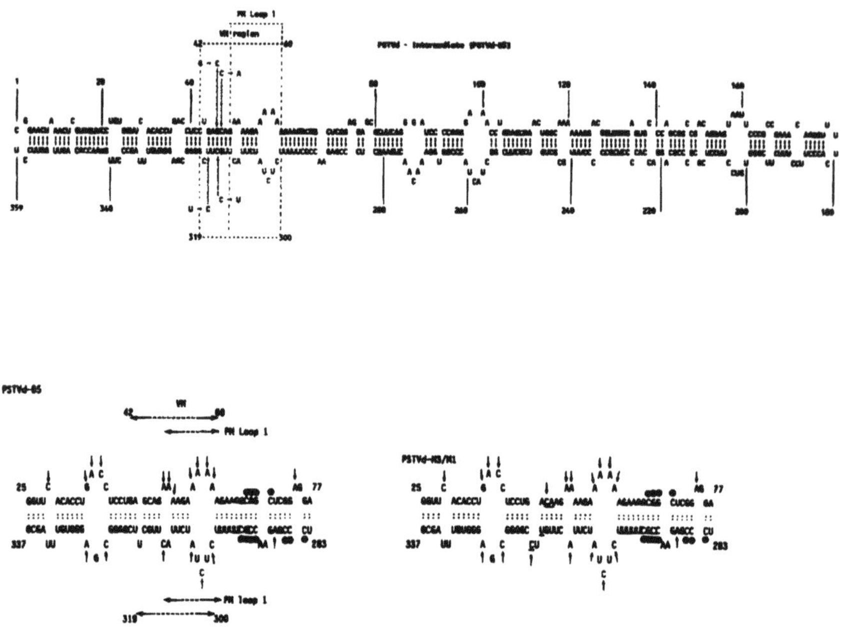

Figure 2. A: The intermediate strain of PSTVd indicating the four nucleotide differences between the intermediate and a severe strain. The P domain is indicated by VM region. Mutagenesis of the intermediate strain by introduction of these four nucleotide differences in various combinations has been described by Hammond (1992). B: Predicted structural effects in the P domain of the four mutations which result in conversion of the intermediate (PSTVd-B5) to severe strain (PSTVd-M3/M1). Sites of chemical modification indicating single-stranded regions are marked by arrows. Solid circles indicate strong stops (Hammond, 1992).

structural rearrangement within the right side of the P domain had no detectable effect upon symptom expression. Mutations that affected base pairing in secondary hairpins I and II were either lethal or rapidly reverted to wild-type. The behaviour of mutants affecting the secondary hairpins point toward an important role in viroid replication.

The functional relevance of a hairpin II-containing structure was also examined by Loss *et al.* (1991). Infectivity trials of a collection of eight single-site mutations in segments which form hairpin II showed that mutations in the core region of the hairpin reverted to the wild-type sequence whereas mutations in the peripheral regions remained stable at the sequence level, suggesting that the core of hairpin II is critical for infectivity. A hypothesis was put forward as to the function of hairpin II as a binding site for host cell transcription factors due to its similarity in sequence and structure to certain 5'-upstream regions in genes.

TASVd: A series of nucleotide substitutions was introduced into the P domain of another member of the PSTVd group of viroids, TASVd (Owens, 1990). The symptoms of TASVd on tomato are similar, but more pronounced than, the symptoms of the intermediate strain of PSTVd; i.e., curling of leaves and apical stunting. The pathogenicity domain of TASVd shows considerable natural sequence variation. Differences between previously described Ivory Coast and Indonesian isolates are concentrated within a particular region of this domain (Candresse *et al.*, 1987).

Infectivity analyses of twelve A → G substitutions were evaluated for their effects upon infectivity and symptom expression (Owens, 1990). None of the mutations abolished infectivity or had a detectable effect upon symptom expression; however, four gave rise to altered progeny. Sequence evolution from mutant to wild-type in some cases involved the passage through other sequence variants. The selection for these molecules may result from a need to conserve a particular native structure or interaction with host components. Similar sequence changes have been observed in infectivity trials with infectious transcripts of cucumber mosaic virus (CMV) satellite RNA's (Collmer and Kaper, 1988; Kurath and Palukaitis, 1990).

In summary, the results of site-directed and random mutagenesis have frustrated attempts to provide a convincing correlation between sequence/structure and pathogenicity, although in some cases these studies have allowed the identification of structural elements which may be necessary for replication and cell-to-cell movement. The next strategy to locate symptom determinants was to construct hybrid viroid genomes from either two strains of a viroid or two different viroids exhibiting contrasting symptoms.

Chimeras: Novel recombinants of plant viruses have been useful in the study of host range and pathogenicity (e.g., brome mosaic virus (Allison *et al.*, 1988), CaMV (Anderson *et al.*, 1991)). This strategy has also been useful in the study of structure/function relationships in the satellite RNA's of CMV (Devic *et al.*, 1990).

In vitro manipulation of the viroid genome has also been performed by the production of chimeric molecules. Initally, infectious intraspecific chimeras between two strains of CEVd revealed that the region important for determining the severity of the viroid is located in the left half of the viroid native structure (Visvader and Symons, 1986). In another study, infectious intraspecific recombinants were constructed by exchanging the left and right sides of two naturally occurring strains of TASVd (Owens *et al.*, 1990). These recombinants appeared to be stable; however, no unusual biological properties were exhibited in a bioassay - the two parent strains, although quite distinct in their nucleotide sequence, were very similar in their biological properties.

Interspecific hybrids between more distantly related viroids have also been constructed in our laboratory, and the results of bioassays with these hybrids have provided us with evidence for the presence of long-range nucleotide sequence interactions between distant parts of the viroid and more definitive evidence for the role of structurally defined domains in pathogenicity.

Initial attempts to construct infectious hybrids by exchanging the left and right halves of PSTVd and TASVd were unsuccessful, probably due to sequence divergence within the secondary hairpin stems I and II (Owens, *et al.*, 1986). Construction of a recombinant between TASVd and CEVd, however, resulted in an infectious, stable hybrid, the symptoms of which were similar to the parent which was the source of the pathogenicity domain (Owens *et al.*, 1990). The ability of this chimera to produce a systemic infection is most likely due to the sequence identity of these two viroids in the stems of the secondary hairpins. These combined results provide additional evidence for the importance of the formation of these secondary hairpins *in vivo*, and, therefore, the existence of alternate viroid conformations.

The experiments described above involved the exchange of the left and right halves of viroids with the main result that of identifying the left half of the viroid as the major contributor in determining the level of severity of symptoms. As stated earlier, comparative sequence analysis of viroids suggests that viroids contain five structural domains. For example, the left half of the viroid contains the left terminal domain (T_L), the pathogenicity domain (P), and a portion of the central conserved region (CCR), while the right half of the viroid contains the right terminal domain (T_R), the variable region (V), and the right portion of the CCR (Figure 1a). The purpose of the experiments completed by Sano *et al.* (1992) and briefly described below was to investigate the role of these individual domains in pathogenicity and replication.

A series of interspecific hybrids were constructed by exchanging the T_L and/or P domains between TASVd and CEVd viroids (Figure 3). Each of the chimeras was able to form hairpins I and II, and the construction of the chimeras was designed to minimize disruption of the native molecule. All six chimeras tested replicated stably in tomato, and the symptoms exhibited by infected plants were intermediate between the two parental types (Figure 4). Quantitative comparisons of symptom development and progeny accumulation revealed that chimeras containing the right half of TASVd accumulated to higher levels early in infection, and the infected plants developed more severe symptoms than those whose right halves were derived from CEVd. The T_L domain of TASVd contains a determinant required for appearance of severe veinal necrosis in tomato, and the severe epinasty and stunting characteristic of TASVd requires the presence of both its T_L and P domains. Examination of the sequence and secondary structure of these domains show that sequence variation is localized in the T_L domain and is not unexpectedly extensive in the P domain (Figure 5). The molecular mechanisms by which these variations influence pathogenesis are unknown; however, more extensive studies are now underway. The results of this study show, however, that regulation of viroid pathogenicity appears to be more complex than previously thought, involving structural elements within at least three of their five proposed structural domains (Figure 5).

Conclusions

Conclusive evidence now exists for the importance of the P domain in symptom induction in tomato. Our latest study, however, also points to a critical role of the left terminal (T_L) domain in determining the severity of stunting, epinasty, and veinal necrosis. Viroid pathogenicity in infected plants is a complex process that is influenced by both the viroid and host plant genomes. Manipulation of the viroid RNA by mutagenesis or construction of chimeras combined with analyses of infectivity, structure, and tissue localization suggests functional roles for regions; and in some cases, specific nucleotides in replication, symptom induction, cell-to-cell movement, and in stabilization of alternate conformations of the viroid molecule. The biological activity of many different RNA's depends upon long-range interactions of nucleotides as well as unconventional interactions such as non-Watson:Crick base pairings. In addition, nucleic acid structures are

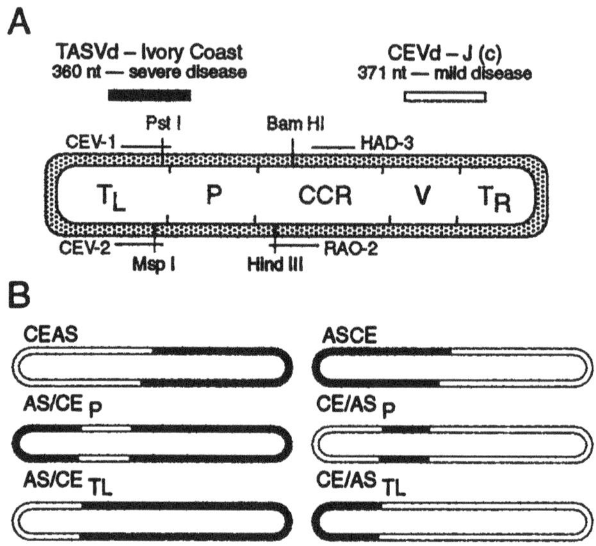

Figure 3. Construction of TASVd-CEVd Chimeras. A: Locations of the five structural domains proposed by Keese and Symons (1985) are shown within a schmematic representation. Boundaries of these domains and locations of restriction sites used for cDNA fragment exchanges are indicated by vertical lines. B: Structures of the six interspecific TASVd-CEVd chimeras. TASVd- and CEVd-derived sequences are indicated by filled and open lines, respectively.

stabilized by interactions with proteins. The results from the experiments described above challenge our previous models of structure/function relationships and viroid pathogenicity and point to a need to determine the structure of the viroid molecule as it occurs *in vivo*. Only by an understanding of the events and interactions which occur between a viroid and host plant that result in disease will we be able to effectively design control strategies.

Acknowledgments

We thank S. M. Thompson, M. Hale, and D. R. Smith for skilled technical assistance. Support for these studies was provided by the USDA Competitive Research Grants Program (Grant 88-37263-3990).

Figure 4. Symtpoms Expressed in Tomato Plants (A) and Leaves (B) Infected With TASVd-CEVd Chimeras. A: The severe stunting characteristic of TASVd infections 6 weeks post-inoculation were seen only after substitution of both its T_L and P domains [i.e., ASCE]. B: Induction of veinal necrosis by viroids containing the T_L domain of TASVd. Severe necrosis is visible in systemically-infected leaves of tomatoes inoculated with either TASVd, ASCEp, or ASCE. Necrosis is absent in plants inoculated with CEVd or much reduced in plants containing ASCE-T_L. Leaves were collected at 4 weeks post-inoculation.

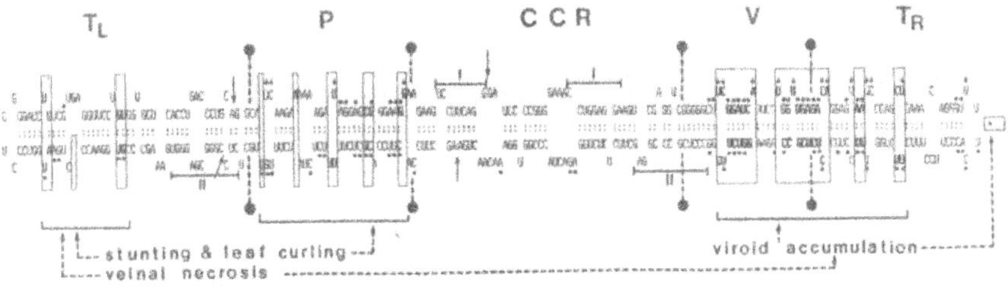

Figure 5. The Nucleotide Sequence and Proposed Secondary Structure of the Ivory Coast Isolate of TASVd; Sequence Differences With CEVd-Strain J(c) are Indicated by Small Dots. Variations in secondary structure between the two viroids have been boxed. Boundaries of the individual structural domains are indicated by vertical dotted lines, and locations of sequences forming the stems of secondary hairpins I and II are shown by horizontal lines. Arrows indicated restriction enzyme sites that were used to construct the chimeras.

References

Allison, R.F., Thompson, C., and Ahlquist, P. 1988. Regeneration of a functional RNA virus genome by recombination between deletion mutants and requirement for cowpea chlorotic mottle virus 3A and coat genes for systemic infection. Proc. Natl. Acad. Sci. U.S.A. 87:1820.

Anderson, E.J., Qiu, S.G., and Schoelz, J.E. 1991. Genetic analysis of determinants of disease severity and virus concentration in cauliflower mosaic virus. Virology 181:647.

Branch, A.D., Benenfeld, B.J., and Robertson, H.D. 1985. Ultraviolet light-induced cross-linking reveals a unique region of local tertiary structure in potato spindle tuber viroid and HeLa 5SRNA. Proc. Natl. Acad. Sci. U.S.A. 82:6590.

Candresse, T., Smith, D., and Diener, T.O. 1987. Nucleotide sequence of a full-length infectious clone of tomato apical stunt viroid. Nucleic Acids Res. 15:10597.

Collmer, C.W. and Kaper, J.M. 1988. Site-directed mutagenesis of potential protein-coding regions in expressible cloned cDNA's of cucumber mosaic viral satellites. Virology 163:293.

Dawson, W.O., Bubrick, P., and Grantham, G.L. 1988. Modification of the tobacco mosaic virus coat protein gene affecting replication, movement, and symptomatology. Phytopathology 78:783.

Devic, M., Jaegle, M., and Baulcombe, D. 1990. Symptom production on tobacco and tomato is determined by two distinct domains of the satellite RNA of cucumber mosaic virus (strain Y). J. Gen. Virol. 71:1443.

Diener, T.O. 1986. Viroid processing: A model involving the central conserved region and hairpin I. Proc. Natl. Acad. Sci. U.S.A. 83:58.

Diener, T.O. (ed.) 1987. *The Viroids*. Plenum Press, New York. 344 pp.

Galindo, A.J., Smith, D.R., and Diener, T.O. 1984. A disease-associated host protein in viroid-infected tomato. Physiol. Plant Path. 24:257.

Haas, B., Klanner, A., Ramm, K., and Sänger, H.L. 1988. The 7S RNA from tomato leaf tissue resembles a signal recognition particle RNA and exhibits a remarkable sequence complementarity to viroids. EMBO J. 7:4063.

Hammond, R.W. 1992. Analysis of the virulence modulating region of potato spindle tuber viroid (PSTVd) by site-directed mutagenesis. Virology 187:654.

Hammond, R.W. and Owens, R.A. 1987. Mutational analysis of potato spindle tuber viroid reveals complex relationships between structure and infectivity. Proc. Natl. Acad. Sci. U.S.A. 84:3967.

Henco, K., Sänger, H.L., and Riesner, D. 1979. Fine structure melting of viroids as studied by kinetic methods. Nucleic Acids Res. 6:3041.

Hiddinga, H.C., Crum, J.H., and Roth, D.A. 1988. Viroid induced phosphorylation of a host protein related to a dsRNA dependent protein kinase. Science 241:451.

Keese, P. and Symons, R.H. 1985. Domains in viroids: Evidence of intermolecular RNA rearrangements and their contribution to viroid evolution. Proc. Natl. Acad. Sci. U.S.A. 82:4582.

Klaff, P., Gruner, R., Hecker, R., Sättler, A., Theissen, G., and Riesner, D. 1989. Reconstituted and cellular viroid-protein complexes. J. Gen. Virol. 70:2257.

Kurath, G. and Palukaitis, P. 1990. Serial passage of infectious transcripts of a cucumber mosaic virus satellite RNA clone results in sequence heterogeneity. Virology 176:8.

Loss, P., Schmitz, M., Steger, G., and Riesner, D. 1991. Formation of a thermodynamically metastable structure containing hairpin II is critical for infectivity of potato spindle tuber viroid RNA. EMBO J. 10:719.

Owens, R.A. 1990. Mutational analysis of virod pathogenicity. Molec. Plant Microbe Interact. 3:374.

Owens, R.A., Candresse, T., and Diener, T.O. 1990. Construction of novel viroid chimeras containing portions of tomato apical stunt and citrus exocortis viroids. Virology 175:238.

Owens, R.A., Hammond, R.W., Gardner, R.C., Kiefer, M.C., Thompson, S.M., and Cress, D.E. 1986. Site-specific mutagenesis of potato spindle tuber viroid cDNA: Alterations within premelting region 2 that abolish infectivity. Plant Mol. Biol. 6:179.

Owens, R.A., Thompson, S.M., and Steger, G. 1991. Effects of random mutagenesis upon potato spindle tuber viroid replication and symptom expression. Virology 185:18.

Riesner, D. 1987. Physical-chemical properties. Structure Formation. In: *The Viroids.* T.O. Diener (ed.), pp. 63. Plenum Press, New York.

Riesner, D. 1990. Structure of viroids and their replication intermediates. Are thermodynamic domains also functional domains? Seminars in Virology 1:83.

Sänger, H.L., Klotz, G., Riesner, D., Gross, H.J., and Kleinschmidt, A.K. 1976. Viroids are single-stranded covalently closed circular RNA molecules existing as highly basepaired rod-like structures. Proc. Natl. Acad. Sci. U.S.A. 73:3852.

Sano, T., Candresse, T., Hammond, R.W., Diener, T.O., and Owens, R.A. 1992. Identification of multiple structural domains regulating viroid pathogenicity. Proc. Natl. Acad. Sci. U.S.A. 89:10104.

Schnölzer, M., Haas, B., Ramm, K., Hofmann, H., and Sänger, H.L. 1985. Correlation between structure and pathogenicity of potato spindle tuber viroid (PSTV). EMBO J. 4:2181.

Schoelz, J.E. and Shepherd, R.J. 1988. Host range control of cauliflower mosaic virus. Virology 162:30.

Schumacher, J., Sänger, H.L., and Riesner, D. 1983. Subcellular localization of viroids in highly purified nuclei from tomato leaf tissue. EMBO J. 2:1549.

Shintaku, M.H., Zhang, L., and Palukaitis, P. 1992. A single amino acid substitution in the coat protein of a cucumber mosaic virus isolate induces chlorosis in tobacco. The Plant Cell 4:751.

Steger, G., Tabler, M., Brüggemann, W., Colpan, M., Klotz, G., Sänger, H.L., and Riesner, D. 1986. Structure of viroid replicative intermediates: Physico-chemical studies on SP6 transcripts of cloned oligomeric potato spindle tuber viroid. Nucleic Acids Res. 14:9613.

Tabler, M. and Sänger, H.L. 1984. Cloned single- and double-stranded DNA copies of potato spindle tuber viroid (PSTV) RNA and co-inoculated subgenomic DNA fragments are infectious. EMBO J. 3:3055.

Visvader, J.E. and Symons, R.H. 1985. Eleven new sequence variants of citrus exocortis viroid and correlation of sequence with pathogenicity. Nucleic Acids Res. 13:5843.

Visvader, J.E. and Symons, R.H. 1986. Replication of *in vitro* constructed viroid mutants: Location of the pathogenicity-modulating domain of citrus exocortis viroid. EMBO J. 5:2051.

RNA Recombination in the Turnip Crinkle Virus System

A. E. Simon, P. J. Cascone, C. Zhang, and C. D. Carpenter
Department of Biochemistry and Molecular Biology and
Program in Molecular and Cellular Biology
University of Massachusetts
Amherst, Massachusetts 01003

RNA recombination has been described for a growing number of viruses in the plant and animal kingdoms and is thought to involve a copy-choice mechanism whereby the viral replicase jumps from one template to an alternate location on the same or different template where polymerization of the nascent strand continues. Recombination between two sub-viral RNA's associated with turnip crinkle virus is not random and can, therefore, be used to study the requirement for specific sequences and structures in the generation of recombinant molecules. Mutagenesis of residues both upstream and downstream of the junction point in the sub-viral RNA which forms the 3'-end of recombinant molecules indicates that single base mutations which disrupt either the stem or the loop of one of the two computer-predicted stem-loop structures in the crossover region eliminate detectable recombinant molecules. However, recombinants are detected if compensatory mutations are generated which reform a stable stem loop structure. These results provide strong evidence for the necessity of specific structures in the formation of recombinant molecules in this system.

Introduction

Recombination between RNA molecules has been described for a number of segmented and non-segmented viruses in both the plant and animal kingdoms. Since RNA recombination results in a mixing of parental genomes, this phenomenon may be an important factor in the evolution of viruses. RNA recombination can be either precise or imprecise involving homologous or heterologous molecules. In a recent review (Lai, 1992), recombination events were divided into three classes.

(1) Homologous recombination: recombinant progeny are precisely derived from highly related parental molecules.

(2) Aberrant homologous recombination: progeny molecules are generated from RNA's sharing some sequence similarity; but nucleotide insertions, deletions, and duplications can be found at the crossover point.

(3) Non-homologous recombination: recombination occurs between unrelated RNA's.

Recombining viral RNA genomes is thought to be a replicase-driven event (Lazzarini *et al.*, 1981; Kirkegaard and Baltimore, 1986; King, 1988). The viral replicase, after ceasing replication on one template, jumps to a second template where polymerization of the nascent strand continues. Irrespective of the type of recombination which occurs, the selection of the crossover sites by the replicase is not yet understood. In some systems, such as mouse hepatitis virus (under non-selective conditions), recombination sites appear to be random (Baric *et al.*, 1987). Alternatively, recombination in picornaviruses seems to occur more frequently in regions predicted by computer analysis to contain strong secondary structure (Romanova *et al.*, 1986; Wilson *et al.*, 1988; Kuge *et al.*, 1989). The seemingly random nature of recombination events, combined with the large sizes of the genomic RNA's which undergo recombination, complicates investigations on the mechanism(s) which generate recombinant molecules.

The discovery that a sub-viral RNA associated with turnip crinkle virus (TCV) contained nearly the entire sequence of a TCV satellite (sat-) RNA (sat-RNA D) joined to two regions of TCV genomic RNA, suggested that recombination events between heterologous templates could occur in this system (Simon and Howell, 1986). Further evidence establishing that discontinuous RNA's are routinely generated was the finding that defective interfering RNA's arise *de novo* from cloned TCV genomic RNA transcripts (Li *et al.*, 1989). Since this earlier work, we have now found a large number of recombinant molecules associated with TCV (Figure 1).

The TCV system is uniquely suited for studying the mechanism of RNA recombination. The TCV genomic RNA (4054 bases) is among the smallest viral RNA genomes (Carrington *et al.*, 1989; Collmer *et al.*, 1991) and contains only four or five open reading frames (Hacker *et al.*, 1992). *In vitro* synthesized transcripts of both the viral genomic RNA (Heaton *et al.*, 1989) and the sub-viral sat- and DI RNA's (Simon and Howell, 1987; Li and Simon, 1991) are infectious in whole plants, and both protoplast (Collmer *et al.*, 1991; Zhang and Simon, unpublished) and *in vitro* (Song and Simon, unpublished) replication systems have been established. The discovery that recombinant molecules are generated in a non-random manner permits the detailed examination of the sequences and structures involved in the event. This chapter describes: (1) the circumstances under which recombinant molecules have been detected; (2) the three different replicase-recognition motifs, one of which is present at nearly every crossover point analyzed to date; and (3) the RNA secondary structure that is required for recombination between two sub-viral RNA's.

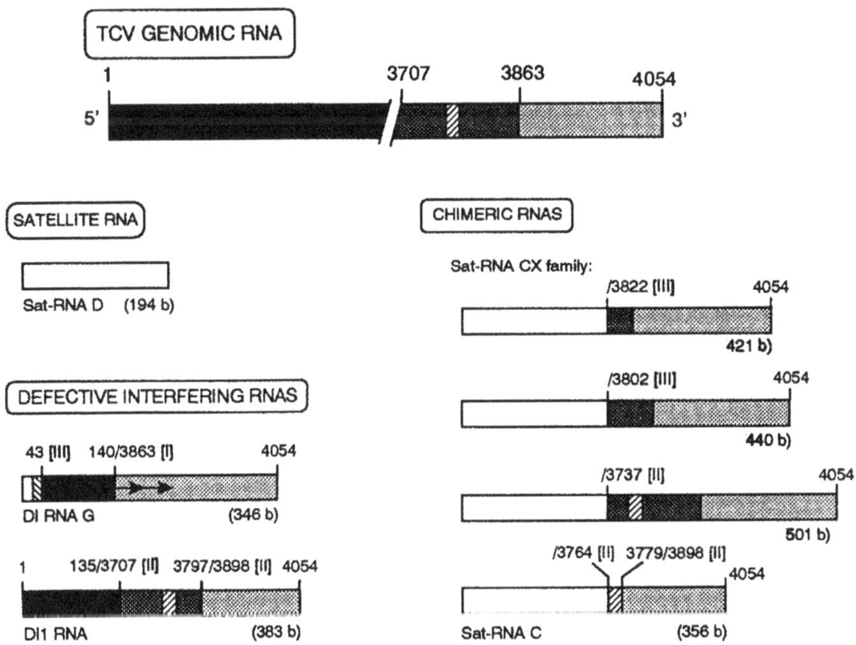

Figure 1. Comparison of Related Sequences Among the Genomic and Selected Sub-Viral RNA's of the TCV System. Similar sequences are shaded alike. The 5'-portion of sat-RNA C shares only 88 percent similarity with the analogous region in sat-RNA D. Numbers indicate positions of TCV-derived sequence. Roman numerals I, II, and III in brackets denote positions of Motifs I, II, or III (see text). Arrows signify a repeated sequence.

Experimental

Plant Growth and Infection Conditions: All of our work to date on recombination has made use of plasmids containing full-length sat-RNA C cDNA downstream from an *Escherichia coli* RNA polymerase promoter. Transcripts synthesized *in vitro* contain exact 5'-ends and five additional bases at the 3'-ends. All transcripts are co-inoculated with a helper virus preparation onto at least six turnip plants. The helper virus preparation consists of total RNA isolated from turnip infected with TCV and sat-RNA D. All sat-RNA C molecules containing insertions, deletions, and single base changes are sequenced by direct dideoxynucleotide techniques (Carpenter and Simon, 1990) to confirm retention of the altered nucleotides.

Detection of Recombinant Molecules: Total RNA extracted from plants possibly containing recombinant molecules is first analyzed on ethidium bromide-stained, denaturing polyacrylamide gels. Sat-RNA's, both natural and derived from recombination events, in general accumulate at levels sufficiently high to be detectable in this manner. Next, Northern blot analysis is used to test if the recombinant sat-RNA's hybridize to probes specific for sat-RNA C and sat-RNA D (only recombinant molecules will hybridize to both probes). Finally, polymerase chain reactions (PCR) using sat-RNA C and sat-RNA D specific primers are used to detect extremely low levels of recombinant molecules. Double-stranded molecules generated by PCR are cloned and sequenced to determine the exact junction points and to detect possible revertants.

Results and Discussion

Recombination Between sat-RNA D and sat-RNA C: Recombinant molecules, derived *de novo* from *in vitro* synthesized sat-RNA C transcripts, were first detected during studies of sat-RNA C transcripts containing either large insertions or deletions in a region about 80-110 bases from the 5'-end (Cascone *et al.*, 1990). Instead of plants accumulating sat-RNA's of 345-382 nucleotides, a new species of about 380 bases was found in nearly all infected plants. Attempts to determine the sequence of this new species by direct dideoxy-sequencing techniques using total RNA preparations (Carpenter and Simon, 1990) were unsuccessful due to sequence heterogeneity which began precisely at position 176, regardless of the original inoculated transcript. Cloning and sequencing the RNA's from several plants revealed that the new species were recombinant molecules, derived from sat-RNA D (which was always included in the helper virus preparation) and the inoculated transcript. By recombining with sat-RNA D, the inoculated transcripts containing insertions and deletions near the 5'-end were "corrected" by replacing mutant sat-RNA C sequence with the sat-RNA C-similar sequence from sat-RNA D (see Figure 1).

Inspection of the junctions of these recombinant molecules revealed that a type of aberrant homologous recombination had occurred which was unique in the field of viral RNA recombination. All of the 20 recombinants cloned had nearly full-length or full-length sat-RNA D sequence at the 5'-end joined to the mid-region of sat-RNA C at one of three consecutive bases (Cascone *et al.*, 1990). Furthermore, about 30 percent of the junctions contained additional non-template encoded nucleotides, mainly uracil residues. Since this work was completed, over 150 recombinant molecules formed from sat-RNA D and sat-RNA C transcripts have been sequenced; with only three exceptions, all are joined to one of five consecutive bases beginning at position 175 with non-template nucleotides at the junctions of about 30 percent of the RNA's (Cascone *et al.*, 1993) (see arrowheads in Figure 4).

Recombination Between sat-RNA D and TCV Genomic RNA: We have also isolated a series of recombinants between sat-RNA D and TCV genomic RNA (Zhang *et al.*, 1991). These RNA's, known collectively as sat-RNA CX, contain nearly full-length sat-RNA D at the 5'-end joined to variable lengths of TCV genomic RNA (Figure 1). Sat-RNA CX molecules are larger than normal sat-RNA D/sat-RNA C recombinants, ranging from 420 to 506 bases. Possibly because of this size difference, they accumulate to a lower level than sat-RNA C or D/C recombinants and are only detectable by Northern blot analysis. Curiously, sat-RNA CX species accumulate to higher levels in plants which have been inoculated with helper virus and uninfectious sat-RNA C transcripts than in plants inoculated with the helper virus alone (Zhang *et al.*, 1991; Cascone *et al.*, 1993).

Recombination Between sat-RNA CX and sat-RNA C: Besides recombinants between sat-RNA D and sat-RNA C or sat-RNA D and TCV genomic RNA, we recently cloned a recombinant between sat-RNA CX and sat-RNA C (Figure 2). This recombinant molecule resembles sat-RNA C in having been generated from two crossover events.

One of Three Different Motifs are Always Found at the Right Sides of Crossover Junctions: The availability of numerous recombinant molecules, including defective interfering RNA's (Li *et al.*, 1989), permitted the determination that one of three dissimilar motifs are

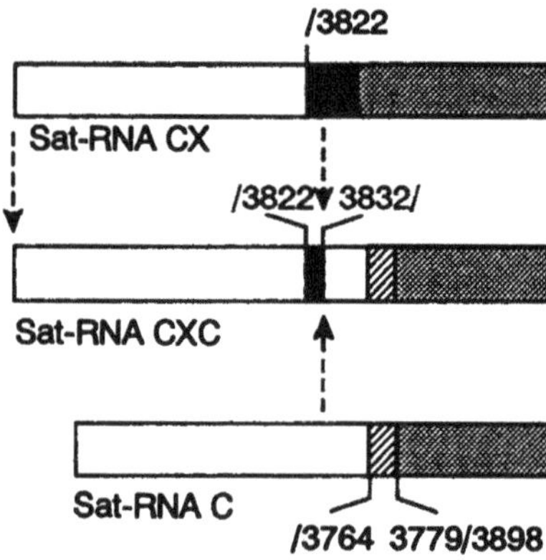

Figure 2. Structure of a Recombinant Between sat-RNA C and sat-RNA CX. Shading is according to Figure 1.

located at the right sides of nearly all junctions analyzed to date (*Cascone et* al., 1990; Zhang *et al.*, 1991) (Figure 1, numbers in brackets). Since recombination is thought to be a replicase-driven event, the replicase might recognize these particular sequences (or structures which include these motifs) as internal promoters. The motifs were, therefore, compared with sequences at the 3'- and 5'-ends of the genomic RNA, sub-viral sat- and DI RNA's, and subgenomic RNA's. As shown in Figure 3, Motif I (the sequence in sat-RNA C found at the junction of sat-RNA D/sat-RNA C recombinants) resembles a sequence 10 bases from the 5'-end of TCV; Motif II resembles the sequence at the 5'-end of sat-RNA D and DI RNA G; Motif III is similar to sequence at the 5'-end of the one mapped TCV subgenomic RNA (1.45 kb).

Model for the Generation of TCV-Associated Discontinuous RNA's: A model for recombination has been proposed based on the finding that the specific motifs, which are at the right side of crossover points, are related to sequences at or near the 5'-ends of the TCV genomic, sub-viral, and subgenomic RNA's (Cascone *et al.*, 1990). We believe that there are three minus (-) strand replicase recognition signals where replication can be initiated or reinitiated. As the replicase copies the (-) strand, it can detach from the template either prematurely or when it reaches the natural end point. Premature termination may not involve a specific sequence since there are no sequence similarities at the left sides of the junctions. Before the replicase releases the nascent (+) strand, it reinitiates synthesis at a recognition signal on the same template, forming DI RNA's, or a different (-) strand template, forming recombinant RNA's.

```
                   10                                               # of Junctions
                   |
  TCV 5' end ...CAAAUCCCUGCACCCGCCUA...
                ********|   ****|**
      Motif I   CAAAUCCCNAGACCCNCCAG                                      2

                1
                |
Sat-RNA D 5'end GGGAUAAAAAUAAGGGAUUC...
                ********** | * ** *
DI RNA G 5'end  GGGAUAAAAAA.GGAGGCUU...
                *********** ** ****
     Motif II   GGGACCAAAAANGGUGGCAG                                      8
                    UA    G     AA U

TCV Subgenomic ...GGGUAAUAUAUGCUUUCUACAACUCUUCUCACU...
                     *****|*||****  ***|||||***   **| *
     Motif III       GUAAUNGNNGCUUAGUANNNNUCUCUCUNUU                      7
                           U        C          UC
```

Figure 3. Comparison of Consensus Motifs and Sequences at the 5'-Ends of TCV-Associated Molecules. The number of junctions where motif-like sequences are found is indicated. All sat-RNA D/sat-RNA C recombinants are counted as a single junction (Motif I).

This copy choice model can also account for the generation of linear sat- and DI RNA multimers without invoking the type of rolling circle mechanism thought responsible for the formation of multimers of circular sat-RNA's and viroids (Branch *et al.*, 1988). After completing synthesis of a full-length satellite or DI RNA, the replicase complexed with the newly synthesized strand initiates a second round of replication either using the same template or a second homologous template (Carpenter *et al.*, 1991). Evidence to support the model, that the formation of TCV sub-viral RNA multimers and RNA recombinants occurs by the same mechanism, comes from determining that junctions between monomeric units of sat-RNA D, and the very similar sat-RNA F, had the same truncations and non-template encoded nucleotides as junctions between recombinant sat-RNA D/sat-RNA C molecules (Carpenter *et al.*, 1991). Furthermore, junctions between monomeric units of other TCV sub-viral RNA's (sat-RNA C and DI RNA G) also contained additional non-template residues.

Still a mystery is why the sat-RNA D sequence at the left side of the junctions in sat-RNA D/sat-RNA C recombinants and sat-RNA D multimers is frequently between one and 11 bases short of full length. Several possibilities exist:

(1) The replicase produces a full length (+) strand which is then exposed to 3'-5'-exonuclease activity resulting in random, minor deletions before re-initiation of replication; and

(2) The replicase terminates polymerization of the (+) strand prematurely due to RNA secondary or tertiary structure.

Analysis of Motif I: The discovery that sat-RNA D and sat-RNA C recombine at a specific site in sat-RNA C affords the unique opportunity to study the sequences and structures involved in this event. A series of insertions, deletions, and single base changes have been engineered both within and external to the Motif I-type sequence in sat-RNA C that is always found at the right side of junctions in sat-RNA D/sat-RNA C recombinants (Cascone *et al.*, 1993) (Figure 4). Sat-RNA C transcripts containing these alterations were all infectious when co-inoculated with helper virus onto turnip. To test the affect of the different alterations on the ability of sat-RNA C to recombine with sat-RNA D, each mutation was transferred into a variant of sat-RNA C which was engineered to contain a 22 b deletion near the 5'-end of the molecule (see Figure 4). Sat-RNA C molecules which contain this deletion are not viable, and recombination with sat-RNA D is required to produce molecules which can accumulate in plants. Detection of recombinant molecules was assayed by both Northern blots and PCR using sat-RNA C- and sat-RNA D-specific probes and primers.

The results of this analysis, summarized in Figure 4, indicate that single base changes within Motif I can eliminate the detection of recombinant molecules even using sensitive PCR assays. Furthermore, sequences that extend well beyond the borders of Motif I are also important for recombination. The possible significance of secondary structure in this region was suggested by the finding that transcripts containing a deletion of four nucleotides, 15 bases downstream of Motif I, could no longer recombine with sat-RNA D, while recombinants were detected in plants inoculated with transcripts containing this deletion and an insertion within Motif I.

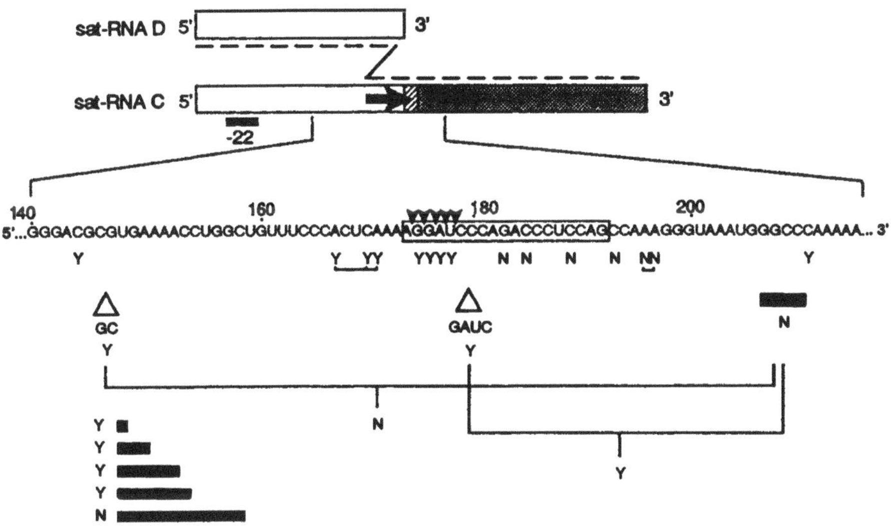

Figure 4. Mutations Generated in and Around Motif I Sequence in sat-RNA C. Arrow denotes Motif I. Arrowheads indicate the various crossover points in over 150 cloned recombinant species. Motif I sequence is boxed. Bars and triangles indicate positions of deletions and insertions, respectively. "Y" indicates alteration still allows for the detection of recombinant molecules. "N" indicates no recombinant species were detected even by PCR. "Y" and "N" below individual bases indicate transcripts containing single base alterations were assayed for recombination. Connected letters indicate double or triple base changes were in a single construct.

The secondary structure for this region of the minus strand of sat-RNA C (the strand proposed to be involved in the recombination event (Cascone *et al.*, 1990)) was determined by chemical and enzymatic modification methods (Carpenter *et al.*, in press) and is shown in Figure 5. The lowest free energy structure contains a stem-loop, where the crossover point is at the 3'-base of the stem. Nucleotide changes in the single-stranded region at the base of the stem still allowed for recombination with sat-RNA D, while alterations in the stem and loop eliminate recombination.

To assess whether single base changes in the stem affect the ability of the replicase to cross over to this sequence by disrupting the structure of the stem, compensatory mutations were generated such that original A:U pairings were replaced with C:G pairings. In the two locations within the stem where these compensatory mutations were generated (bases 205 + 182 and 201 + 184), recombination was re-established between sat-RNA D and sat-RNA C. This data strongly supports both the existence of the hairpin and its importance in the recombination event.

Secondary Structure at Other Recombination Sites in the TCV Genomic RNA: As discussed earlier, Motif I is one of three dissimilar sequence motifs found at the right side junctions of discontinuous RNA's associated with TCV. Computer folding of 72 residues of the TCV genomic RNA minus strand encompassing the Motif I sequence found at the right side junction of one of the DI RNA's (DI RNA G) also produced a stem-loop structure with the crossover site at the base of the stem. Recently, a similar mutational analysis as desbribed above for Motif I was conducted on the Motif III sequence at the junction in Sat-RNA CX species (Carptenter *et al.*, in press). A hairpin similar to that described in Figure 5 was also found to be important in targetting recombination to the base of the stem.

Future Directions: The analysis of junction sequences in sat-RNA D/sat-RNA C recombinants represents the first detailed analysis of a recombination site in any plant or animal viral system. However, we hope to address many remaining questions:

(1) What is the relationship between the 19 b Motif I sequence and the larger 63 base structure required for recombination?

(2) Why are the sat-RNA D sequences at the left-side junction truncated and what is the origin of the non-template encoded nucleotides?

(3) Are there structures required for recombination at Motif II sequences?

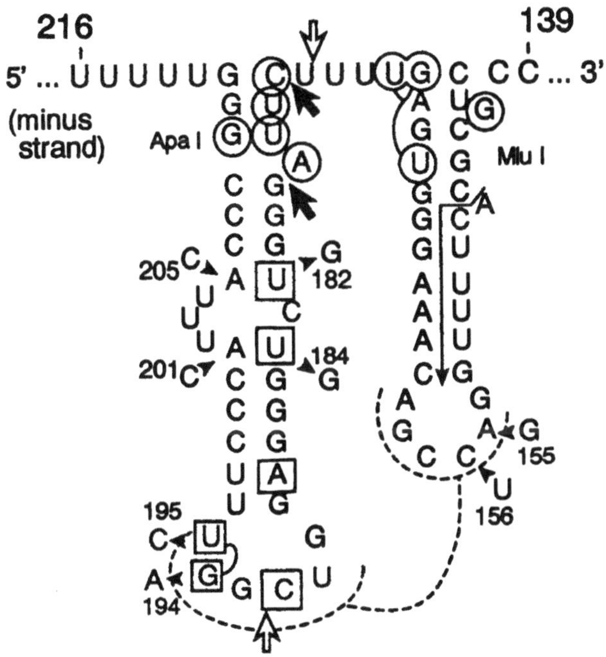

Figure 5. Secondary Structure in the Region of sat-RNA C Involved in Recombination. Circles indicate change still allows for recombination; squares indicate no recombinants are detected in plants inoculated with transcripts containing alterations in these nucleotides. Motif I sequence is between open arrows (note that this is the compliment to the strand shown in Figure 3). Recombination sites are between closed arrows. Compensatory mutations are denoted by arrowheads towards nucleotides in the stem-loop.

(4) What is the nature of the TCV replicase that results in the large number of recombinant species in this system?

The recent availability of protoplast and *in vitro* replication systems will expand the tools available for the further analysis of recombination in the TCV system.

Acknowledgments

This work was supported by National Science Foundation Grants MCB 9315948 and DMB 9105890 to A. E. Simon and Sigma Xi Grant-in-Aid of Research to Pamela J. Cascone and Chunxia Zhang.

References

Baric, R.S., Shieh, C.K., Stohlman, S.A., and Lai, M.M.C. 1987. Analysis of intracellular small RNA's of mouse hepatitis virus: Evidence for discontinuous transcription. Virology 156:342-354.

Branch, A.D., Benenfeld, B.J., and Robertson, H.D. 1988. Evidence for a single rolling circle in the replication of potato spindle tuber viroid. Proc. Natl. Acad. Sci. U.S.A. 85:9128-9132.

Carpenter, C.D., Cascone, P.J., and Simon, A.E. 1991. Formation of multimers of linear satellite RNA's. Virology 183:586-594.

Carpenter, C.D., Oh, J.-W., Zhang, C., and Simon, A.E. Involvement of a stem-loop structure in the location of junction sites in viral RNA recombination. J. Mol. Biol. In Press.

Carpenter, C.D. and Simon, A.E. 1990. Simplified RNA sequencing using dideoxy chain termination. BioTechniques 8:26-27.

Carrington, J.C., Heaton, L.A., Zuidema, D., Hillman, B., and Morris, T.J. 1989. The genome structure of turnip crinkle virus. Virology 170:219-226.

Cascone, P.J., Carpenter, C.D., Li, X.H., and Simon, A.E. 1990. RNA recombination between satellite RNA's of turnip crinkle virus. EMBO J. 9:1709-1715.

Cascone, P.J., Haydar, T.F., and Simon, A.E. 1993. Sequences and structures required for recombination between virus-associated RNA's. Science 260:801-805.

Collmer, C.W., Stenzler, L., Fay, N., and Howell, S.H. 1991. Nonmutant forms of the avirulent satellite D or turnip crinkle virus are produced following inoculation of plants with mutant forms synthesized *in vitro*. Virology 183:251-259.

Hacker, D.L., Petty, I.T.D., Wei, N., and Morris, T.J. 1992. Turnip crinkle virus genes required for RNA replication and virus movement. Virology 186:1-8.

Heaton, L.A., Carrington, J.C., and Morris, T.J. 1989. Turnip crinkle virus infection from RNA synthesized *in vitro*. Virology 170:214-218.

King, A.M.Q. 1988. Preferred sites of recombination in poliovirus RNA: An analysis of 40 intertypic cross-over sequences. Nuc. Acids. Res. 29:921-928.

Kirkegaard, K. and Baltimore, D. 1986. The mechanism of RNA recombination in poliovirus. Cell 47:433-443.

Kuge, S., Kawamura, N., and Nomoto, A. 1989. Genetic variation occurring on the genome of an *in vitro* insertion mutant of poliovirus type I. J. Virol. 63:1069-1075.

Lai, M.M.C. 1992. RNA recombination in animal and plant viruses. Microbiological Reviews 56:61-79.

Lazzarini, R.A., Keene, J.D., and Schubert, M. 1981. The origins of defective interfering particles of the negative-strand RNA viruses. Cell 26:145-154.

Li, X.H., Heaton, L.A., Morris, T.J., and Simon, A.E. 1989. Turnip crinkle virus defective interfering RNA's intensify viral symptoms and are generated *de novo*. Proc. Natl. Acad. Sci. U.S.A. 86:9173-9177.

Li, X.H. and Simon, A.E. 1991. *In vivo* accumulation of a turnip crinkle virus defective interfering RNA is affected by alterations in size and sequence. J. Virol. 65:4582-4590.

Romanova, L.I., Blinov, V.M., Tolskaya, E.A., Viktorova, E.G., Kolesnikova, M.S., Guseva, E.A., and Agol, V.I. 1986. The primary structure of crossover regions of intertypic poliovirus recombinants: A model of recombination between RNA genomes. Virology 155:202-213.

Simon, A.E. and Howell, S.H. 1986. The virulent satellite RNA of turnip crinkle virus has a major domain homologous to the 3'-end of the helper virus genome. EMBO J. 7:3423-3428.

Simon, A.E. and Howell, S.H. 1987. Synthesis *in vitro* of infectious RNA copies of the virulent satellite of turnip crinkle virus. Virology 156:146-152.

Wilson, V., Taylor, P., and Desselberger, U. 1988. Crossover regions in foot-and-mouth disease virus (FMDV) recombinants correspond to regions of high local secondary structure. Arch. Virol. 102:131-139.

Zhang, C., Cascone, P.J., and Simon, A.E. 1991. Recombination between satellite and genomic RNA's of turnip crinkle virus. Virology 184:791-794.

Self-Cleavage and Replication of Barley Yellow Dwarf Virus Satellite RNA

S. L. Silver*, L. Rasochova, S.P. Dinesh-Kumar* and
W. A. Miller*#
Plant Pathology Department and
**Molecular, Cellular and Developmental Biology Program*
Iowa State University
Ames, Iowa 50011

#Presenting and corresponding author.

The satellite RNA of barley yellow dwarf virus (sBYDV) contains a hammerhead self-cleavage structure which differs from the consensus hammerhead in several ways. These differences reduce the self-cleavage rate when compared to mutants designed to resemble the consensus structure. The sequence comprising the hammerhead folds into an alternative pseudoknot-like conformation. We propose that this sequence performs two functions, one being self-cleavage via the hammerhead, the other being some other aspect of the satellite life-cycle that requires the pseudoknot. To test this, we constructed a permuted dimeric cDNA clone of sBYDV RNA, from which infectious monomeric RNA was generated by *in vitro* transcription and self-cleavage. sBYDV RNA required helper BYDV genomic RNA for replication in oat protoplasts; but the helper did not require sBYDV, proving that sBYDV RNA is a true satellite RNA. In the future, we can use wild-type and mutant infectious transcripts to determine the role(s) of the alternative structure and the biological properties of the satellite RNA in general.

Introduction

A satellite RNA by definition requires a helper virus for replication but is not required by the helper virus (reviewed by Roossinck *et al.*, 1992). Satellite RNA's have no significant sequence similarity with their helper viruses. An RNA molecule (sBYDV RNA) associated with the RPV serotype of barley yellow dwarf virus (BYDV) was reported to have many features of a satellite RNA (Miller *et al.*, 1991). It (1) is 322 nucleotides long, (2) has no significant sequence homology to the helper BYDV RNA, (3) is found in BYDV particles, (4) is not found in all RPV isolates, and (5) like BYDV itself, is not mechanically transmissible. Circular and multimeric linear forms of the RNA are found *in vivo*, with the linear monomer predominating in the virion, suggesting that sBYDV RNA replicates by a rolling circle mechanism

(Branch and Robertson, 1984; Symons, 1989). Like satellite RNA's of the nepo- and sobemoviruses (virusoids), and avocado sunblotch viroid (ASBV), sBYDV RNA is capable of self-cleavage via hammerhead structures. (See reviews by Bruening, 1989 and Symons, 1991.)

The sBYDV RNA (+) strand self-cleavage structure differs from other naturally occuring hammerheads (Bruening, 1989; Symons, 1991; Forster and Symons, 1987) as follows (Figure 1): (1) the trinucleotide at the 5'-side of the cleavage site is AUA rather than GUC, (2) unpaired bases are found at the top (proximal end) of helix II, and (3) base-pairing between loops L1 and L2a may result in an unusual pseudoknot-like structure, rather than a hammerhead. We showed previously that the pseudoknot-like structure is indeed favored and that single base mutations that disrupt this structure and favor a hammerhead increase the self-cleavage rate by two orders of magnitude (Miller and Silver, 1991). Here we summarize those results and examine the effects of the unpaired bases at the proximal end of helix II that are absent in more efficient hammerheads.

Our goals are to determine the role(s) of this unusual structure in self-cleavage and other aspects of the satellite life cycle, as well as to understand the biology of the three-way relationship between satellite, virus and host. We describe the first step in studying satellite RNA replication: construction of a full-length infectious transcript of sBYDV RNA. We also show for the first time that sBYDV RNA fits the definition of a true satellite RNA.

Experimental

Nomenclature: Nucleotides are referred to by the base followed by the number of its position in the full-length monomeric satellite RNA (Miller *et al.*, 1991).

Clone Construction and in vitro Transcription: Site-directed mutagenesis was performed as described by Herlitze and Koenen (1990). *In vitro* transcription and self-cleavage assays were as described previously (Miller and Silver, 1991). Plasmid pSAT6 contained permuted dimeric sBYDV RNA sequence flanked by T7 and SP6 RNA polymerase promoters (Figure 3). It was derived by site-directed mutagenesis from pBWS30 (Miller *et al.*, 1991) which lacked the 3'-terminal base (A322) of both copies of the monomeric RNA. Details have been described in Silver *et al.* (1994). Following *in vitro* transcription of linearized plasmid, monomeric RNA was gel-purified (Sambrook *et al.*, 1989) and used for innoculation of protoplasts.

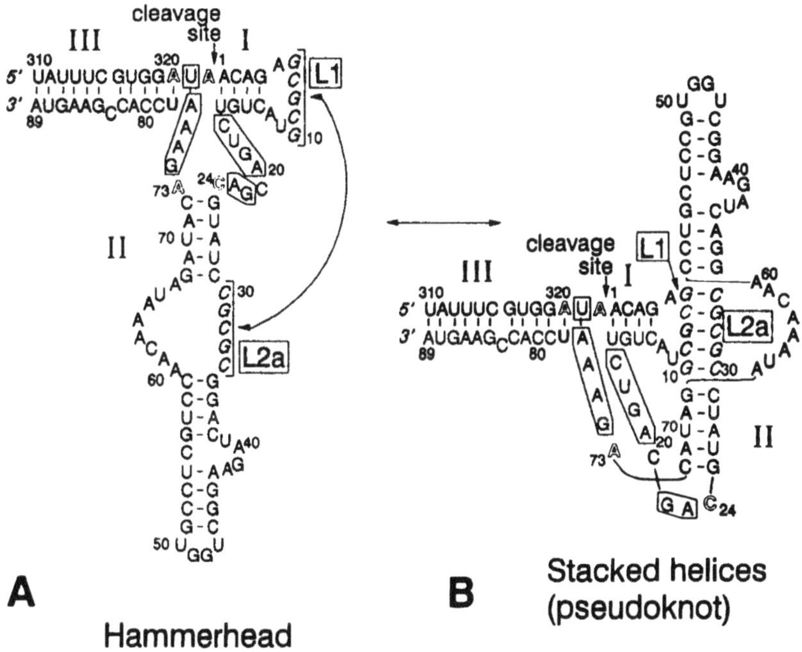

Figure 1. Alternative Conformations of the Sequence Comprising the Hammerhead Self-Cleavage Structure on the (+) Strand of sBYDV RNA. Bases are numbered as in the full-length sequence (Miller *et al.*, 1991). Boxed bases are conserved in all known hammerheads. Outlined bases differ from consensus. Helical domains I, II, and III are numbered as in Hertel *et al.*, 1992. Bases (bracketed and in italics) in loops in domains I (L1) and II (L2a) that base-pair to form proposed pseudoknot-like structure (B) are connected by the curved arrow in structure A. To facilitate two-dimensional representation, lines are added to indicate some phosphodiester bonds in structure B.

Inoculation of Protoplasts with Satellite RNA Transcript: BYDV RNA (a mixture of PAV and RPV serotypes) was purified as in Waterhouse *et al.* (1986). Protoplasts were prepared and inoculated, and Northern hybridizations were performed as described previously (Dinesh-Kumar *et al.*, 1992).

Results and Discussion

Alternative Base-Pairing Inhibits Formation of an Efficient Hammerhead Structure: Several base changes were introduced into a cDNA clone spanning only the sBYDV RNA self-cleavage sequence (Figure 1) adjacent to a T7 polymerase promoter. Mutant RNA's were then obtained by *in vitro* transcription. The self-cleavage rate was defined as the half-life of the purified full-length transcript after addition of Mg^{2+} as monitored by gel electrophoresis and radioanalytical quantification of the RNA fragments. Transcripts M1, M2, M1/2, M19, M20, and M19/20 contain mutations expected to disrupt or restore base-pairing between loops L1 and L2a (Figure 2). Disruptions of this base-pairing should favor hammerhead formation. As expected, transcripts containing mutations in either L1 or L2a cleaved more rapidly than wild-type (Figure 2). Double mutants in both L1 and L2a, that were expected to restore base-pairing, cleaved more slowly than the single mutants. Mutant M1 self-cleaved more slowly than the other single mutants because it had a tendency to form yet another nonhammerhead conformation as revealed by partial digestion with structure-sensitive nucleases under self-cleavage conditions (Miller and Silver, 1991). The cleavage rates of the wild-type and double mutant transcripts expected to form the L1-L2a base-pairing were proportional to the predicted stability of the L1-L2a helix. For example, M19/20 contains an A-U pair in this helix, weakening it relative to wild-type (SS1) or M1/2 which contain only G-C pairs. This allows M19/M20 to self-cleave more rapidly, presumably because it can form a hammerhead more readily than SS1 or M1/2. Thus, the sequence likely exists in a dynamic equilibrium between the inactive pseudoknot-like structure and the hammerhead which self-cleaves irreversibly (Figure 1).

Effects of Unpaired Bases at the Proximal End of Helix II: In almost all other hammerheads, helix II begins at the equivalent positions of bases C24 and A73; i.e., at the positions immediately 3' of the conserved CUGANGA and 5' of the conserved GAAA sequences (Bruening, 1989; Symons, 1991). It is usually a G-C basepair. The hammerheads of two other satellite RNA's have an apparently unpaired U at the position equivalent to C24. Ruffner *et al.* (1990) showed that disruption of the basepair that normally forms, between bases at the positions equivalent to C24 and A73, greatly reduced *trans*-cleavage by an ASBV hammerhead ribozyme. Sheldon and Symons (1989) obtained similar results with self-cleavage of satellite lucerne transient streak virus RNA. Thus, we tested whether deletion of one or both of bases C24 and A73 would increase self-cleavage. Deletion of both would most resemble the consensus structure.

Mutant M3 lacks base C24, mutant M4 lacks base A73, and the double mutant, M3/4 lacks both. The self-cleavage rates of the single mutants did not differ significantly from wild-type (Figure 2). However, the double mutant (M3/4), which most closely fits the consensus hammerhead structure, cleaved much more rapidly (Figure 2). This was somewhat surprising because the potential base-pairing between loops L1 and L2a was not disrupted in mutant M3/4. Thus, efficient formation of the active hammerhead was not expected. Perhaps the unpaired C24 and A73 provide the length and flexiblity needed to join the long horizontal helical region with the long vertical helix II in the pseudoknotted conformation as drawn in Figure 1.

The above studies dealt only with the 102 bases comprising the (+) strand cleavage structure. Perhaps in the full-length context, the RNA folds into a hammerhead more readily. However, the base-pairing between L1 and L2a is the strongest continous helix in the entire satellite genome; conformations which favor the hammerhead only in the full-length or dimeric contexts are not apparent. A permuted dimeric clone has been constructed from which full-length sBYDV RNA can be transcribed (Figure 3). We are in the process of measuring the self-cleavage rate of wild-type and mutant sBYDV RNA's in this dimeric context.

A Molecular Switch? We propose that formation of alternative structures by one sequence serves act as a molecular switch in which the hammerhead provides the self-cleavage function, and the pseudoknot-like structure serves some other function. That one sequence may serve more than one function is not surprising because one-half of the 322 base satellite RNA encodes either the (+) strand hammerhead sequence or the complement of the (-) strand sequence. If the self-cleavage structure-encoding sequences did not serve more than one role, then only 161 bases divided into two portions of the satellite RNA would have to provide all the other functions, including replication origin, origin of assembly, and other undefined functions. Formation of alternative conformations by circular RNA's may be a general rule as this has been observed or predicted with several satellite RNA's (Davies *et al.*, 1990), viroids (Hernandez and Flores, 1992), and hepatitis delta virus (Rosenstein and Been, 1990). In many viroids, the potential to form an alternative stem-loop structure

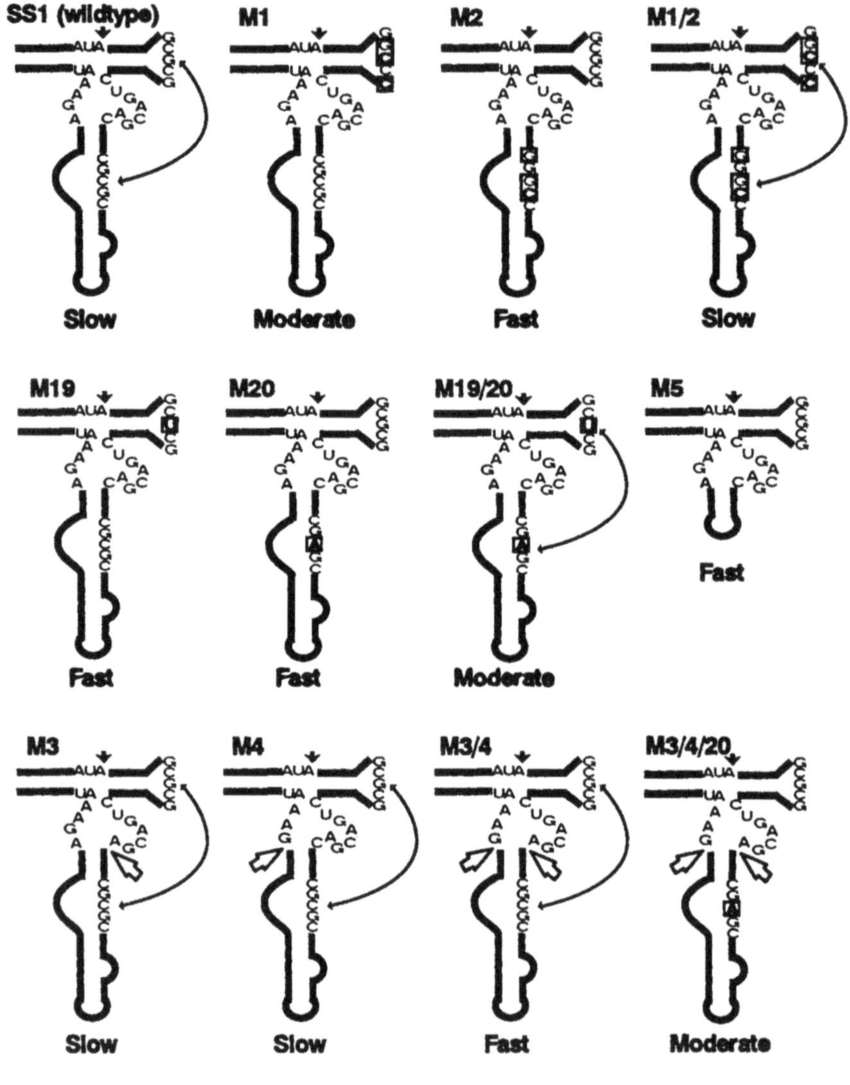

Figure 2. Relative Rates of Self-Cleavage of Mutant Transcripts Spanning sBYDV RNA Self-Cleavage Structure. Curved, double-headed arrow indicates expected base-pairing between L1 and L2a (Figure 1). Substitutions are boxed, deletions indicated by open arrows. Small vertical arrow indicates cleavage site. Cleavage rates of fast, moderate and slow correspond to half-lives of <50 min., 100-600 min., and >1000 min., respectively.

is conserved, even though this structure is not part of the well-characterized rod-shaped conformation of purified viroid RNA (Riesner, 1991).

Infection of Protoplasts with a Full-Length Transcript of sBYDV RNA Requires the Presence of BYDV Genomic RNA: To test the biological significance of the alternative base pairing within the sBYDV RNA self-cleavage structure, our next goal was to construct a full-length cDNA clone from which infectious sBYDV RNA could be transcribed. To generate an infectious transcript, we constructed a permuted, dimeric cDNA clone by ligation at a convenient *Bgl*II site within sBYDV RNA (Miller *et al.*, 1991). This dimer was adjacent to an SP6 RNA polymerase promoter. Upon transcription, a monomer with correct termini would be generated by self-cleavage. For unknown reasons, cDNA clones from which the dimeric clone was derived lacked the base (A322) representing the 3'-terminus of sBYDV RNA.

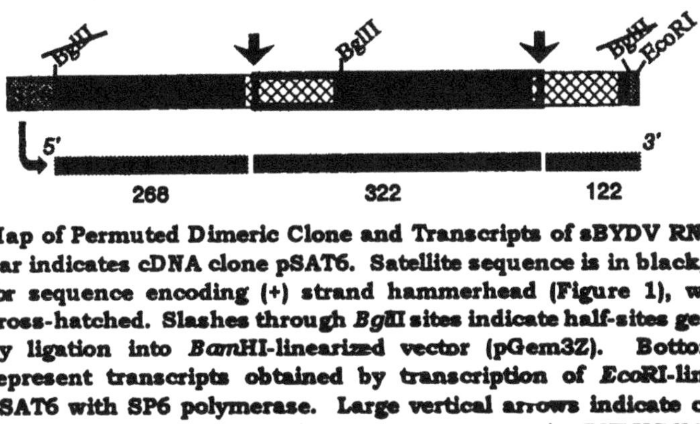

Figure 3. Map of Permuted Dimeric Clone and Transcripts of sBYDV RNA. Top bar indicates cDNA clone pSAT6. Satellite sequence is in black, except for sequence encoding (+) strand hammerhead (Figure 1), which is cross-hatched. Slashes through *Bgl*II sites indicate half-sites generated by ligation into *Bam*HI-linearized vector (pGem3Z). Bottom bars represent transcripts obtained by transcription of *Eco*RI-linearized pSAT6 with SP6 polymerase. Large vertical arrows indicate cleavage sites, which upon transcription generate monomeric sBYDV RNA (black, 322 base transcript cleavage product), and partial-length RNA's (grey 268 base and 122 base fragments).

Thus, the transcripts were incapable of self-cleavage (Miller *et al.*, 1991). A322 was introduced into both cleavage sites of the dimer by site-directed mutagenesis to generate pSAT6 (Figure 3).

Upon transcription of *Eco*RI-linearized pSAT6, the expected 322 base monomeric RNA and flanking fragments and partial cleavage products were evident (data not shown). The monomer was gel-purified and quantitated before electroporation into oat protoplasts. Oat protoplasts were inoculated with a mixture of BYDV RPV and PAV RNA's with or without satellite RNA. Northern blot analysis after 36 and 72 hours revealed an accumulation of large quantities of full-length RPV genomic RNA (Figure 4, upper panel). Lower molecular weight bands may be subgenomic RNA's, although only the largest one was predicted (Vincent *et al.*, 1991). Alternatively, they may represent degradation products or defective interfering (DI) RNA's. The sizes of these RNA's differ from the satellite multimers in the lower panel.

To test for sBYDV RNA replication, all radioactivity was removed from the blot by boiling and the blot was rehybridized with radiolabeled, (+) sense sBYDV RNA transcript. This would detect only (-) strand satellite RNA, which could be produced only if at least the first step of sBYDV RNA replication [(-) strand synthesis] occurred. Indeed, large quantities of (-) sBYDV RNA were detected by 36 hrs. after inoculation. The presence of multimeric forms and the accumulation of large quantities, that were not present at 0 hr., verifies that complete replication occurred. Furthermore, the presence of multimeric and monomeric (-) strand indicates that sBYDV replicates by a rolling circle mechanism in both strands. Some other satellites and viroids form only large multimers of (-) strand which do not self-cleave but do serve as templates for production of multimeric (+) strand which does self-cleave (Davies *et al.*, 1990). Plus strand sBYDV RNA probe did not hybridize to itself (Figure 4, lane 11). As expected, the sBYDV probe did not hybridize to any RNA from cells infected by BYDV in the absence of satellite RNA (Figure 4, lanes 2-5).

These results demonstrate that the sBYDV RNA transcript is indeed infectious and that the same BYDV preparation which can support the satellite also replicates efficiently in the absence of sBYDV RNA. This fulfills an important part of the definition of the satellite RNA, ruling out the remote possibility that sBYDV RNA was a second genomic RNA unique to the particular BYDV isolate with which it was discovered.

To prove that the satellite RNA depends on BYDV genomic RNA and that it is incapable of independent replication like a viroid, protoplasts were inoculated with the infectious monomeric transcript in the presence or absence of helper virus RNA. A northern blot similar to that in Figure 4 revealed accumulation of multimeric and monomeric sBYDV RNA only in cells inoculated with sBYDV RNA transcript *and* helper virus RNA (data not shown). This shows the dependence of sBYDV RNA on helper BYDV RNA, fulfilling the definition of satellite RNA.

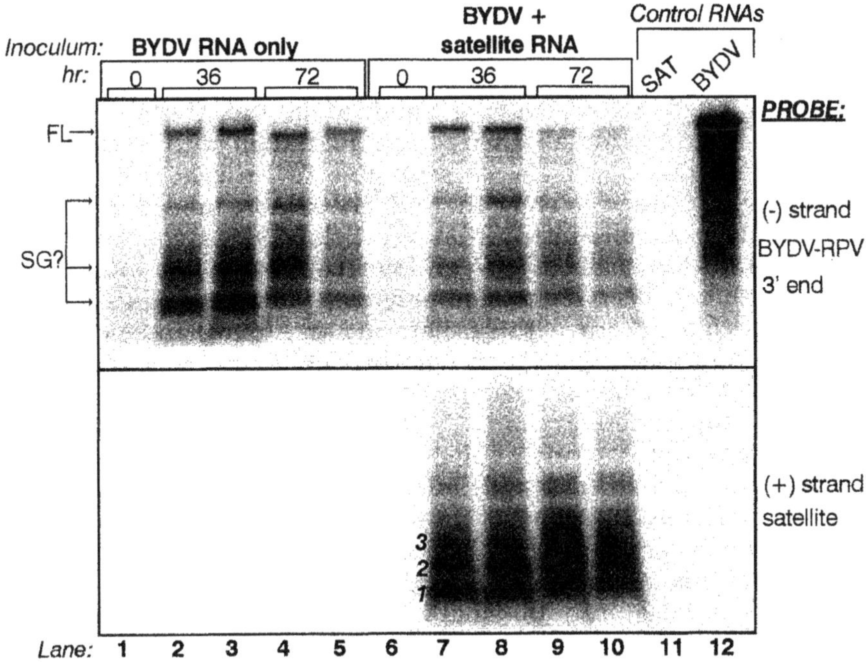

Figure 4. Northern Hybridization of Total RNA from Infected Oat Protoplasts. Total RNA from protoplasts was run on a 1 percent agarose gel, blotted to Genescreen (duPont/NEN) nylon membrane. RNA was isolated at indicated times (duplicate samples of 36 and 72 hr. extracts were used) after inoculation. Top panel shows Phosphorimager (Molecular Dynamics) image after probing with a ^{32}P-labeled (-) sense transcript comprising the 3'-terminal 250 bases of the BYDV-RPV genome (Beckett, R., unpublished data). Probe was stripped by boiling for 15 min. in 0.1xSSC, 0.1 percent SDS and the blot was reprobed with a ^{32}P-labeled form of the same (+) sense sBYDV RNA transcript that was used to inoculate cells (Figure 3), to give the pattern in the bottom panel. Times after inoculation with 0.5 µg BYDV genomic RNA (lanes 1-10) and monomeric sBYDV RNA transcript (lanes 6-10) are indicated. Lanes 11 and 12 contain 0.2 µg of monomeric sBYDV RNA transcript, and 0.2 µg of purified BYDV genomic RNA, respectively. Mobilities of full-length BYDV RNA (FL), possible subgenomic or DI RNA's (SG?), and monomeric, dimeric, and trimeric-sized sBYDV RNA's (*1,2,3* adjacent to lane 7 in bottom panel) are indicated.

Separation of viral RNA components from a virus preparation, including satellite from helper is an arduous task and can never be achieved with certainty. With *in vitro* transcription of a cDNA clone, infectious satellite RNA can be obtained in the absence of viral infection, assuring the absence of contamination with helper virus RNA. Using this approach we were able to perform the separate inoculations required to show that sBYDV RNA is, indeed, a satellite RNA. It is also noteworthy that the helper virus consisted of a mixture of the NY isolate of RPV, and the IL isolate of PAV. Until now, the satellite had been associated only with an Australian isolate of RPV and with a mixture of this isolate and an Australian PAV isolate. Thus, the Australian and North American BYDV isolates are similar enough to allow them both to support replication of the same satellite RNA. Although the Australian RPV isolate alone can support sBYDV RNA, we have not rigorously ruled out the ability of PAV to support the satellite. Using full-length infectious transcripts of the helper virus and satellite RNA in this protoplast system we will be able to determinine the serotype-specificity of sBYDV RNA and the sequences that control this. This system also allows us to introduce mutations that affect cleavage into full-length sBYDV RNA and determine the role(s), if any, of the alternative structure in the satellite and the biological reason, if any, for such a slow rate of natural self-cleavage. This system also opens the way for study of the effect of sBYDV RNA on disease symptoms and virus replication. Ultimately we hope to dissect the complex three-way relationship between plant, virus, and satellite.

Acknowledgements

This work was funded by the Consortium for Plant Biotechnology Research and the Quaker Oats Company. This is Journal Paper No. J-15086 of the Iowa Agriculture and Home Economics Experiment Station; Project No. 3127.

References

Branch, A.D. and Robertson, H.D. 1984. A replication cycle for viroids and other small infectious RNA's. Science 223:45.

Bruening, G. 1989. Compilation of self-cleaving sequences from plant virus satellite RNA's and other sources. Methods in Enzymol. 180:546-558.

Davies, C., Haseloff, J., and Symons, R.H. 1990. Structure, self-cleavage, and replication of two viroid-like satellite RNA's (Virusoids) of subterranean clover mottle virus. Virology 177:216-224.

Dinesh-Kumar, S.P., Brault, V., and Miller, W.A. 1992. Precise mapping and *in vitro* translation of a trifunctional subgenomic RNA of barley yellow dwarf virus. Virology 187:711-722.

Forster, A.C. and Symons, R.H. 1987. Self-cleavage of plus and minus RNA's of a virusoid and a structural model for the active sites. Cell 49:211-220.

Herlitze, S. and Koenen, M. 1990. A general and rapid mutagenesis method using polymerase chain reaction. Gene 91:143-147.

Hernandez, C. and Flores, R. 1992. Plus and minus RNA's of peach latent mosaic viroid self-cleave *in vitro* via hammerhead structures. Proc. Natl. Acad. Sci. U.S.A. 89:3711-3715.

Hertel, K.J., Pardi, A., Uhlenbeck, O.C., Koizumi, M., Ohtsuka, E., Uesugi, S., Cedergren, R., Eckstein, F., Gerlach, W.L., Hodgson, R., and Symons, R.H. 1992. Numbering system for the hammerhead. Nucleic Acids Res. 20:3252.

Miller, W.A., Hercus, T., Waterhouse, P.M., and Gerlach, W.L. 1991. A satellite RNA of barley yellow dwarf virus contains a novel hammerhead structure in the self-cleavage domain. Virology 183:711-720.

Miller, W.A. and Silver, S.L. 1991. Alternative tertiary structure attenuates self-cleavage of the ribozyme in the satellite RNA of barley yellow dwarf virus. Nucleic Acids Res. 19:513-520.

Riesner, D. 1991. Viroids: From thermodynamics to cellular structure and function. Molec. Plant-Microbe Interact. 4:122-131.

Roossinck, M.J., Sleat, D., and Palukitis, P. 1992. Satellite RNA's of plant viruses: Structures and biological effects. Microbiol. Rev. 56:265-279.

Rosenstein, S.P. and Been, M.D. 1990. Self-cleavage of hepatitis delta virus genomic strand RNA is enhanced under partially denaturing conditions. Biochemistry 29:8011-8016.

Ruffner, D.E., Stormo, G.D., and Uhlenbeck, O.C. 1990. Sequence requirements of the hammerhead RNA self-cleavage reaction. Biochemistry 29:10695-10702.

Sambrook, J., Fritsch, E.F., and Maniatis, T. 1989. Molecular Cloning: A Laboratory Manual. Second Edition. Cold Spring Harbor Laboratory. Cold Spring Harbor, NY.

Sheldon, C.C. and Symons, R.H. 1989. Mutagensis analysis of a self-cleaving RNA. Nucleic Acids Res. 17:5679-5685.

Silver, S.L., Rasochova, L., Dinesh-Kumar, S.P., and Miller, W.A. 1994. Replication of barley yellow dwarf virus satellite RNA transcripts in oat protoplasts. Virology 199:331-335.

Symons, R.H. 1989. Self-cleavage of RNA in the replication of small pathogens of plants and animals. Trends Biochem. Sci. 14:445-450.

Symons, R.H. 1991. Ribozymes. Critical Reviews in Plant Sciences 10:189-234.

Vincent, J.R., Lister, R.M., and Larkins, B.A. 1991. Nucleotide sequence analysis and genomic organization of the NY-RPV isolate of barley yellow dwarf virus. J. Gen. Virol. 72:2347-2355.

Waterhouse, P.M., Gerlach, W.L., and Miller, W.A. 1986. Serotype-specific and general luteovirus probes from cloned cDNA sequences of barley yellow dwarf virus. J. Gen. Virol. 67:1273-1281.

Parasitic Nucleic Acids Competing for Genetic Expression: A Conceptual Framework for the Study of Satellite-Mediated Viral Symptom Modulation and its Implications for the Epidemiology and Control of Plant Viruses

J. M. Kaper
Microbiology and Plant Pathology Laboratory
Plant Sciences Institute
Agricultural Research Service, USDA
Beltsville, Maryland 20705

Satellite-mediated viral symptom modulation has been responsible for a rapidly rising interest in the use of satellites for the practical control of viral crop disease. Cucumber mosaic virus (CMV) satellites have been at the leading edge of this emerging technology due to their striking ability to completely ameliorate viral symptoms in certain crops. However, CMV satellites can cause new diseases in other crops and, therefore, are also major plant pathogens. This paper presents a conceptual framework that defines the dynamic behavior of satellites and their helper viruses in a susceptible host plant, with the secondary agent depending upon the primary, and the primary agent upon the host in a nested parasitic relationship. It shows that satellite-mediated viral symptom modulation, although genetically controlled by their nucleotide sequences, can be influenced by several other factors that affect the replication dynamics of the two competing infectious nucleic acids of the triad.

Abbreviations

CMV:	cucumber mosaic virus
CARNA 5:	CMV-associated RNA 5
SBE:	satellite biological expression
VBE:	virus biological expression
SRS:	satellite replication support
PAGE:	polyacrylamide gel electrophoresis
RFP:	replication footprint profile
ss:	single-stranded
ds:	double-stranded
3'nDetm:	3'-half necrogenic determinant

Introduction

The notion of parasitic nucleic acids competing for genetic expression forms the basis for studying a biological phenomenon which is gaining increasing attention from a standpoint of plant protection. In the past 15 years satellite-mediated viral symptom modulation has begun to impact practical agriculture, insofar as it concerns the effects of an economically important and widespread plant pathogen, the cucumber mosaic virus (CMV) [see Tomlinson, 1987 for a listing of several important crop viruses worldwide]. This impact is being felt in two areas: that of virus disease epidemiology and virus control. The disease-causing potential of certain CMV satellites has been known since their discovery (Kaper *et al.*, 1976; Kaper and Waterworth, 1977) and has driven much of the basic research effort on these small parasitic RNA molecules. However, it is the exploitation of viral satellites as molecular parasites of viruses for the protection of thousands of hectares of different crops against CMV disease in China for several years (Tien and Wu, 1991), in addition to the imminent commercialization of this novel method of biocontrol in other countries, that is becoming a prime example of how spin-offs from basic research and application of biotechnological advances can rapidly benefit agriculture.

Trilateral Interactions: Satellite-mediated plant viral symptom modulation is basically determined by the trilateral interactions among satellite, virus, and host plant during an infection. Its biological outcome depends upon the genetic specificity of the three interaction partners and the competitive replication dynamics of the infecting virus/satellite combination, which in turn is decisively influenced by the satellite's parasitic relationship with the virus (Kaper, 1992). Although the symptom modulating signal is initiated at the satellite level, where its nucleotide sequence encodes information for satellite biological expression (SBE), it is the pathogenic response of the host plant to virus infection which in the perception of the observer becomes modified. Since the latter is encoded in the viral RNA, the nucleotide sequences of each helper virus/satellite combination in the aggregate will determine the ultimate biological response of a given host plant to their joint infection.

Experimental Approaches: "Reverse genetics" has made it possible to manipulate the primary structure of pathogenic RNA molecules with the aim of mapping their "functional domains." Traditionally (with viral RNA or viroids) this is done by correlating existing structural differences, or changes introduced *in vitro*, with concurrent changes in the disease symptoms of the infected host plant. However, when this approach is followed in attempts to map SBE domains of viral satellites, complications quickly arise. This is because instead of the traditional bilateral parasite - host interaction, here one deals with at least a set of trilateral interactions, due to the obligate dependence of the satellite on the replicative presence of its helper virus. Unfortunately, at the present time there exists insufficient knowledge concerning any of these interactions individually. For this reason structure - biological function studies of viral satellites should be pursued on an "everything else being equal" basis. This means that in bioassays following structural manipulation aimed at identifying SBE domains in satellites not only should host plant cultivar and helper virus be the same, but environmental conditions should also be kept scrupulously constant. Bioassays should be quantified to the extent possible and the rate of accumulation (or "titer") of the infectious RNA's in the host plant carefully monitored.

Chapter's Outline: In this paper first some of the methodology, developed in our laboratory in the past 15 years, for experimentation with satellite mediated viral symptom modulation in the laboratory and in the field will be described. Then a conceptual framework for such studies will be presented, focusing on the lethal necrosis syndrome and on symptom attenuation in tomato known to be induced by certain CMV satellite variants ever since they were first discovered (Kaper *et al.*, 1976; Kaper and Waterworth, 1977). Finally, this framework will also be linked to the epidemiology of CMV satellites (Kaper *et al.*, 1990) and their application as biocontrol agents in the field (Tien and Wu, 1991).

Dilution Endpoint Bioassay for Quantifying Satellite Biological Expression

An essential requirement in the experimental study of any molecule with a biological function is that a bioassay be developed that detects the presence of this specific molecule in the system under study as quantitatively as possible. Such a need first arose with CMV satellites when a particular variant linked to the lethal tomato necrosis disease (Kaper and Waterworth, 1977) could be shown to also ameliorate the viral symptoms in many other plant species (Waterworth *et al.*, 1979).

This quantitative bioassay is illustrated in Figure 1A for a newly identified CMV satellite, which our laboratory has designated as J876-CARNA 5 (CARNA 5 = CMV-Associated RNA 5). This satellite was of particular interest because it induced tomato necrosis at the same dilution endpoint as the prototype necrogenic CMV satellite D-CARNA 5 (Collmer and Kaper, 1986), in spite of a large structural deviation in the 5'-half of the molecule (Tousignant and Kaper, 1993; Wu and Kaper, 1992). The bioassay demonstrates unequivocally that lethal necrosis in tomato is linked to the J876-CARNA 5 inoculum concentration, since any necrogenicity attributable to a possible minor contamination of the purified material by another necrogenic satellite would have diluted out sooner. Moreover, polyacrylamide gel electrophoretic (PAGE) analysis of total nucleic acid extracts of randomly sampled tissues from each different infection (Figure 1B) showed a direct correlation between the score of dead plantlets, the J876-CARNA 5 inoculum concentration, and the presence of single-stranded (ss) and double-stranded (ds) CARNA 5 in the tissue samples visualized by PAGE analysis.

This type of dilution endpoint bioassay can basically be used in all viral satellite investigations where satellite replication leads to modulation of a viral symptom that can be measured or scored, such as with the stunting phenomenon caused by certain satellites in infections of the Ixora strain of CMV in tomato (Kaper *et al.*, 1990). However, as with any attempt at quantification of a bioassay, it is imperative that such an assay is performed under scrupulously controlled growth chamber conditions, with plants that have been cultivated or selected for optimal uniformity, using standardized inoculums and inoculation technique.

Replication Footprint Analysis for Monitoring Satellite Replication Support

Satellite replication support (SRS) is a <u>conditio</u> <u>sine</u> <u>qua</u> <u>non</u> for satellite-mediated viral symptom modulation to occur. Thus, it is very important that in such studies a way be found to quantify SRS. As a first approximation this can be done by estimating overall satellite versus virus accumulation during infection. However, in addition, it is important to obtain an insight into the qualitative aspects of the satellite replication process, since both quantitative and qualitative differences in SRS can influence the symptom modulation process significantly.

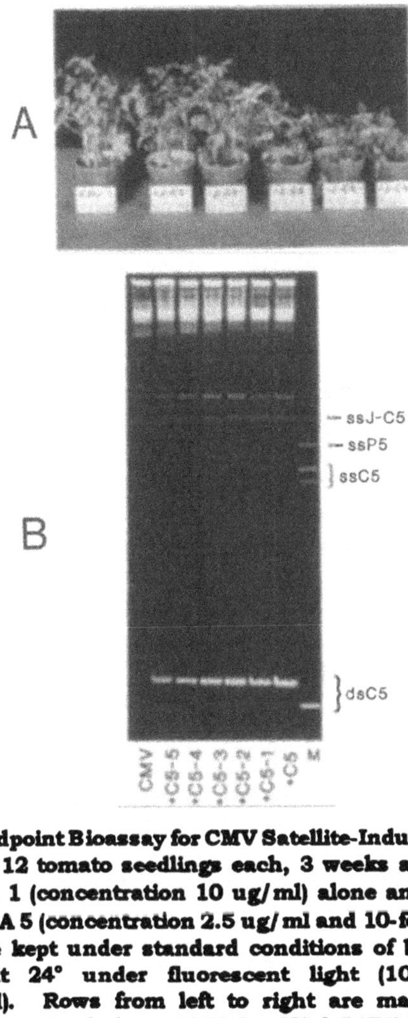

Figure 1. **Dilution Endpoint Bioassay for CMV Satellite-Induced Tomato Necrosis. A: Rows of 12 tomato seedlings each, 3 weeks after inoculation with CMV strain 1 (concentration 10 ug/ml) alone and in the presence of J876-CARNA 5 (concentration 2.5 ug/ml and 10-fold dilutions thereof). Plants were kept under standard conditions of bioassay in a growth chamber at 24° under fluorescent light (10,000 Lux, 16-hour photoperiod). Rows from left to right are marked with inoculum designations: CMV-1 alone; CMV-1 + J876-CARNA 5 at dilution of 10^{-5}; 10^{-4}; 10^{-3}; 10^{-2}; 10^{-1}. Plants inoculated with J876-CARNA 5 at 2.5 ug/ml are not shown. Concentration of J876-CARNA 5 lethal to 50 percent of test plants (LD_{50}) is 2.5 X 10^{-4} ug/ml. B: EtBr-visualized semidenaturing PAGE analysis of total nucleic acid extracts from tissues randomly sampled 14 days after inoculation from the rows of plants shown in (A). Gel lanes are marked with inoculum designations in the same left to right order as the plant rows. Lane M contains a mixture of single-stranded and double-stranded satellite RNA markers (Kaper *et al.*, 1990). Total nucleic acid extraction, PAGE, and other experimental details are in White and Kaper, 1989; Smith *et al.*, 1992a; Tousignant and Kaper, 1993.**

In our laboratory several approaches have been taken to delve into the problem of the interrelated processes of CMV satellite and helper virus replication. For each approach the appropriate experimental technique to estimate the titer of the satellite and viral RNA's was developed concurrently. Thus, the replicative processes of CARNA 5 and CMV can now be assessed *in vivo*, after infection with the appropriate helper virus/satellite combination, at the whole plant level (White and Kaper, 1989) and at the protoplast level (Smith *et al.*, 1992b; Saunders *et al.*, 1989). They can also be studied via helper virus challenge infection of transgenic plants that express satellite sequences (McGarvey *et al.*, 1990). Recently it has also become possible to study the replication of CMV satellites *in vitro* at the enzymatic level (Wu *et al.*, 1991).

For the purpose of this paper a technique that has been named "replication footprint analysis" (Kaper *et al.*, 1990; Smith *et al.*, 1992b) shall be discussed briefly, since it is relevant to experimental attempts at understanding how satellites can modulate the symptoms caused by viruses in their host plants. The technique consists of a microscale total nucleic acid extraction of plant tissue (White and Kaper, 1989) followed by semidenaturing PAGE analysis (Kaper *et al.*, 1981) of the extract, which separates the viral RNA, the ss, and the dsCARNA 5 and also often differentiates ssCARNA 5 sequence variants of essentially equal lengths (see Figure 6). In addition, a PAGE method capable of separating CARNA 5 strands of opposite polarity has also been developed (Kaper and Tousignant, 1983). The virus/satellite infection-related RNA's are visualized in autoradiographs after Northern transfer and hybridization with suitable sequence-specific radioactive probes. In this way a semiquantitative estimate of the amounts of viral and satellite RNA is obtained. Because this type of analysis estimates the sum-total of accumulation and processing of these RNA's at the time the tissue is extracted, and only provides evidence of past replication events, the visualized result of such an analysis is called a "replication footprint." A series of replication footprints taken at different time intervals in an infection is referred to as a "replication footprint profile" or RFP. RFP analysis, therefore, conveys an impression, albeit imperfect, of the replication dynamics of satellite vis-a-vis viral RNA. In experiments where differences in necrogenic expression of CMV satellite variants are correlated with their structural differences, RFP analysis serves to ascertain that such differences in SBE are indeed attributable to structural differences and not to differences in SRS (Wu and Kaper, 1992).

RFP analysis can also be of considerable practical value because it shows the time-course of satellite and viral RNA accumulation *in planta*. Figure 2 shows an RFP of an infection of tomato with the combination CMV-S/S-CARNA 5.

Here it appears that ssS-CARNA 5 reaches an optimal titer 7 days post-inoculation. Determining the time of optimal satellite titer in infected plants is important in preparing for large-scale preinoculation of crop seedlings for protection against CMV in the field (Tien and Wu, 1991). This particular experiment was used to obtain an inoculum for a field test that determined the efficacy of satellite-mediated control of CMV during a recent tomato necrosis epidemic in southern Italy (Gallitelli *et al.*, 1991).

A Framework for the Analysis of Satellite-Mediated Viral Symptom Modulation

Because satellite-mediated viral symptom modulation clearly involves a large number of factors, a framework that accounts for these factors, and that shows schematically how they are interrelated and influence the biological symptoms that are observed, is presented in Figure 3. It shows how a set of "nested" parasitic viral and satellite nucleic acids is amplified by the virus-encoded replicase and how their competing replication determines the proportions by which their expression functions VBE (viral biological expression) and SBE (satellite biological expression) impact and interact with the host plant's physiological functions to yield the symptoms perceived by the observer. It also shows how extraneous factors (here exemplified by the environmental temperature) may affect the symptom modulation at both the replicative and the biological expression level. Some of these factors will be discussed in more detail below.

Host Plant Response to Satellite Presence: Symptoms Observed

Satellite-mediated viral symptom modulation takes two basic forms; it either aggravates or attenuates viral disease symptoms. Symptom aggravation in its most extreme form results in death, symptom attenuation in a healthy-looking plant. Both extremes have been recorded for CMV satellites in tomato (Kaper and Waterworth, 1977; Mossop and Francki, 1979) (Figure 4). Although less dramatic, these symptom modulation effects by CMV satellites are equally conspicuous in tobacco (Takanami, 1981; Gonsalves *et al.*, 1982) and several other host plants (Waterworth *et al.*, 1979).

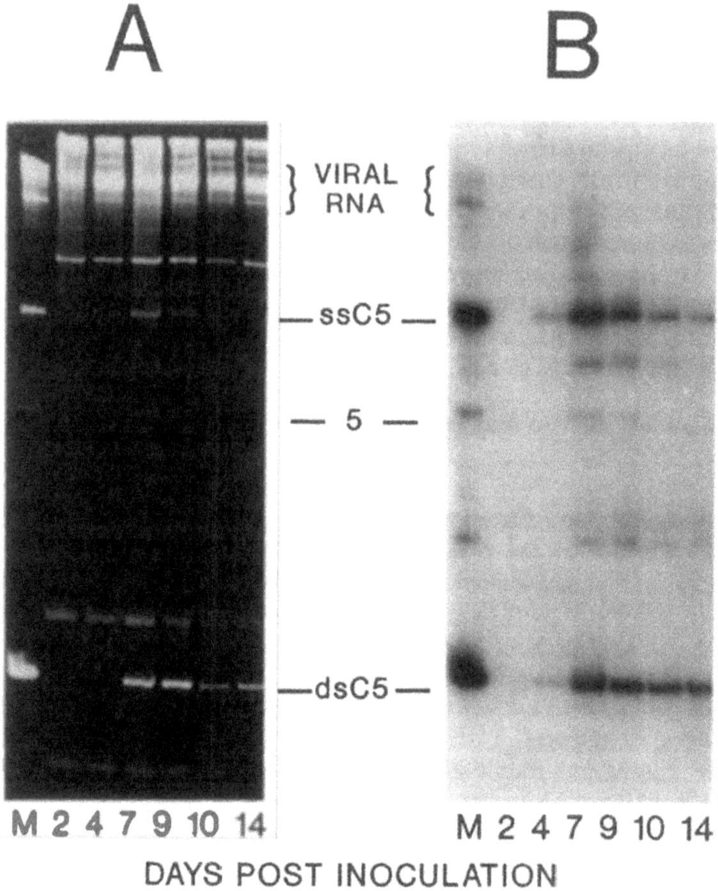

Figure 2. Replication Footprint Profile (RFP) of Ameliorative CMV Satellite Accumulation in Infected Tomato. A: EtBr-visualized semidenaturing PAGE analysis of total nucleic acid extracts of tissues randomly sampled at different times after inoculation of tomato plants with the combination CMV-S/S-CARNA 5. Gel lanes are marked with numbers designating the number of days after inoculation that tissues were sampled for extraction. Lane M is a marker lane and represents a mixture of total RNA extracted from CMV-S + dsS-CARNA 5. B: Autoradiograph of the PAGE analysis shown in (A) after Northern transfer, and hybridization with ^{32}P-labeled sequence-specific riboprobes for CMV-S RNA and CARNA 5. Experimental details for nucleic acid extraction, PAGE, Northern transfer, and hybridization with riboprobes are in White and Kaper, 1989; Smith *et al.*, 1992a.

Basically three routes can be envisioned (Figure 3) by way of which in the triad satellite - helper virus - host plant, satellite, and helper virus sequences codetermine the pathogenic response of a given host plant to their replicative presence in its cells.

(1) After symptom modulation is initiated in specific regions of the satellite's nucleotide sequence, such regions (SBE-donor domains) could interact with SBE-acceptor domains in the viral RNA, which could be located within viral biological expression (VBE) sequences. The interaction could also be with the gene products that control the pathogenic response of the host plant to the virus. This type of symptom modulation in essence would represent a modification of the VBE mechanism in response to interference by an SBE signal.

(2) SBE-donor domains on the satellite, like VBE sequences on the helper virus RNA, could activate their own mechanism for symptom formation directly in the host plant. Parallel SBE and VBE signals would result in mixed symptoms and be perceived as a symptom modulating effect of the satellite.

(3) At the SRS level replicative interference by the satellite in itself, through competition for virus-encoded replicase (Wu *et al.*, 1991; Kaper, 1982) could generate a symptom modulation effect. In the absence of, or with a silent SBE signal, this would result in a diminished VBE signal and an ameliorative response on the part of the host plant as a consequence of diminished viral replication.

Satellite Replication Support: The Indispensable Helper Virus Function

The delivery of satellite replication support (SRS) is of primary importance in studies of satellite-mediated viral symptom modulation. Without SRS there can be no satellite replication and, thus, no symptom modulation. Even if satellite symptoms are expressed directly in the host plant, via a separate SBE mechanism, independently of the helper virus [Figure 3, symptom type (2)], the SRS function in this process cannot be ignored since it controls the minimal replicative threshold level of the satellite required for symptom expression.

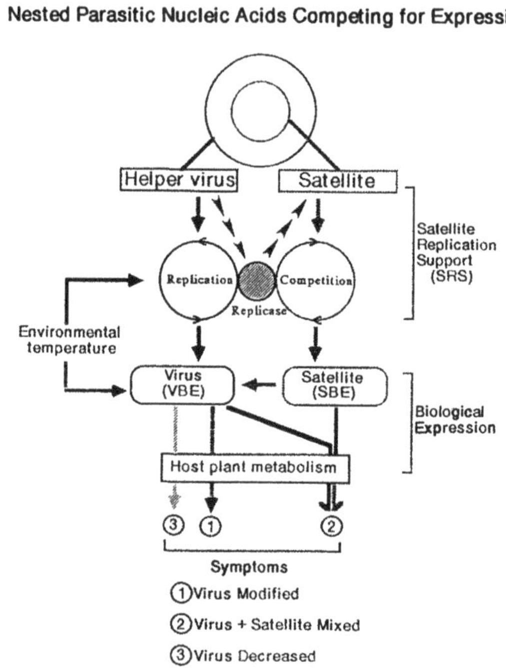

Figure 3. Schematic Representation of the Multiple Factors Affecting Satellite-Mediated Viral Symptom Modulation. Viral and satellite sequences which determine the satellite replication support (SRS) and viral/satellite biological expression (VBE/SBE) functions, are shown as concentric circles symbolizing their "nested parasitism" (satellite parasitizes virus which parasitizes the host plant). Virus and satellite functions are shown as circles, boxes, and arrows that are drawn (virus) or shaded (satellite), respectively. In the SRS phase, the competing nature of virus versus satellite amplification by the replicase is indicated by two circles revolving in opposite directions (see arrows). The one-way flow of arrowheads from the virus box to the replicase circle to the satellite box shows the satellite's replication dependence upon virus-encoded replicase. Arrows emanating from the virus and satellite boxes represent biological expression signals that interact with the plant's metabolism which is shown in the lower box. After emergence from this box they result in the symptom types designated (1) or (2), depending upon what expression route was followed. Symptom (3) results from a weakened VBE signal designated by a lightly shaded arrow. The influence of environmental temperature is shown as differently shaded arrows impacting both the SRS and the VBE/SBE phases during satellite-mediated symptom modulation.

Figure 4. **Extreme Forms of Viral Symptom Modulation in Tomato by Sequence Variants of CMV Satellites. From left to right pots marked H, +S, CMV-1, +D, contain tomato plants that were uninfected; inoculated with CMV-1 + S-CARNA 5; inoculated with CMV-1 alone; and inoculated with CMV-1 + D-CARNA 5, respectively. Plants were kept under standard conditions of bioassay (see Figure 1) and photographed 3 weeks after inoculation. S-CARNA 5 (339 nt) is a prototype ameliorative and D-CARNA 5 (335 nt) a prototype necrogenic satellite (Collmer and Kaper, 1986) with sequences differing in more than 40 positions.**

SRS is encoded in the viral genome where it may occupy one or more regions here designated SRS-donor domains. The satellite accepts replication support; and its sequence, therefore, contains one or more SRS-acceptor domains. In a simple structural model such SRS-acceptor sequences might be involved in binding the virally encoded RNA-dependent RNA polymerase or replicase. For CMV such an RNA-dependent RNA polymerase recently was shown to replicate CMV satellites *in vitro* with high template specificity (Wu *et al.*, 1991). In addition, other interacting factors, each of which presumably under some form of genetic control, must exist. These would include the auxiliary host factors that complement the virally encoded subunit to make up the replicase enzyme complex. However, any genetic analysis of the factors that contribute to the satellite's replication efficiency would be subject to additional constraints imposed upon the SRS function by the competitive, yet virus-dependent, nature of satellite replication. In a highly simplified fashion this one-way dependence is also schematically shown in Figure 3. The dynamics of these replicative interactions, which presumably are governed by well-established biochemical principles of enzyme competition and feed-back regulation, have previously been discussed in detail (Kaper, 1984). They are best analyzed with the SBE function silent, when symptom modulation is mainly regulated at the SRS level, and symptom type (3) is expressed (Figure 3).

Multiple Sequence Domains Control the Necrogenic Expression of CMV Satellites in Tomato

After a sufficient number of CMV satellites had been sequenced and their strikingly different symptoms in CMV-infected tomato (Figure 4) shown to correlate with differences in their primary structures (Kaper *et al.*, 1988), several laboratories provided evidence that a small (ca. 15-nucleotide) conserved region in the 3'-half of these molecules could be linked to their ability to induce necrotic symptoms in tomato (Jacquemond and Lauquin, 1988; Masuta and Takanami, 1989; Devic *et al.*, 1990; Sleat and Palukaitis, 1990). This seemed to identify a specific SBE-donor domain in the sequence of these satellites as principal "necrogenic determinant" (abbreviated 3'nDetm for the purpose of this paper). Thus far, all CMV satellites capable of inducing the lethal tomato necrosis disease (Figures 1 and 4) have been shown to possess the 3'nDetm. However, the presence of the 3'nDetm alone appears not to be sufficient for inducing tomato necrosis. A recent investigation has shown that other sequence elements need to be present in the CMV satellite structure and placed in the proper sequence context for the molecule to induce the disease (Wu and Kaper, 1992). This is best illustrated in Figure 5A, in which the most important structural features of a CMV satellite associated with CMV strain Y are displayed schematically. Although in possession of the 3'nDetm, the variant Y-CARNA 5 does not cause necrosis in tomato under conditions where the prototype necrogenic D-CARNA 5 (Collmer and Kaper, 1986) would kill tomato seedlings within 3 weeks after inoculation (Figure 5B). Only under certain stress-related conditions, and with a more sensitive tomato cultivar, has the satellite Y-CARNA 5 been able to induce a limited form of necrosis (Wu *et al.*, 1993). Figure 5A shows that structurally Y-CARNA 5 differs significantly from D-CARNA 5 in its 5'-half where a block of about 70 nucleotides is replaced by 100 others with these nucleotide blocks essentially unrelated in sequence. Within the replaced 70-nt region of D-CARNA 5 there must, therefore, reside a second important SBE donor domain for tomato necrosis. This idea was put to the test by constructing a chimera in which the 5'-half of Y-CARNA 5 was replaced by the 5'-half of D-CARNA 5. RNA transcripts of the chimeric clone DY were only partially necrogenic, however (Figure 5B). In a typical bioassay at least 80 percent of the plants would escape death from tomato necrosis (Wu and Kaper, 1992). Further restoration to full necrogenicity apparently depends upon further alteration of some of the remaining sequence differences in the 3'-halves of Y-CARNA 5 and D-CARNA 5 (Figure 5A). This notion was confirmed when site-directed mutation of Y-CARNA 5 of a nucleotide in close proximity to the 3'nDetm restored practically all necrogenicity of the satellite (Wu and Kaper, 1992).

Extraneous Factors and Satellite-Mediated Viral Symptom Modulation

The preceding shows how in the triad satellite - helper virus - host plant the genetically controlled "intrinsic" SRS/SBE functions basically determine the biological response of the host to infection by a given helper virus/satellite combination. In addition a variety of extraneous factors can obviously influence this response. In Figure 3 only the influence of environmental temperature has been indicated. Temperature clearly is an important factor and could play a major role in viral/satellite pathogenesis and biocontrol under field conditions. Tomato necrosis is the first reported lethal plant disease of satellite etiology (Kaper *et al.*, 1976; Kaper and Waterworth, 1977), and ameliorative CMV satellites have been used successfully in field tests for protection of tomato against lethal necrosis (Gallitelli *et al.*, 1991; Montasser *et al.*, 1991). Temperature also is a desirable parameter from an experimental standpoint as it can be manipulated fairly simply and reproducibly. Our laboratory has recently directed its focus towards the effects of environmental temperature on the necrotic response of tomato to infections of different CMV/satellite combinations. Although not completely unanticipated, the preliminary findings of this study have shown a bewildering variety of ways in which different environmental temperatures affect this response to infections by four specific CMV strains in combination with a prototype necrogenic satellite D-CARNA 5 (Collmer and Kaper, 1986). At standard bioassay temperature (24°) necrosis was observed in all infections. However, at higher temperature (32°) the response

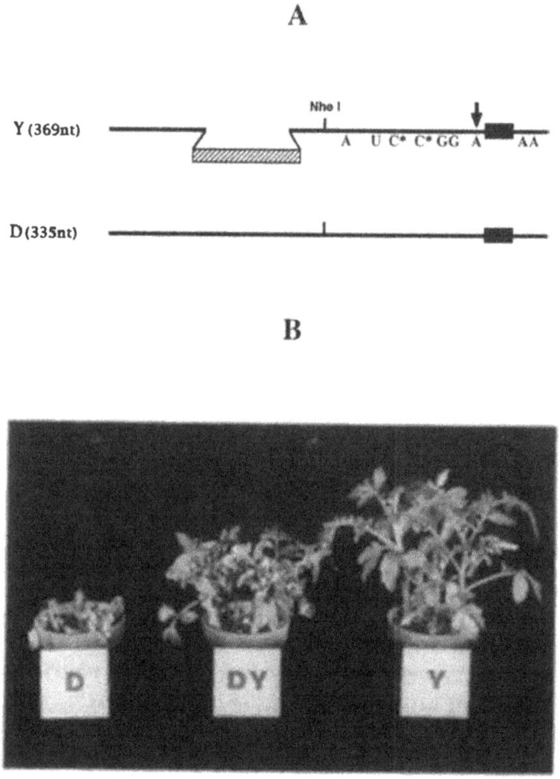

Figure 5. Nucleotide Sequence Elements of CMV Satellites Involved in Eliciting Lethal Necrosis in Tomato. A: Schematic representation of the sequences of two CMV satellites: Y-CARNA 5, which is incapable of inducing lethal necrosis, and D-CARNA 5, which is lethal to tomato within 3 weeks, as shown in (B). The two satellites have large identical stretches of sequence (drawn line) which includes the 3'nDetm domain (solid blocks). In the 5'-half this sequence homology is interrupted where ca. 70 nt of D-CARNA 5 are replaced by a stretch of ca. 100 nt in Y-CARNA 5 (shaded block) of which the sequences have only minimal homology. Other sequence differences that may influence necrogenicity differences between D- and Y-CARNA 5 are shown in the 3'-half of Y-CARNA 5. The arrow points at a nucleotide position where site-directed mutation had a significant effect upon satellite necrogenic expression (Wu and Kaper, 1992). B: Necrogenic expression of *in vitro* RNA transcripts of cloned recombinant CMV satellite cDNA in tomato 3 weeks after inoculation in the presence of helper virus CMV-1. From left to right pots marked D, DY, and Y, represent plants inoculated with D-CARNA 5 cDNA transcripts; DY (= 5' D- 184/185-Y 3')-CARNA 5 cDNA transcripts; and Y-CARNA 5 transcripts, respectively.

varied from total absence of necrosis to accelerated necrosis. These different responses could be linked to the specific nature of the helper CMV strain. With two CMV strains the profound changes in necrotic response were independent of the coinfecting satellite. Here, the changes in temperature affected viral replicative ability in tomato, which, therefore, may have been directly responsible for the changes in pathogenic response. With two other strains the SRS function seemed to have been affected, since RFP analysis revealed significant differences in D-CARNA 5 accumulation patterns of such infections (White *et al.*, 1995; Kaper *et al.*, 1995).

Viral Satellites as Causal Agents of Crop Epidemics and as Agents of Virus Biocontrol

The first unequivocal identification of a viral satellite as principal etiological agent in a major crop epidemic came after the tomato crop in the Basilicata region of southern Italy was lost to tomato necrosis in 1988 (Kaper *et al.*, 1990). Figure 3 suggests that such a tomato necrosis outbreak in the field could be determined by the genetically controlled SRS/SBE functions, provided they were proportioned in favor of optimal necrogenic expression of the satellite and combined with favorable extraneous influences, such as an abnormal intensity of aphid movement and others as yet not understood. Such lethal necrosis would probably be expressed via type (1) or type (2) symptoms. In the opposite case, when a CMV satellite functions as biocontrol agent, as is currently widely practiced in the field in China (Tien and Wu, 1991), the most straightforward interpretation in terms of Figure 3 is that SBE remains silent in the presence of a strong SRS function. This means that viral replication is outcompeted by satellite and that, consequently, viral symptoms are expressed as type (3), resulting in a visibly decreased host response.

There has been only one field-tested case where an ameliorative variant of the CMV satellite was used successfully to protect a crop against necrogenic satellite-induced disease (Gallitelli *et al.*, 1991; Montasser *et al.*, 1991). In this test evidently a situation existed where two satellites supported by the same type helper virus competed for expression and where the SRS function may have been proportioned in favor of the protecting satellite. Recently this type of field test was mimicked in a simple model experiment where the replicative competition of prototype benign and necrogenic CARNA 5, inoculated simultaneously in the presence of a given helper CMV, was monitored in tomato seedlings by means of replication footprint

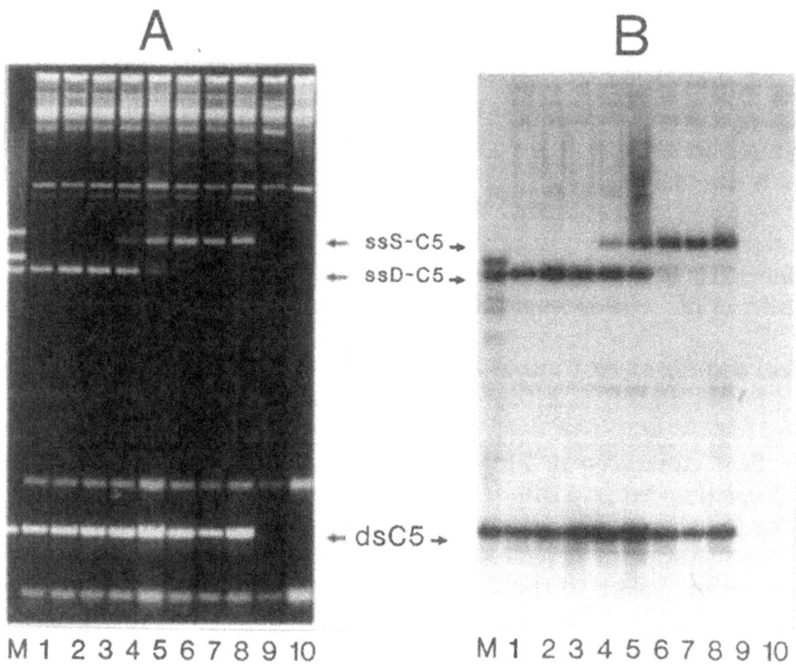

Figure 6. Replication Footprint Analysis Showing Competition of Prototype Ameliorative and Necrogenic CMV Satellites in Tomato Seedlings in the Presence of Helper Virus CMV Strain 1. A: EtBr-visualized semidenaturing PAGE analysis of total nucleic acid extracts of tissues sampled 10 days after inoculation with CMV-1 (10 ug/ml) + S-CARNA 5 (2.5 ug/ml) + D-CARNA 5 (at 2.5 ug/ml and a series of 10-fold dilutions thereof). From left to right: Lane M, a mixture of single- and double-stranded satellite RNA markers (Kaper *et al.*, 1990); lane 1, CMV-1 + D-CARNA 5 (2.5 ug/ml); lane 2, CMV-1 + S-CARNA 5/D-CARNA 5 mixture, each at 2.5 ug/ml; lanes 3-7, CMV-1 + S-CARNA 5/D-CARNA 5 mixture with D-CARNA 5 diluted 10^{-1}, 10^{-2}, 10^{-3}, 10^{-4}, 10^{-5}, respectively; lane 8, CMV-1 + S- CARNA 5 (2.5 ug/ml); lane 9, CMV-1 alone; lane 10, extract from uninfected tomato plants. B: Autoradiograph of the PAGE analysis shown in (A) after Northern transfer and hybridization with 32P labeled riboprobes specific for CMV satellites. Experimental details for nucleic acid extraction, PAGE, Northern transfer and hybridization with riboprobes are in White and Kaper, 1989; Smith *et al.*, 1992a.

analysis (Smith *et al.*, 1992a). It was surprising to learn that in this situation necrogenic D-CARNA 5 was preferentially replicated over S-CARNA 5, even when at a one thousand-fold disadvantage in inoculum concentration, presumably due to a strong SRS advantage over S-CARNA 5 (Figure 6). However, this replication advantage of D-CARNA 5 rapidly disappeared in a closer imitation of the field situation, when the necrogenic satellite was inoculated 3 days later than the protecting S-CARNA 5 (Smith *et al.*, 1992a). Clearly, to explain the complexity of real world virus/satellite infections, which can be controlled by benign satellites, more sophisticated models are needed than current technology can provide.

The interplay of satellite-mediated symptom modulation and viral symptom expression becomes even more complicated when viral disease is strongly expressed as well. In the previous section the influence of environmental temperature on the necrogenic response of specific CMV strains in combination with necrogenic D-CARNA 5 was briefly mentioned. One of the combinations used, CMV-Y/D-CARNA 5, is of specific interest in the context of this discussion, since tomato also responds necrotically to infection by the helper virus CMV-Y in the absence of satellite. At standard bioassay temperature (24°) Rutgers tomato seedlings infected with CMV-Y alone, within 1 week will display a severe necrotic response on the leaves from which the plantlets barely recover after 3 weeks. An increase in environmental temperature of merely 2-3° even becomes lethal to the plants during the second week of such an infection. When the biological response to an infection of CMV-Y in the presence of prototype necrogenic D-CARNA 5 at 24° is carefully monitored, however, several observations demonstrating the complexity of the situation can be made. First the plantlets appear to go through an initial stage of symptom attenuation where helper virus induced necrosis is clearly suppressed. However, in the infection's third week D-CARNA 5 induced necrosis rapidly surfaces and is followed by epinasty and death of the plant. On the basis of Figure 3 one can deduce that initially, with high levels of SRS activity, D-CARNA 5 replication outcompetes CMV-Y replication and suppresses helper virus-induced necrosis. Then, after the satellite accumulates to high levels, D-CARNA 5's SBE function takes over and is responsible for rapid necrogenic expression of the satellite, resulting in lethal necrosis and death of the plant.

Many other examples of this complicated interplay of CMV and its satellites undoubtedly remain to be identified. Due to the occurrence of several massive CMV outbreaks among vegetable crops in southern

Italy and in eastern Spain, the Mediterranean region in the past 4 years has become a focal point for several field studies of this virus and its associated satellites (Gallitelli *et al.*, 1988; Gallitelli *et al.*, 1988; Gallitelli *et al.*, 1988; Crescenzi *et al.*, 1993b; Jorda *et al.*, 1992). Because of the enormous economic importance of the tomato processing industry to southern Italy, and of the fresh market tomato production in eastern Spain, several research groups have now focused their efforts on CMV epidemiology in these regions, as well as on attempts towards control of such epidemics. Several types of CMV disease in tomato were recognized in these surveys. In addition to the now well-known lethal necrosis some of these diseases were also suspected to be of satellite etiology (Jorda *et al.*, 1992). One deceptive and economically particularly destructive disease is that of "tomato fruit necrosis." In Italy the disease is characterized by the presence of variously extended internal browning of fruits borne by vigorous plants that are virtually symptomless, or show only a mild mottling of the leaves. Fruit necrosis correlated invariably with the presence of a satellite named Tfn-CARNA 5, which is 390 nt in length and does not possess the 3'nDetm responsible for necrosis (Crescenzi *et al.*, 1992). However, in Spain tomato fruit necrosis failed to correlate with satellite presence and appeared to be caused by CMV alone (Jorda *et al.*, 1992). Thus, combining these two observations it would seem quite possible that the main role of Tfn-CARNA 5 would not be as a pathogen but to function as an agent containing the unlimited spread of CMV; a kind of natural biocontrol agent. It would prevent full expression of the necrogenicity of the fruit necrosis CMV strain and the plants would have a healthy appearance. However, unable to prevent viral necrogenic expression entirely, it would allow some form of partial necrosis to occur which the observer detects as the fruit necrosis syndrome (Crescenzi *et al.*, 1993a).

Unresolved Central Questions

The concept of parasitic nucleic acids competing for expression (Figure 3) has left unresolved most questions directed at problems of more central importance. Examples are: When and where in the replication cycle does what satellite-related component or gene product interact with which virus or virus-related components, or gene products, so as to signal the host plant to display a modified biological response? Is such a satellite-virus interaction absolutely needed to result in symptom modulation; or could each agent, satellite and virus individually, activate its own mechanism for symptom formation and then interact to result in a modulated symptom? Finally the most central question: Where and how would a satellite agent and/or a viral agent or their interaction product interfere with the host plant's metabolism to elicit a disease symptom?

Clearly Figure 3 does not provide answers to these fundamental questions. Its principal merit is that it does show how in infections of specific combinations of helper virus and satellite systematic consideration has to be given to all components that are interlinked in this multifaceted system. These include the genetic determinants and extraneous factors that can influence the symptomatology of the virus individually, or in combination with the satellite, via their replicative interactions or via their individual capacities to induce disease.

Acknowledgments

The experimental contributions upon which much of the above discussion is based were produced by many coworkers and colleagues over a 15-year time period. However, the excellence of technical execution exemplified by several of the illustrations was only possible thanks to the expert assistance of Lynn Geletka.

References

Collmer, C.W. and Kaper, J.M. 1986. Infectious RNA transcripts from cloned cDNA's of cucumber mosaic viral satellites. Biochemical and Biophysical Research Communications 135:290-296.

Crescenzi, A., Barbarossa, L., Cillo, F., DiFranco, A., Volvas, N., and Gallitelli, D. 1993a. Role of cucumber mosaic virus and its satellite RNA in the etiology of tomato fruit necrosis in Italy. Arch. Virol. 131:321-334.

Crescenzi, A., Barbarossa, L., Gallitelli, D., and Martelli, G.P. 1993b. A survey of cucumber mosaic cucumovirus populations in Italy under natural epidemic conditions and after a satellite-mediated protection test. Plant Dis. 77:28-33.

Crescenzi, A., Grieco, F., and Gallitelli, D. 1992. Nucleotide sequence of a satellite RNA of a strain of cucumber mosaic virus associated with tomato fruit necrosis. Nucl. Acids Res. 20:2886.

Devic, M., Jaegle, M., and Baulcombe, D. 1990. Cucumber mosaic virus satellite RNA (strain Y): Analysis of sequences which affect systemic necrosis on tomato. J. Gen. Virol. 71:1443-1449.

Gallitelli, D., DiFranco, A., and Vovlas, C. 1988. Epidemie del virus del mosaico del cetriolo e di potyvirus in Italia meridionale repertorio dei virus e aspetti ecologici. L'Informatore Agrario 44:40-45.

Gallitelli, D., DiFranco, A., Vovlas, C., Crescenzi, A., and Ragozzino, A. 1988. Una grave virosi del pomodoro in Italia meridionale. L'Informatore Agrario 44:67-70.

Gallitelli, D., DiFranco, A., Vovlas, C., and Kaper, J.M. 1988. Infezioni miste del virus del mosaico del cetriolo (CMV) e di potyvirus in colture ortive di Puglia e Basilicata. Informatore Fitopatologico 38:57-64.

Gallitelli, D., Vovlas, C., Martelli, G.P., Montasser, M.S., Tousignant, M.E., and Kaper, J.M. 1991. Satellite-mediated protection of tomato against cucumber mosaic virus. II. Field test under natural epidemic conditions in Southern Italy. Plant Dis. 75:93-95.

Gonsalves, D., Provvidenti, R., and Edwards, M.C. 1982. Tomato white-leaf: The relation of an apparent satellite RNA and cucumber mosaic virus. Phytopathology 72:1533-1538.

Jacquemond, M. and Lauquin, G. 1988. The cDNA of cucumber mosaic virus-associated satellite RNA has *in vivo* biological properties. Biochemical and Biophysical Research Communications 151:388-395.

Jorda, C., Alfaro, A., Aranda, M.A., Moriones, E., and Garcia-Arenal, F. 1992. An epidemic of cucumber mosaic virus plus satellite RNA in tomatoes in eastern Spain. Plant Dis. 76:363-366.

Kaper, J.M. 1982. Rapid synthesis of double-stranded cucumber mosaic virus-associated RNA 5: Mechanism controlling viral pathogenesis? Biochemical and Biophysical Research Communications 105:1014-1022.

Kaper, J.M. 1984. Plant disease regulation by virus-dependent satellite like replicating RNA's. In: *Control of Virus Diseases*. E. Kurstak (ed.), pp. 317-343. Marcel Dekker, Inc., New York and Basel.

Kaper, J.M. 1992. Satellite-induced viral symptom modulation in plants: A case of nested parasitic nucleic acids competing for genetic expression. Res. Virol. 143:5-10.

Kaper, J.M., Gallitelli, D., and Tousignant, M.E. 1990. Identification of a 334-ribonucleotide viral satellite as principal aetiological agent in a tomate necrosis epidemic. Res. Virol. 141:81-95.

Kaper, J.M., Geletka, L.M., Wu, G.S., and Tousignant, M.E. 1995. Effect of temperature on cucumber mosaic satellite-induced lethal tomato necrosis is helper strain dependent. Arch. Virol. In press.

Kaper, J.M. and Tousignant, M.E. 1983. Separation of the complementary strands of double-stranded cucumber mosaic virus-associated RNA 5 and peanut stunt virus-associated RNA 5. Biochemical and Biophysical Research Communications 116:1168-1175.

Kaper, J.M., Tousignant, M.E., and Geletka, L.M. 1990. Cucumber mosaic virus-assoicated RNA 5. XII. Symptom modulating effect is codetermined by satellite replication support function of helper virus. Res. Virol. 141:487-503.

Kaper, J.M., Tousignant, M.E., and Lot, H. 1976. A low molecular weight replicating RNA associated with a divided genome plant virus: Defective or satellite RNA? Biochemical and Biophysical Research Communications 72:1237-1243.

Kaper, J.M., Tousignant, M.E., and Steen, M.T. 1988. Cucumber mosaic virus-associated RNA 5. XI. Comparison of 14 CARNA 5 variants relates ability to induce tomato necrosis to a conserved nucleotide sequence. Virology 163:284-292.

Kaper, J.M., Tousignant, M.E., and Thompson, S.M. 1981. Cucumber mosaic virus-associated RNA 5. VIII. Identification and partial characterization of a CARNA 5 incapable of inducing tomato necrosis. Virology 114:526-533.

Kaper, J.M. and Waterworth, H.E. 1977. Cucumber mosaic virus-associated RNA 5: Causal agent for tomato necrosis. Science 196:429-431.

Masuta, C. and Takanami, Y. 1989. Determination of sequence and structural requirements for pathogenicity of a cucumber mosaic virus satellite RNA (Y-satRNA). Plant Cell 1:1165-1173.

McGarvey, P.B., Kaper, J.M., Avila-Rincon, M.J., Pena, L., and Diaz-Ruiz, J.R. 1990. Transformed tomato plants express a satellite RNA of cucumber mosaic virus and produce lethal necrosis upon infection with viral RNA. Biochemical and Biophysical Research Communications 170:548-555.

Montasser, M.S., Tousignant, M.E., and Kaper, J.M. 1991. Satellite-mediated protection of tomato against cucumber mosaic virus. I. Greenhouse experiments and simulated epidemic conditions in the field. Plant Dis. 75:86-92.

Mossop, D.W. and Francki, R.I.B. 1979. Comparative studies on two satellite RNA's of cucumber mosaic virus. Virology 95:395-404.

Saunders, J.A., Smith, C.R., and Kaper, J.M. 1989. Effects of electroporation profiles on the incorporation of viral RNA into tobacco protoplasts. BioTechniques 7:1124-1131.

Sleat, D.E. and Palukaitis, P. 1990. Site-directed mutagenesis of a plant viral satellite RNA changes its phenotype from amerliorative to necrogenic. Proc. Natl. Acad. Sci. U.S.A. 87:2946-2950.

Smith, C.R., Tousignant, M.E., Geletka, L.M., and Kaper, J.M. 1992a. Cucumber mosaic virus satellite competition in tomato seedlings and electroporated protoplasts. Plant Dis. 76:1270-1274.

Smith, C.R., Tousignant, M.E., and Kaper, J.M. 1992b. Replication footprint analysis of cucumber mosaic virus electroporated into tomato protoplasts. Anal. Biochem. 200:310-314.

Takanami, Y. 1981. A striking change in symptoms on cucumber mosaic virus-infected tobacco plants induced by a satellite RNA. Virology 109:120-126.

Tien, P. and Wu, G-S. 1991. Satellite RNA for the biological control of plant disease. In: *Advances in Research*. Vol. 39. K. Maramorosch, F. Murphy, and A. Shatkin (eds.), pp. 321-338. Academic Press, New York.

Tomlinson, J.A. 1987. Epidemiology and control of virus diseases of vegetables. Ann. Appl. Biol. 110:661-681.

Tousignant, M.E. and Kaper, J.M. 1993. Cucumber mosaic virus associated RNA 5. XIII. Two variants with similar 5'-half insertion/deletion elements produce opposite biological responses in tomato. Res. Virol. 144:349-360.

Waterworth, H.E., Kaper, J.M., and Tousignant, M.E. 1979. CARNA 5, the small cucumber mosaic virus-dependent replicating RNA, regulates disease expression. Science 204:845-847.

White, J.L. and Kaper, J.M. 1989. A simple method for detection of viral satellite RNA's in small plant tissue samples. J. Virol. Meth. 23:83-94.

White, J.L., Tousignant, M.E., Geletka, L.M., and Kaper, J.M. 1995. The replication of a necrogenic cucmber mosaic virus satellite is temperature-sensitive in tomato. Arch. Virol. In press.

Wu, G-S. and Kaper, J.M. 1992. Widely separated sequence elements within cucumber mosaic virus satellite contribute to their ability to induce lethal tomato necrosis. J. Gen. Virol. 73:2805-2812.

Wu, G-S., Kaper, J.M., and Jaspars, E.M.J. 1991. Replication of cucumber mosaic virus satellite RNA *in vitro* by an RNA-dependent RNA polymerase from virus infected tobacco. FEBS Letters 292:213-216.

Wu, G-S., Kaper, J.M., Tousignant, M.E., Masuta, C., Kuwata, S., Takanami, Y., Pena, L., and Diaz-Ruiz, J.R. 1993. Tomato necrosis and the 369-nucleotide Y satellite of cucumber mosaic virus: Factors affecting satellite biological expression. J. Gen. Virol. 74:161-168.

Domains in the Coat Protein and HC-Pro Protein Involved in Aphid Transmission of a Potyvirus

C. D. Atreya, P. L. Atreya, D. W. Thornbury, and T. P. Pirone
Department of Plant Pathology
University of Kentucky
Lexington, Kentucky 40546

Aphid transmission of potyviruses requires virions with an appropriate amino acid sequence near the N-terminus of the coat protein (CP) as well as a virus-encoded nonstructural protein"helper component" (HC). We are attempting to define amino acid residues or sequences in the CP and HC that are essential for aphid transmission by making site-directed mutations in tobacco vein mottling virus (TVMV). Our studies have confirmed the hypothesis that the sequence asp-ala-gly near the N-terminus of the CP is involved in aphid transmission. Substitutions in the position immediately following the gly of this triplet also affect transmissibility. Mutations in the N-terminal portion of the HC-Pro protein alter HC activity. A lys to glu change at residue 307 of the TVMV polyprotein (residue 51 of TVMV HC-Pro) abolishes HC activity. This and several other mutations in this region also reduce virus accumulation and symptom expression, indicating function(s) in addition to helper component and protease.

Introduction

Aphid transmission of potyviruses involves an interaction among aphid, virus, and "helper component," a virus-encoded nonstructural protein which is an absolute requirement for potyvirus transmission (Pirone and Thornbury, 1984; Pirone, 1991). The efficiency, or level, of transmission is dependent upon properties of the viral coat protein (Pirone and Thornbury, 1984; Pirone, 1991).

Potyvirus virions are flexous rods 680-900 nm long and ca 11 nm wide comprised of a single strand of (+) sense RNA of approximately 10 kb. The RNA is encapsidated by some 2000 copies of a single species of coat protein, Mr ca 30 kDa (Shukla and Ward, 1989). The gene products involved in the aphid transmission process have been characterized for a number of potyviruses. In the case of tobacco vein mottling virus (TVMV), the coat protein has a Mr of 29 kDa and HC-Pro is approximately 50 kDa (Pirone and Shaw, 1988; Thornbury *et al.*, 1985). The location of the genes encoding the helper component and the coat protein is shown in Figure 1. The gene encoding the helper component also encodes one of the proteases needed for cleavage of the polyprotein translational product and is referred to as HC-Pro (Carrington *et al.*, 1989).

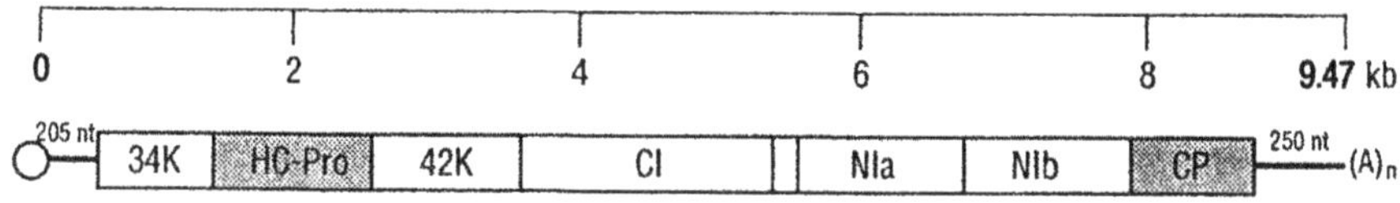

Figure 1. Gene Map of TVMV. Open boxes depict the positions of proteins in the polyprotein. Gene products involved in aphid transmission are shaded.

A major focus of recent research in our laboratory has been the identification of amino acid residues in the coat protein and the HC-Pro protein that affect aphid transmissibility. We have compared amino acid sequences of the coat protein and HC-Pro of aphid transmissible (AT) and non-transmissible (NAT) isolates of potyviruses to identify differences that might be correlated with transmissibility. Site-specific mutagenesis of TVMV cDNA was then used to substitute amino acid residues typical of NAT isolates into the coat protein or HC-Pro of an AT isolate of TVMV (Atreya *et al.*, 1991; Atreya *et al.*, 1992). Run-off transcripts were then mechanically inoculated onto tobacco plants, and the effect of the mutations was tested by aphid transmission assays. Details of these protocols have been published previously (Atreya *et al.*, 1991; Atreya *et al.*, 1992).

N-Terminal Coat Protein Sequences and Aphid Transmissibility

The N-terminus of the coat protein is exposed on the surface of the virus particle (Allison *et al.*, 1985), and enzymatic removal of the N-terminal domain results in loss of aphid transmissibility (Salomon, 1989). As first pointed out by Harrison and Robinson (1988) an amino acid triplet, Asp-Ala-Gly (DAG, in single letter code), is typically conserved near the N-terminus of the coat protein of aphid-transmissible potyviruses; but substitutions for one or more of these amino acid residues occur in non-transmissible isolates. A comparison of sequences of naturally-occurring AT and NAT isolates of several potyviruses is shown in Table 1.

Table 1. Coat Protein N-Terminal Amino Acid Sequence of Some Naturally-Occurring Potyvirus Isolates and Aphid Transmissibility of the Virus.

Virus/ Isolate[a]	N-Terminal Amino Acid Sequence	Aphid Transmissibility
TVMV-AT	SDTV**DAGK**DK	Yes (Atreya *et al.*, 1990)
TVMV-NAT	SDTV**DAEK**DK	No
TEV-AT	SGTV**DAGA**DA	Yes (Allison *et al.*, 1985)
TEV-NAT	GGTV**DASA**DV	No
TuMV #1	AGETL**DAGL**TD	Yes (Nakashima *et al.*, 1991)
TuMV #31	AGETL**DANL**TD	No
PRSV-P	SKNEAV**DAGL**NE	Yes (Quemada *et al.*, 1991)
PRSV-W	SKNEAV**DTGL**NE	No
ZYMV-AT	SGTQPTVA**DAGA**TK	Yes (Gal-on *et al.*, 1992)
ZYMV-NAT	SGTQPTVA**DTGA**TK	No
SMV-G2	SGKEKEGDM**DAGK**DP	Yes (Jayaram *et al.*, 1991)
SMV-N	SGKEKEGDM**DADK**DP	No
PPV-AT	ADEREDEEEV**DAGK**PL	Yes (Maiss *et al.*, 1988)
PPV-NAT	ADEREDEEEV**DALQ**PP	No

[a] Abbreviations: TVMV, tobacco vein mottling virus; TEV, tobacco etch virus; TuMV, turnip mosaic virus; PRSV, papaya ringspot virus; ZYMV, zucchini yellow mosaic virus; SMV, soybean mosaic virus; PPV, plum pox virus.

Our experiments with TVMV have provided direct evidence that substitutions or deletions in the DAG region can result in decrease or loss of aphid transmissibility. The results of some of these experiments are summarized in Table 2.

These data, combined with data from comparisons given in Table 1 and elsewhere (Atreya *et al.*, 1991) as well as unpublished results from our laboratory suggest the following predictions. Within the DAG triplet, an acidic residue seems to be required at the first position. A neutral residue may be tolerated, but a basic residue at this position results in loss of aphid transmissibility. In the second position, a small non-polar residue is preferred; an increase in size

Table 2. Effect of Amino Acid Deletions or Substitutions in the N-Terminal Sequence (SDTVDAGKDK) of TVMV Coat Protein on Aphid Transmissibility.

Mutation	Sequence change after mutation	% Aphid transmission	Mutation status in progeny[a]
Deletion of of DAG	SDTV---KDK	0	Retained
D --> N	SDTV**N**AGKDK	100	Retained
D --> K	SDTV**K**AGKDK	?[b]	Reverted
A --> T	SDTVD**T**GKDK	3	Retained
G --> E	SDTVDA**E**KDK	0	Retained
K --> E	SDTVDAG**E**DK	0.5	Retained
Q^{14} --> P^{14}[c]	SDTVDAGKDK...**P**	100	Retained

[a] Presence of mutation confirmed by nucleotide sequence analysis.
[b] Effect on transmission could not be determined due to reversion to wild-type.
[c] Fourteenth residue from coat protein N-terminus

and/or polarity reduces the transmissibility of the virus. Substitutions in the third position occur most often in NAT isolates; and even the replacement of gly with the next smallest amino acid, ala, results in a drastic decrease in transmissibility (Atreya, unpublished data). The apparent essentiality of gly in this position may suggest a function as a hinge in the N-terminal structure. Although the residue immediately following the DAG is not conserved among AT isolates of potyviruses, the change from lys to glu in this position results in a drastic decrease in the transmissibility of TVMV, indicating that the context within which the DAG occurs is also important.

Amino Acid Substitutions that Affect Helper Component Activity

Potato virus C (PVC), a strain of potato virus Y (PVY), produces HC-Pro protein similar in molecular weight to PVY HC-Pro but PVC-HC-Pro is non-functional as helper component in aphid transmission assays. Comparison of the derived amino acid sequence of PVC HC-Pro with that of HC-Pro of PVY and several other potyviruses revealed two PVC-specific amino acid changes within areas of conserved sequence; one is a change from lys to glu and the other from ile to val (Thornbury *et al.*, 1990). We proposed that one or both of these changes might be responsible for lack of helper component activity of PVC HC-Pro (Thornbury *et al.*, 1990).

In the TVMV polyprotein, HC-Pro represents a 457 residue polypeptide from amino acid position 257 to 713 (Mavankal and Rhoads, 1991). A lys to glu and an ile to val change at amino acid positions 307 and 482, respectively, of the TVMV polyprotein were made to mimic the PVC HC-Pro-specific amino acid residues. Additional mutations, which included depeptide changes thr-ser to ile-asp and thr-ala to leu-glu at amino acid positions (283/284) and (368/369) respectively, were made to further test the effects of mutations in this region on helper component activity. A diagrammatic representation of the location of these mutations is shown in Figure 2.

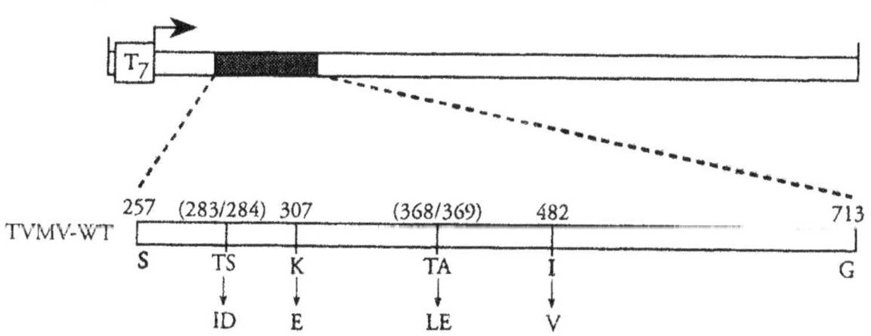

Figure 2. Location and Nature of the Amino Acid Substitutions made in the HC-Pro Region of TVMV. The numbers represent amino acid positions in the TVMV polyprotein; the N- and C-terminal amino acids of TVMV HC-Pro are shown according to Mavankal and Rhoads (1991).

One of the two mutations made to mimic the PVC-specific differences, the lys to glu change at residue 307 (residue 51 of TVMV HC-Pro), resulted in the production of HC-Pro that lacked helper component activity, while the ile to val change at residue 482 had no effect (Atreya *et al.*, 1992). The demonstration that the lys to glu change in TVMV abolishes helper component activity suggests that the presence of a glu rather than a lys residue at the corresponding position in PVC is responsible for lack of PVC helper component activity. Other recently tested substitutions at position 307 have also inactivated helper component activity, suggesting that the lys residue at this position in TVMV is critical for helper component activity. The lys residue is highly basic and might be involved in a charge interaction by which it maintains its activity. The recent report (Grumet *et al.*, 1992) that a similar lys/glu difference occurs between an AT and an NAT isolate of zucchini yellow mosaic virus (ZYMV) strengthens this conclusion. A comparison of conserved amino acid residues in the N-terminal region of HC-Pro of several potyviruses is shown in Table 3.

Table 3. Conserved Amino Acid Residues (Single Letter Code) in the N-Terminal Region of HC-Pro, Beginning with the Serine at the N-Terminus. The K-->E Substitution Occurs in PVC and ZYMV-CT, Both of Which Produce Defective Helper Component; Other Potyviruses Listed Here Produce Active Helper Component.

PVC (Thornbury *et al.*, 1990)	S-x_{21}-H-x-C-x_8-C-G-x_2-A-A-x_7-P-x_2-**E**-x_2-C-x_2-C
PVY-O (Thornbury *et al.*, 1990	S-x_{21}-H-x-C-x_8-C-G-x_2-A-A-x_7-P-x_2-**K**-x_2-C-x_2-C
PVY-N (Robaglia *et al.*, 1989)	S-x_{21}-H-x-C-x_8-C-G-x_2-A-A-x_7-P-x_2-**K**-x_2-C-x_2-C
PeMV (Vance *et al.*, 1992)	S-x_{21}-H-x-C-x_8-C-G-x_2-A-A-x_7-P-x_2-**K**-x_2-C-x_2-C
PPV (Maiss *et al.*, 1989)	S-x_{23}-H-x-C-x_8-C-G-x_2-A-A-x_7-P-x_2-**K**-x_2-C-x_2-C
TEV-HAT (Allison *et al.*, 1986)	S-x_{24}-H-x-C-x_8-C-G-x_2-A-A-x_7-P-x_2-**K**-x_2-C-x_2-C
TVMV-AT (Domier *et al.*, 1986)	S-x_{22}-H-x-C-x_8-C-G-x_2-A-A-x_7-P-x_2-**K**-x_2-C-x_2-C
ZYMV-ISR (Gal-on *et al.*, per com)	S-x_{23}-H-x-C-x_7-C-G-x_2-A-A-x_7-P-x_2-**K**-x_2-C-x_2-C
ZYMV-CT (Grumet *et al.*, 1992)	S-x_{23}-H-x-C-x_8-C-G-x_2-A-A-x_7-P-x_2-**E**-x_2-C-x_2-C

Of the other mutants in the HC-Pro region of TVMV characterized so far, the (368/369) mutation, which is outside the conserved N-terminal HC-Pro region had no adverse effect on helper component activity, while the (283/284) mutant produced insufficient HC-Pro for analysis of helper component activity (Atreya *et al.*, 1992).

Other Effects of Coat Protein and HC-Pro Mutations on Virus Phenotype

None of the stable N-terminal coat protein mutations have resulted in additional effects such as on the rate or severity of symptom expression or virus concentration in infected plants, and with one exception the mutations have thusfar not reverted to wild-type. The asp to lys mutation in the first residue of the DAG triplet (Table 1) proved to be unstable; sequencing of PCR-amplified RNA revealed only the wild-type DAG sequence in systemically infected leaves (Atreya *et al.*, 1991). This suggests that there is strong selection pressure against either the nucleotide or the amino acid change made at this location, and we are currently attempting to determine at what stage this selection pressure is manifested.

In contrast, the mutations in HC-Pro at residues 307, (283/284) and (368/369) all had significant effects, resulting in delayed and reduced expression of initial symptoms followed by the development of a symptomless systemic infection (Atreya *et al.*, 1992). Virus and viral RNA concentrations initially were substantially lower than those of wild-type TVMV, and declined to barely detectable levels over time (Atreya *et al.*, 1992).

Role of the Coat Protein and Helper Component in the Transmission Process

The process of transmission of potyviruses by aphids is very rapid; virus can be acquired by the insect in a matter of seconds and can be transmitted in a similar amount of time (Pirone, 1991). Transmissibility is lost rapidly, however, typically in a matter of minutes. These characteristics, as well as other lines of experimental evidence, indicate that virus of importance in transmission is retained in the anterior parts of the aphid alimentary tract; i.e., stylets and/or foregut, and does not enter living cells.

In order to effect aphid transmission of a potyvirus, helper component protein must be acquired prior to, or simultaneously with, virus particles (Govier and Kassanis, 1974). This suggests that helper component acts by allowing virus uptake or retention. Studies with ^{125}I-labelled potyviruses have shown no effect of helper component on the ability of aphids to acquire virus; but that only virus acquired along with helper component is retained on the stylets and foregut while in the absence of helper component virus passes into the midgut, without retention (Berger and Pirone, 1986).

The data on site of retention support the hypothesis that helper component acts by binding to virus and to sites in the aphid. Transmission-competent helper component, in its native state, is a dimer; and our current working hypothesis is that one subunit binds to the aphid and the other to the virus. The requisite sequence DAG in the surface-exposed N-terminus of the coat protein is presumably involved in the binding complex, and further molecular analysis hopefully will provide an indication of the type of interactions that is involved.

Since AT and NAT forms of TVMV virions, as well as transmission-competent and -incompetent forms of both TVMV-HC and PVY-HC are available and we have produced antibodies for each, we are currently attempting to acquire more detailed information about the location and nature of the putative binding complex, through the use of gold-labelled antibodies and electron microscopy.

Conclusions

Our studies have resulted in the delineation of domains in the coat protein and HC-Pro that are involved in the aphid transmission process and have begun to identify amino acid sequences or residues that are critical to the process. Further mutational analysis will be required, however, to refine our current understanding of how these sequences act in allowing potyviruses to be transmitted by aphids.

The identification of mutations that result in loss of aphid transmissibility may have some practical applications. The possibility has been raised that coat protein expressed in transgenic plants might encapsidate heterologous RNA's and result in aphid transmission of a heretofore non-aphid transmissible genotype (de Zoeten, 1991). Deletion of one or all of the codons for DAG would result in expression of an NAT coat protein and obviate objections based upon this possibility. Another potential application is based upon our finding that certain mutations in the HC-Pro result in dramatic decreases in virus titer and symptom expression. We have found that prior infection with one of these mutants results in effective cross-protection against wild-type TVMV (Atreya and Pirone, unpublished). We are currently testing a version of TVMV, made doubly defective for aphid transmission by the deletion of key amino acids in the coat protein as well as mutation of HC-Pro sequences that also result in the mild phenotype. Such a virus might have the potential to be used for cross-protection under field conditions.

Acknowledgements

The research reported here was supported by the USDA Competitive Grant Program, by USDA-ARS Contracts, and by grants for the R.J. Reynolds Corporation.

References

Allison, R.F., Doughtery, W.G., Parks, T.D., Willis, L., Johnston, R.E., Kelly, M., and Armstrong, F.B. 1985. Biochemical analysis of the capsid protein gene and capsid protein of tobacco etch virus: N-Terminal amino acids are located on the virion's surface. Virology 147:309-316.

Allison, R.F., Johnston, R.E., and Dougherty, W.G. 1986. The nucleotide sequence of the coding region of tobacco etch virus genomic RNA: Evidence for synthesis of a single polyprotein. Virology 154:9-20.

Atreya, P.L., Atreya, C.D., and Pirone, T.P. 1991. Amino acid substitutions in the coat protein result in loss of insect transmissibility of a plant virus. Proc. Natl. Acad. Sci. U.S.A. 88:7887-7891.

Atreya, C.D., Atreya, P.L., Thornbury, D.W., and Pirone, T.P. 1992. Site-directed mutations in the potyvirus HC-PRO gene affect helper component activity, virus accumulation, and symptom expression in infected tobacco plants. Virology 191:106-111.

Atreya, C.D., Raccah, B., and Pirone, T.P. 1990. A point mutation in the coat protein abolishes aphid transmissibility of a potyvirus. Virology 178:161-165.

Berger, P.H. and Pirone, T.P. 1986. The effect of helper component on the uptake and localization of potyviruses in Myzus persicae. Virology 153:256-261.

Carrington, J.C., Cary, S.M., Parks, T.D., and Dougherty, W.G. 1989. A second proteinase encoded by a plant potyvirus genome. EMBO J. 8:365-370.

de Zoeten, G.A. 1991. Risk assessment: Do we let history repeat itself? Phytopathology 81:585-586.

Domier, L.L., Franklin, K.M., Shahabuddin, M., Hellman, G.M., Overmeyer, J.H., Hiremath, S.T., Siaw, M.F.E., Lomonossoff, G.P., Shaw, J.G., and Rhoads, R.E. 1986. The nucleotide sequence of tobacco vein mottling virus RNA. Nucleic Acids Res. 14:5417-5430.

Gal-on, A., Antignus, Y., Rosner, A., and Raccah, B. 1992. A zucchini yellow mosaic virus coat protein gene mutation restores aphid transmissibility but has no effect on multiplication. J. Gen. Virol. 73:2183-2187.

Gal-on, A., Huet, H., Chejanovsky, N., and Raccah, B. Personal communication.

Govier, D.A. and Kassanis, B. 1974. Evidence that a component other than the virus particle is needed for aphid transmission of potato virus Y. Virology 57:285-286.

Grumet, R., Bada, R., and Hammar, S. 1992. Analysis of the zucchini yellow mosaic virus (ZYMV) potyviral helper component, possible identification of an aphid-interaction domain. Phytopathology 82:1176.

Harrison, B.D. and Robinson, D.J. 1988. Molecular variation in vector-borne plant viruses: Epidemiological significance. Phil. Trans. Roy. Soc. London B321:447-462.

Jayaram, C., Hill, J.H., and Miller, W.A. 1991. Nucleotide sequences of the coat protein genes of two aphid-transmissible strains of soybean mosaic virus. J. Gen. Virol. 72:1001-1003.

Maiss, E., Timpe, U., Brisske, A., Jelkmann, W., Casper, R., Himmler, G., Mattanovich, D., and Katinger, H.W.D. 1989. The complete nucleotide sequence of plum pox virus RNA. J. Gen. Virol. 70:513-524.

Maiss, E., Timpe, U., Jelkmann, W., and Casper, R. 1988. Comparison of two different plum pox virus isolates on nucleic acid level. Acta Hort. 235:313-317.

Mavankal, G. and Rhoads, R.E. 1991. *In vitro* cleavage at or near the N-terminus of the helper component protein in the tobacco vein mottling virus polytprotein. Virology 185:721-731.

Nakashima, H., Sako, N., Joh, K., Hori, K., and Nonaka, F. 1991. Nucleotide sequences of the coat protein genes of aphid transmissible and non-transmissible isolates of turnip mosaic virus. Ann. Phytopath. Soc. Japan 57:549-557.

Pirone, T.P. 1991. Viral genes and gene products that determine insect transmissibility. Seminars in Virology 2:81-87.

Pirone, T.P. and Shaw, J.G. 1988. Tobacco vein mottling virus. CMI/AAB Descriptions of Plant Viruses No. 325.

Pirone, T.P. and Thornbury, D.W. 1984. The involvement of a helper component in nonpersistent transmission of plant viruses by aphids. Microbiol. Sci. 1:190-193.

Quemada, H., L'Hostis, B., Gonsalves, D., Reardon, I.M., Heinrikson, R., Hiebert, E.L., Sieu, L.C., and Slightom, J.L. 1990. The nucleotide sequences of the 3'-terminal regions of papaya ringspot virus strans W and P. J. Gen. Virol. 71:203-210.

Robaglia, C., Durand-Tardif, M., Tronchet, M., Boudazin, G., Astier-Manifacier, S., and Casse-Delbart, F. 1989. Nucleotide sequence of potato virus Y (N strain) genomic RNA. J. Gen. Virol. 70:935-947.

Salomon, R. 1989. Res. Virol. 140:453-455.

Shukla, D.D. and Ward, C.W. 1989. Structure of potyvirus coat proteins and its application in the taxonomy of the potyvirus group. Adv. Virus Res. 36:273-314.

Thornbury, D.W., Hellmann, G.M., Rhoads, R.E., and Pirone, T.P. 1985. Purification and characterization of potyvirus helper component. Virology 144:260-267.

Thornbury, D.W., Patterson, C.A., Dessens, J.T., and Pirone, T.P. 1990. Comparative sequence of the helper component (HC) region of potato virus Y and a HC-defective strain, potato virus C. Virology 178:573-578.

Vance, V.B., Moore, D., Turpen, T.H., Bracker, A., and Hollowell, VC. 1992. The complete nucleotide sequence of pepper mottle virus genomic RNA: Comparison of the encoded polyprotein with those of other sequenced potyviruses. Virology 191:19-30.

Precise Interactions Between Tobamoviruses and Plants

W. O. Dawson, A. G. C. Lindbeck, and D. J. Lewandowski
University of Florida
Citrus Research and Education Center
Lake Alfred, Florida 33850

J. N. Culver
Department of Botany
University of Maryland
College Park, Maryland 20742

M. E. Hilf
United States Department of Agriculture
Horticultural Laboratory
Orlando, Florida 32803

G. Stubbs
Department of Molecular Biology
Vanderbilt University
Nashville, Tennessee 37212

Tobamovirus gene products interact precisely with plant gene products at several levels. Efficient systemic infection of a plant requires that the virus replicate, move cell-to-cell, and move long distances effectively. Each of these processes appears to require precise viral-plant interactions. The efficacy of these primary interactions of the viral gene products with different plants defines the host range of viruses. Another level of interaction occurs when specific plant gene products recognize specific viral gene products and mount active defense responses that result in resistance. Additionally, a level of interactions, perhaps accidental interactions, occurs when viral gene products interfere with necessary plant processes, resulting in disease symptoms. Each of these interactions can be extraordinarily precise with as little as a single nucleotide alteration in the viral genome differentiating between systemic infection and immunity, susceptibility and resistance, or disease and tolerance.

Introduction

Tobamoviruses constitute a group of plus-sense RNA viruses whose 6-7 kb genomes reside on single RNA molecules that are encapsidated into rigid rod-shaped virions approximately 18 X 300 nm. The type member is tobacco mosaic virus (TMV). The genome of tobamoviruses consists of three open reading frames (ORF's) that express four or five proteins. The 5'-ORF encodes a replicase protein of approximately 130 kDa; and, by occasional read-through of the amber stop codon, a protein with an additional polymerase domain is produced (180 kDa). The capsid protein (17.5 kDa) and the cell-to-cell movement protein (30 kDa) are produced via 3'-co-terminal subgenomic mRNA's. An additional mRNA has been found that could produce a 54 kDa protein which would be the read-through region of the 180 kDa protein. These viruses occur throughout the world causing considerable crop losses. They accumulate to high concentrations in infected tissues and virions are unusually stable outside of infected cells. Transmission from plant to plant normally occurs mechanically by contaminated materials coming in contact with plant surfaces. Tobamoviruses tend to have unusually broad experimental host ranges (Holmes, 1946) but tend to naturally occur only in a limited number of plant species (Bald, 1960).

When considering random viruses and plants, it is only rare combinations that allow infections. Even viruses with broad host ranges infect only a small proportion of plant species. If a particular plant species is surveyed, it is probable that a number of viruses will be found that can infect it; but most viruses will not be able to infect that plant. When plants become infected, sometimes the infections result in severe disease and growth reduction, whereas other combinations result in minor or no symptoms.

The theme of this chapter is that precise interactions between virus-encoded gene products and plant-encoded gene products are required for a virus to infect and invade a plant. There are different levels of interactions between the virus and the plant. We define primary levels of interactions as those that are required for the virus to systemically infect a plant. These include interactions required for replication and movement cell-to-cell and long distances. There also appears to be secondary interactions that determine whether the viral infection causes disease symptoms and reductions in growth and yield of infected plants or allow the plant to recognize the viral infection and mount a resistance response. The degree of vulnerability of the plant is defined as susceptibility or resistance which usually is a measurement relative to related virus-plant interactions. There is a wide range of virus-plant interactions that allow some degree of infection, ranging from the virus being able to infect and replicate only in the initially inoculated cell to the infection being confined to inoculated leaves to the virus invading and replicating in all cells of the plant. For purposes of this review, we consider plants as susceptible only if the virus is able to infect initial cells, replicate, move cell-to-cell throughout that leaf, move to other leaves, and invade and replicate in most of the cells of the plant. We will use the term systemic infection to describe this process. Plants that do not allow this level of infection have some degree of resistance to the virus.

It appears that for a plant to be susceptible it must provide precise functions to assist the virus in all necessary functions for viral invasion: the infection process, replication of viral RNA, movement to adjacent cells across the cell wall barrier, and entry into phloem elements and transport from leaf to leaf. Viral encoded gene products do not solely effect these processes. A plant that completely fails to assist in any one of these functions will be insusceptible. However, the process of viral systemic infection is even more complex. There appears to be a required balance of these processes. Each specific function must be performed above a specific threshold. Thus, reduction of any one function may effectively prevent systemic infection and susceptibility.

Thus, for a virus to be virulent in a particular plant it must be able to replicate, move cell-to-cell through plasmodesmata, and move long distances through phloem. For a plant to be susceptible, it must provide the virus with the appropriate environment for each of these processes. This series of virus-plant interactions appears to involve precise protein-protein interactions between all viral proteins and several plant proteins. These interactions are so precise that one or a few amino acid substitutions in a single protein can determine whether the virus can systemically infect a particular plant or whether that plant is resistant to the virus.

Primary Interactions

Replication: Replication is the process which produces many copies, estimated as approximately 10 (Ding *et al.*, 1992) per cell, of the parental infecting RNA. However, this is a complex process that also regulates gene expression via production of subgenomic mRNA's that are differentially synthesized and capped. The enzymes involved in replication are thought to specifically recognize different RNA sequences and to cap and methylate certain 5'-termini. The viral-encoded 130 kDa and 180 kDa proteins are required for RNA replication and are components of a "replicase complex." However, there is evidence that additional proteins encoded by the plant genome are essential to the complete replicase complex.

Although plant proteins constituting the tobamoviral replicase complex have not been characterized, one series of experiments indirectly demonstrate that viral proteins must precisely interact with plant proteins for viral replication. Tomato plants homozygous for the Tm-1 gene are resistant to tomato mosaic tobamovirus (ToMV). Mutants of tomato mosaic virus have been found that overcome this resistance (Meshi *et al.*, 1988). These mutants have two amino acid exchanges in the replicase proteins. This suggests that a protein in tomato plants has a secondary function that is required for viral replication. This protein in Tm-1/Tm-1 plants is slightly altered, still providing its primary function required for the plant, but now not able to function in viral replication. The two amino acid alterations in replicase are required for the viral protein to precisely interact with the Tm-1 protein and allow viral replication. Thus, the ability of this plant protein to interact with the viral replicase proteins appears to be one determinant of the viral host range.

Cell-to-Cell Movement: Once the virus infects and replicates in an individual plant cell, the next step is to move to adjacent cells. However, plant cells have cell walls that can provide an effective barrier to viral passage. Plasmodesmata are complex openings across cell walls connecting adjacent cells. The early literature assumed that viruses moved passively cell-to-cell through the plasmodesmatal openings during normal cytoplasmic streaming. However, an event that triggered substantial interest in this area was the demonstration that cell-to-cell movement required a specific viral function. Isolation of a TMV mutant that was able to replicate in infected cells but was deficient for cell-to-cell spread at high restrictive temperatures demonstrated that some viral function must be involved in the process of moving to adjacent cells (Nishiguchi *et al.*, 1978). This mutant was shown to have an altered 30 kDa protein responsible for this phenotype (Meshi *et al.*, 1987). Since then it has been shown that the viral 30 kDa protein specifically associates with secondary plasmodesmata and increases the exclusion limit of their openings (Ding *et al.*, 1992). Also, the 30 kDa protein appears to bind and encapsidate tobamovirus genomic RNA's serving as a chaperone to move the RNA through the plasmodesmata (Citovsky *et al.*, 1990).

The virus-plant interactions that are involved in cell-to-cell movement are often the block that prevents the virus from systemically infecting a plant. In many plant species in which the virus is able to replicate in infected cells, the virus is not able to effect cell-to-cell movement. Plants of this type have been defined as "subliminal hosts" (Cheo, 1971; Sulzinski and Zaitlin, 1982).

The viral 30 kDa protein appears also to precisely interact with the plant to effect viral cell-to-cell movement. The relative effectiveness of different 30 kDa proteins in particular plants has been examined by creating chimeric viruses in which the only alteration was the substitution of the 30 kDa gene ORF. Both (U5) and *Odontoglossum* ringspot tobamovirus (ORSV) 30 kDa proteins effect cell-to-cell movement of chimerics of TMV in tobacco plants less effectively than virus with the TMV 30 kDa protein, suggesting that these proteins do not interact optimally with tobacco plasmodesmata (Neijdat *et al.*, 1991; Hilf and Dawson, 1992).

In plants where the virus can move cell-to-cell, it has been shown that the type, amount, and timing of 30 kDa protein affect how effectively the virus moves cell-to-cell. Changing the timing of production of the correct 30 kDa protein from early to late reduced the rate of cell-to-cell movement (Lehto *et al.*, 1990). Additionally, the level of expression of 30K gene can be altered by changing its position relative to the 3'-terminus. Removal of the coat protein ORF increases the level of 30 kDa protein production 10-20 fold, whereas positioning further from the 3'-terminus causes decreases in 30 kDa protein production (Culver *et al.*, 1992). Increases in 30 kDa protein production did not increase the rate of cell-to-cell movement suggesting that levels of 30 kDa protein above a certain threshold had no effect (Dawson, 1990). Moderate reductions in 30 kDa protein, up to one-tenth of wild-type levels, did not reduce cell-to-cell movement. However, production of even lower amounts of 30 kDa protein by positioning of the 30 K gene increased distances from the 3'-terminus did result in substantially reduced rates of cell-to-cell movement (Donson *et al.*, 1991).

The precision of the 30 kDa protein-plant interaction is better seen from the ability of the virus to replicate in resistant plants. Tomato plants with the Tm-2 or Tm-2^2 alleles allow ToMV to replicate in inoculated cells but do not allow the virus infection to move to adjacent cells. Thus, the wild-type 30 kDa protein cannot effect viral cell-to-cell movement in these plants. Virulent mutants that can overcome this resistance develop in fields where resistant tomatoes are grown. Genetic analysis of such mutants have found a few amino acids substitutions within the 30K ORF (Meshi *et al.*, 1989;

Calder and Palukaitis, 1992). The resistance in Tm-2 or Tm-2^2 tomato plants appears to be a slight change in a tomato protein, perhaps associated with secondary plasmodesmata, that prevents the interaction required for the virus to move cell-to-cell.

Efficiency of cell-to-cell movement also affects systemic infection. The TMV hybrid containing the substituted ORSV 30K ORF moves cell-to-cell, albeit at a reduced rate, and produced substantial amounts of virus in inoculated leaves of tobacco plants. However, the systemic infection of these plants was greatly retarded and upper leaves never became uniformly infected similar to the situation with TMV-infected tobacco plants (Hilf and Dawson, 1992). These data suggest that a 30 kDa protein with a sequence that does interact precisely enough with the corresponding plant component greatly affects systemic infection. However, production of the optimal 30 kDa protein does not guarantee successful systemic infection. Alteration of the timing or amounts of the wild-type 30 kDa protein has a similar effect. The mutants with delayed or reduced production of 30 kDa protein also were deficient in systemic infection (Lehto *et al.*, 1990; Donson *et al.*, 1991).

A remarkable observation is that several of the mutants that exhibited greatly reduced ability to systemically infect tobacco, mutants with the wrong 30 K gene, mutants with altered timing of 30K gene expression, or mutants with reduced production of 30 kDa protein, were all able to systemically infect *Nicotiana benthamiana* plants similarly to wild-type TMV. Thus, the viral-plant interactions required for systemic infection of these two plants must be different.

Long-distance Movement: Long-distance movement, movement from the initially infected area, often an inoculated leaf, to other leaves and organs throughout the plant, occurs via the phloem, which is a tissue consisting of living cells that conduct metabolites throughout the plant. Once the virus enters the phloem, it generally moves in the direction of metabolites within the phloem which usually is from photosynthetically active areas that export sugars to areas that import sugars, most frequently to roots and shoots that are not yet photosynthetically self-sufficient. Thus, virus can first be found to move from infected leaves to roots and the youngest leaves. Compared to cell-to-cell movement that occurs in microns per hour, long-distance movement in phloem can occur in cms/hr.

The coat protein is the only structural protein. It had been known for a considerable time that coat protein was involved in long-distance movement because TMV mutants that fail to form virions were always deficient in long-distance movement (Siegel *et al.*, 1962; Dawson *et al.*, 1988; Saito *et al.*, 1990). Thus, it was interpreted that virions were the vehicle of long-distance movement. However, the function of coat protein appears to be more complex. We compared the long-distance movement of two tobamoviruses in tobacco plants. TMV efficiently systemically infects tobacco plants. The virus replicates to high levels in inoculated cells, spreads to adjacent cells within inoculated leaves, and quickly moves to roots and the upper leaves, 3-4 days after the lower leaves are inoculated. The virus continues to spread throughout the plant until most cells are infected and have high titers of virus. In contrast, ORSV, a virus normally found in *Cymbidium* orchids, does not noticeably systemically infect tobacco plants. The virus replicates to high titers in infected cells, moves cell-to-cell (albeit at reduced rates), and accumulates substantial amounts of virus in inoculated leaves. However, the virus rarely moves into upper leaves and then only small amounts of virus are found.

A series of hybrids of TMV and ORSV were constructed to determine which ORSV gene products did not effectively interact with tobacco plants to provide systemic infection (Hilf and Dawson, 1992). The definitive hybrid was TMV except for the substitution of the ORSV coat protein ORF (TMVcpO). This virus replicated and moved cell-to-cell almost like the wild-type virus, but was defective in long-distance movement. The production of coat protein for the formation of large numbers of stable virions was not sufficient for efficient long-distance movement. These data suggest that the "correct" tobamovirus coat protein is required, perhaps in a precise interaction with the plasmodesmata that allow entry to and exit from the phloem, in order for the virus to move long distances to systemically infect tobacco plants.

Secondary Interactions

There are virus-plant interactions that are not involved in the systemic infection. Since these interactions are not required for viral replication, movement, and systemic infection of the plant, we define them as "secondary interactions." At least some of these interactions also appear to be precise in that single amino changes can determine whether a certain response occurs. Some of these interactions may be accidental and inconsequential to the virus.

Induction of Disease Symptoms: In general, disease symptoms caused by tobamoviruses result from chlorosis of leaves, with yellowing of developed leaves and/or the development of mosaic patterns of dark green (normal cells) mixed with light green areas (cells with abnormal chloroplasts) in newly forming leaves. We have shown that these two symptoms arise from two different processes.

Mosaic Symptoms: The most common symptom of plants systemically infected with tobamoviruses consists of a mosaic of light green and normal dark green areas in leaves that develop after the systemic infection. However, there is no direct correlation between amount of virus produced and systemic infection and whether symptoms are produced. Tobamovirus mutants have been isolated and characterized that fail to induce symptoms in infected leaves. Also, it is common to find mutants that induce more severe symptoms, often a mosaic of light green and dark green areas mixed with bright yellow or necrotic areas. We still have little information concerning the interactions between virus and plant that cause abnormal plant development and disease. However, it appears that these interactions are secondary, perhaps accidental, interactions.

Chlorotic symptoms induced by tobamoviruses on plants is usually due to the production of abnormal chloroplasts. The common mosaic symptoms occur in leaves that develop after infection by the virus. Mosaic symptoms result from prevention of normal chloroplast development and occur when small leaves become infected. Although coat protein and occasionally virions can be found in such chloroplasts, it is clear that viral coat protein is not involved in the cause of normal light green/dark green mosaic patterns. Mutants of TMV that were constructed not to be able to produce coat protein still induced normal mosaic patterns in developing leaves (Culver and Dawson, 1989; Lindbeck *et al.*, 1992).

Attenuated Symptoms: The severity of mosaic symptoms varies with different tobamovirus mutants or isolates. Genetic characterizations of several such mutants have identified amino acid substitutions in the replicase proteins to be responsible for reduced symptoms. However, each of the replicase alterations resulted in what appears to be different approaches to affecting the induction of mosaic symptoms.

A mutant of ToMV that fails to induce mosaic symptoms differs from the wild-type virus by 10 nucleotides that map to the replicase genes (Nishiguchi *et al.*, 1985). This mutant appears not to be deficient in replication but is deficient in production of 30K subgenomic mRNA, suggesting that cell-to-cell movement might somehow be involved in this interaction. A similar mutant of TMV behaved identically to wild-type virus in protoplasts and inoculated leaves, but moved into upper leaves slightly slower suggesting the ability to move long distances was reduced causing the milder symptoms (Nelson *et al.*, 1992). Another symptomless TMV mutant with only one amino acid alteration in replicase moved cell-to-cell and into upper leaves identically to wild-type virus but replicated to a lower level, accumulating only about one-tenth as much virus in infected leaves as the wild-type virus (Lewandowski and Dawson, 1992). At this time, it is not understood what specific interactions between viral and plant gene products induce mosaic symptoms. Since minor alterations in replicase proteins caused changes in expression of these symptoms, it is possible that the replicase proteins interact specifically with a plant component to prevent normal chloroplast development. However, the replicase alteration also could result in delayed timing of infection of cells in young leaves giving reduced symptoms.

Bright Yellow Symptoms: A common symptom caused in plants by tobamoviruses is a yellowing on both inoculated and systemically-infected leaves. These symptoms are distinct from, but often occur concurrently with, normal mosaic symptoms. Several isolates and mutants of tobamoviruses cause mosaic symptoms that are a combination of dark green, light green, and bright yellow areas of tissue (Lindbeck *et al.*, 1992; McKinney, 1935).

In directly inoculated leaves, where chloroplasts are already fully developed at the time of infection, coat-protein-related polypeptides containing specific deletions in coat protein caused the degradation of fully developed chloroplasts (Lindbeck *et al.*, 1991). Wild-type TMV and other non-yellowing mutants did not cause any significant ultrastructural changes in chloroplasts, whereas yellowing mutants caused chloroplast degradation resulting in interconnected tubular networks which appeared to be derived from rearrangements of existing photosynthetic membranes. Early in the development of these symptoms, chloroplast grana appeared to be unstacking, forming convex structures which evolve into tubules. We since have examined a series of mutants with single amino acid substitutions in the coat protein that cause similar effects on the chloroplasts (Table

1; Figure 1). Many of these coat proteins are still functional for assembly.

Table 1. Summary of Coat Protein Point Mutants.

Mutant	Amino Acid From	Change To	Symptoms on Leaves Inoculated	Systemic	Virions
cp20PL	Pro	Leu	none	mosaic	+
cp29ND	Asn	Asp	yellow spots	none	+
cp33ND	Asn	Asp	yellow spots	none	+
cp42TR	Thr	Arg	necrosis	none	+
cp66DN	Asp	Asn	yellow spots	mosaic	+
cp66DA	Asp	Ala	yellow spots	mosaic	+
cp67FL	Phe	Leu	white spots	mosaic	+
cp67FI	Phe	Ile	white spots	mosaic	+
cp88DI	Asp	Ile	none	none	disks
cp133IP	Ile	Pro	yellow spots	none	-

Bright yellow mosaic symptoms appear to be due to the addition of chlorotic symptoms caused by the coat protein mutation to the classic light green/dark green mosaic symptoms caused by some viral product other than the coat protein (Lindbeck *et al.*, 1992). Normal development of chloroplasts was inhibited in bright yellow areas. The chloroplasts in this tissue were immature and not photosynthetically competent.

Electron microscope examination of chlorotic tissue from both inoculated and systemically-infected tissue showed a correlation betweeen the presence of coat protein aggregates in the cytoplasm of infected cells and chloroplast disorganization (Lindbeck *et al.*, 1992; Lindbeck *et al.*, 1991). The presence of these aggregates suggests that similar mechanisms may be involved in the development of bright yellow symptoms in these leaves. A number of important proteins required for chloroplast development and maintenance are synthesized in the cytoplasm and transported into the chloroplast. It is possible that the induction of bright yellow symptoms results from the coat protein aggregates interfering with the synthesis or translocation of the cytoplasmically-synthesized chloroplast proteins and, thus, preventing chloroplast maintenance.

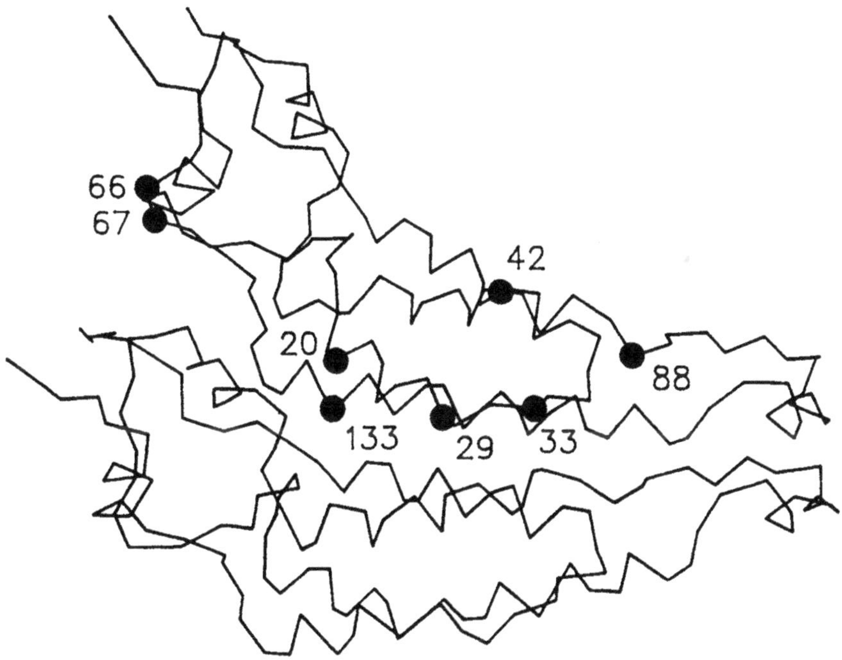

Figure 1. Schematic Diagram of Two Adjacent Coat Protein Molecules Showing the Locations of the Mutations Listed in Table 1.

Necrotic Symptoms: Another symptom type is necrosis. The most studied type of necrosis is the hypersensitive response (HR) which results from a gene-for-gene type interaction in which specific plants recognize specific viruses and initiate an active defense reaction that usually confines the virus to the area of initial infection. However, there are other non-specific necrotic responses. Several isolates and mutants induce necrosis in a range of different plant species, often giving a mosaic of light green and dark green mixed with necrosis. We initially became interested in this phenomenon when we created a series of mutations with deletions in the coat protein ORF's (Dawson *et al.*, 1988). Two of these mutants which have truncated, non-functional coat proteins induced a generalized necrosis in all infected areas in almost all plant species that were examined. Similar mutants of ToMV have also been created and examined (Saito *et al.*, 1989). Since only specific coat protein mutants induce these necrotic symptoms, this cell pathology appears to result from a precise interaction between certain coat proteins and some cell component.

Active Resistance: Virus-plant interactions that allow viral replication, cell-to-cell movement, and long-distance movement do not guarantee susceptibility. A precise interaction resulting in host recognition of the virus can trigger an active defense response that confines the virus to the initial infection area. This recognition consists of an interaction between plants of a specific genotype, perhaps only one specific gene of the plant, and a specific virus or viral gene product.

The best understood system of active resistance is the hypersensitive response (HR) in which a series of plant genes are turned on resulting in a confined necrotic area containing anti-pathogen compounds. The HR provides effective resistance in plants to different pathogens including bacteria, fungi, nematodes, and viruses. This appears to be a generalized defense response in which a series of induced reactions to limit general pathogens and pests are produced. The major difference in this defense mechanism toward different pathogens appears not to be the cascade of induced events that limit the pathogen, but the mechanisms of induction of that cascade of events. However, this response does not occur in all plants against all pathogens. In many cases this is a gene-for-gene interaction in which a specific plant gene product appears to recognize a specific pathogen gene product. Thus, an individual plant is resistant to only a limited number of pathogens which it has evolved to specifically recognize. One model for the induction of the active defense response is that gene products of HR-type resistance are constitutively produced receptors that specifically recognize pathogen components, termed elicitors. Activation of the receptor would trigger the cascade of anti-pathogen processes. There is no information concerning what the plant gene product is, but based on animal signal transduction models a membrane-bound glycoprotein would not be unexpected.

Two *Nicotiana* genes that control an HR against tobamoviruses have been characterized. *N. glutinosa* carries the N gene (bred into tobacco varieties) that reacts to and confines most tobamoviruses.

Other species of *Nicotiana*, for example *N. sylvestris* and some varieties of tobacco (Java and Burley), carry the N' gene that localizes most tobamoviruses, with the exception of TMV. Both N and N' phenotypes segregate as single dominant genes. When the N and N' genes are crossed into the same plant, their effects are additive (Weber, 1951). Also, each of these genes responds to a different viral elicitor. Plants with the corresponding recessive alleles (nn and n'n' genotype) are susceptible to most tobamoviruses and become systemically infected.

The Interaction of Tobamoviruses with the N' Gene of N. Sylvestris: Upon infection, plants with the N' gene are able to recognize most tobamoviruses other than TMV and respond with HR defense mechanisms to prevent further invasion by the virus. Wild-type TMV may have evolved to evade this recognition. The involvement of the coat protein gene in this interaction was first shown by Saito et al. (1987). Substitution of the coat protein gene of TMV-OM, which does not induce the HR, into the genome of the HR-inducing ToMV resulted in a hybrid virus that failed to induce the HR. Knorr and Dawson (1988) demonstrated that a single nucleotide alteration within the coat protein ORF was sufficient to change the phenotype of TMV from disease causing to HR-inducing. Thus, the precision of this interaction is such that N' gene plants are able to distinguish between viruses that have only one nucleotide difference in their coat protein genes.

Culver and Dawson (1989) found that alterations throughout the coat protein gene can induce the HR. Each of the nucleotide alterations responsible for the induction of the N' gene HR resulted in an amino acid substitution in the coat protein, but it was not known whether the coat protein was required for induction of the HR. Removal of the coat protein ORF start codon prevented HR-inducing mutants from inducing the HR in *N. sylvestris*, even though the mutant replicated efficiently (Culver and Dawson, 1989). We did not know whether the altered coat proteins acted alone to induce the HR. To determine if any viral replication processes or components other than coat protein were involved in the induction of the N' gene HR, transgenic *N. sylvestris* plants were produced that expressed only the coat protein ORF of either HR-inducing mutants or wild-type TMV (Culver and Dawson, 1991). Only transgenic *N. sylvestris* plants expressing HR-inducing coat proteins displayed necrosis.

One possibility was that some subset of the coat protein was the elicitor of the HR. However, Saito *et al.* (1989) demonstrated that creation of deletions within the elicitor coat protein caused it to no longer induce the HR. Additionally, we have a coat protein deletion mutant that assembles into virions, but even this deletion mutation causes the loss of ability to induce the HR (Culver, unpublished). Also, amino acid substitutions that cause an alteration in the overall structure of the coat protein by altering the position of alpha helices or loops prevent induction of the HR. These data suggest that only limited structural alterations that allow certain parts of native structure to remain are required for induction of the HR.

Interestingly, the HR-inducing mutants gave distinctly different patterns of lesion development. In general, there are four phenotypic groups based upon the developmental pattern of necrotic lesions induced in N' plants. "Strong elicitors" induce small lesions (1-2 mm in diameter) within 2-3 days after inoculation that do not increase further in size. "Moderate elicitors" induce lesions after 4-5 days (2-3 mm) but increase in size to 5-10 mm diameters. "Weak elicitors" induce necrotic lesions that occur 5-7 days after inoculation and initially appear as large lesions, 4-7 mm in diameter and continually expand, eventually collapsing large portions of the inoculated leaves. "Non-elicitors" do not induce necrotic lesion formation.

Mapping the Elicitor-Positive Region of the Coat Protein: The coat protein three-dimensional structure has been resolved to 2.8 A for disk aggregates, which consist of two stacked rings, each containing 17 coat protein subunits (Bloomer *et al.*, 1978). TMV virion structure has been resolved to 2.9 A (Namba *et al.*, 1989). The protein contains a high proportion of secondary structure. The core (middle radius) of the subunit is composed of a right-handed, four antiparalleled alpha-helix bundle, with left and right slewed helices (LS and RS) and left and right radial helices (LR and RR) (Figure 2). The inner radius of the subunit consists of a short loop connecting the LS and RS helices and a long inner loop containing the RNA-binding site and connecting the LR and RR helices. At the outer radius of the core helices is a region of beta-structure and hydrophobic residues that serve to stabilize the orientation of the four helices.

Our initial objectives were to define positions within the coat protein that, when substituted with other amino acids, would cause the wild-type coat protein to change from virulent to avirulent and then to correlate these positions with coat protein structure. Based

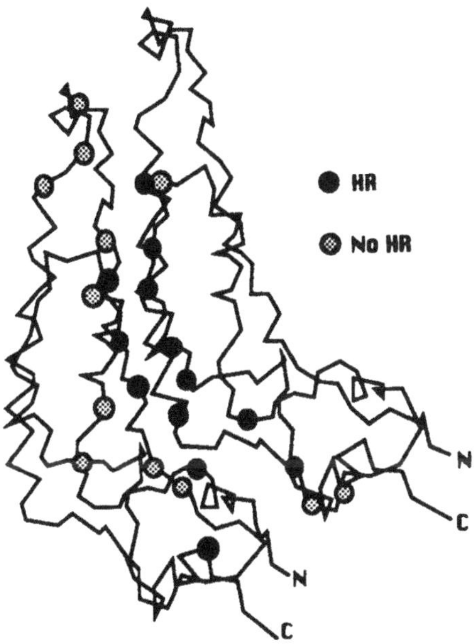

Figure 2. Schematic Representation of Two Tobacco Mosaic Virus Coat-Protein Subunits. Locations of HR-inducing and non HR-inducing amino acids substitutions are shown as indicated.

on computer generated structural models of coat protein aggregates (Namba and Stubbs, 1986), one common feature of HR-inducing amino acid substitutions is that all are located at a surface interface region between adjacent subunits (Figure 2). Amino acid substitutions that do not elicit the HR are either conservative changes or occur outside of the interface region between subunits. Thus, substitutions that were predicted to interfere with subunit to subunit interactions elicited the HR. Overall, these mutations define a region at the interface of protein subunits where most alterations change the phenotype of the virus from virulent to avirulent. These data suggest that mutations that cause a "looser" interaction between adjacent protein subunits are associated with the induction of the HR.

Virions are not required for induction of the HR. Transgenic plants expressing only the elicitor coat protein became necrotic. Also, several of the mutants that induce the HR fail to assemble into virions. This suggests that coat protein in the form of monomers or some type of aggregate would be the molecule recognized by the plant. Perhaps the proportion of monomers to aggregates of coat protein are the key to the induction of the HR. Under normal physiological conditions of the cell, individual coat protein subunits of wild-type TMV preferentially form aggregates and few monomers are thought to be present (Poschka *et al.*, 1988). However, the structural locations of the HR-inducing substitutions suggest that properties of subunit to subunit contact lead to induction of the HR. The possibility exists that coat protein monomers or small aggregates ("A" protein which is mostly side-by-side trimers) are active in HR induction. If so, the HR-inducing amino acid substitutions might shift the equilibrium toward less aggregation. However, it does not appear that the HR is induced merely by large amounts of monomeric coat protein. The correlation instead appears to be between "loosely associated" aggregates of structurally intact protein and induction of the HR. This correlation also holds when different tobamoviruses are examined. Tobamoviruses that induce the HR are less heat stable (Fraser, 1983) and are more susceptible to UV inactivation (Siegel, 1959).

It is possible that there are two interactive sites in the coat protein: the interference with coat protein aggregation by HR-inducing amino acid substitutions at the surface regions allows for the exposure of a second site, the N' gene recognition site, that in non-HR-inducing proteins is buried within aggregates of the molecule and is not accessible to the putative N' gene receptor molecule. The TMV coat protein can induce the HR when amino acid substitutions that reduce the tightness of coat protein interactions are introduced. The wild-type amino acids at the protein interfaces that allow "tight" interactions could function as a suppressor region to prevent binding at the putative receptor-binding area. This may be what allows the wild-type virus to evade detection by the plant. The present combined data suggest that host recognition requires the overall folded structure of the coat protein to be maintained, the affinity for subunit-subunit interaction to be reduced, and that at least one specific structural area of the molecule cannot be altered.

References

Bald, J.G. 1960. Forms of tobacco mosaic virus. Nature 188:645-647.

Bloomer, A.C., Champness, J.N., Bricogne, G., Staden, R., and Klug, A. 1978. Protein disk of tobacco mosaic virus at 2.8 A resolution showing the interactions within and between subunits. Nature 276:362-368.

Calder, V.L. and Palukaitis, P. 1992. Nucleotide sequence analysis of the movement genes of resistant breaking strains of tomato mosaic virus. J. Gen. Virol. 73:165-168.

Cheo, P.C. 1971. Subliminal infection of cotton by tobacco mosaic virus. Phytopathology 60:41-46.

Citovsky, V., Knorr, D.A., Schuster, G., and Zambryski, P. 1990. The P30 movement protein of tobacco mosaic virus is a single-stranded nucleic acid binding protein. Cell 60:637-647.

Culver, J.N. and Dawson, W.O. 1989. Tobacco mosaic virus coat protein: An elicitor of the hypersensitive reaction but not required for the development of mosaic symptoms in *Nicotiana sylvestris*. Virology 173:755-58.

Culver, J.N. and Dawson, W.O. 1989. Point mutations in the coat protein gene of tobacco mosaic virus induce hypersensitivity in *Nicotiana sylvestris*. Mol. Plant-Microbe Interact. 2:209-13.

Culver, J.N. and Dawson, W.O. 1991. Tobacco mosaic virus elicitor coat protein genes produce a hypersensitive phenotype in transgenic *Nicotiana sylvestris* plants. Molec. Plant-Microbe Inter. 4:458-463.

Culver, J.N., Lehto, K., Close, S., Hilf, M., and Dawson, W.O. 1992. Genomic position affects the expression of tobacco mosaic virus movement and coat protein genes. Proc. Natl. Acad. Sci. U.S.A. In press.

Dawson, W.O. 1990. Relationship of tobacco mosaic virus gene expression to movement within the plant. In: *Viral Genes and Plant Pathogenesis*. T. P. Pirone, J. G. Shaw (eds.), pp 39-50. Springer-Verlag, New York.

Dawson, W.O., Bubrick, P., and Grantham, G.L. 1988. Modifications of the tobacco mosaic virus coat protein gene affect replication, movement, and symptomatology. Phytopathology 78:783-789.

Ding, B., Haudenshield, J., Hull, R.J., Wolf, S., Beachy, R.N., and Lucas, W.J. 1992. Secondary plasmodesmata are specific sites of localization of the tobacco mosaic virus movement protein in transgenic plants. The Plant Cell 4:915-928.

Donson, J., Kearney, C., Hilf, M.E., and Dawson, W.O. 1991. Systemic expression of a bacterial gene by a tobacco mosaic virus-based vector. Proc. Natl. Acad. Sci. U.S.A. 88:7204-7208.

Fraser, R.S.S. 1983. Varying effectiveness of the N' gene for resistance to tobacco mosaic virus in tobacco infected with virus strains differing in coat protein properties. Physiol. Plant Pathol. 22:109-19.

Hilf, M.E. and Dawson, W.O. 1992. The tobamovirus capsid protein functions as a host specific determinant of long-distance movement. Virology In press.

Holmes, F.O. 1946. A comparison of the experimental host ranges of tobacco-etch and tobacco-mosaic viruses. Phytopathology 36:643-659.

Knorr, D.A. and Dawson, W.O. 1988. A point mutation in the tobacco mosaic capsid protein gene induces hypersensitivity in *Nicotiana sylvestris*. Proc. Natl. Acad. Sci. U.S.A. 85:170-74.

Lehto, K., Grantham, G.L., and Dawson, W.O. 1990. Insertion of sequences containing the coat protein subgenomic RNA promoter and leader in front of the tobacco mosaic virus 30 K ORF delays its expression and causes defective cell-to-cell movement. Virology 174:145-157.

Lewandowski, D.J. and Dawson, W.O. 1992. A single amino acid change in tobacco mosaic virus replicase prevents symptom production. Molec. Plant-Microbe Inter. In press.

Lindbeck, A.G.C., Dawson, W.O., and Thomson, W.W. 1991. Coat protein-related polypeptides from *in vitro* tobacco mosaic virus coat protein mutants do not accumulate in the chloroplasts of directly inoculated leaves. Molec. Plant-Microbe Inter. 4:89-94.

Lindbeck, A.G.C., Lewandowski, D.J., Culver, J.N., Thomson, W.W., and Dawson, W.O. 1992. Mutant coat protein of tobacco mosaic virus can induce chlorosis in expanded and developing tobacco leaves. Molec. Plant-Microbe Interact. 5:235-241.

McKinney, H.H. 1935. Evidence of virus mutation in the common mosaic of tobacco. J. Agric. Res. 51:951-959.

Meshi, T., Motoyoshi, F., Adachi, A., Watanabe, Y., Takamatsu, N., and Okada, Y. 1988. Two concomitant base substitutions in the putative replicase genes of tobacco mosaic virus confer the ability to overcome the effects of a tomato resistant gene, Tm-1. EMBO J. 7:1575-1581.

Meshi, T., Motoyoshi, F., Maeda, T., Yoshiwoka, S., Wantanabe, H., and Okada, Y. 1989. Mutations in the tobacco mosaic virus 30 kd protein gene overcome Tm-2 resistance in tomato. Plant Cell 1:515-22.

Meshi, T., Watanabe, Y., Saito, T., Sugimoto, A., Maeda, T., and Okada, Y. 1987. Function of the 30 kd protein of tobacco mosaic virus: Involvement in cell-to-cell movement and dispensability for replication. EMBO J. 6:2557-2563.

Namba, K., Pattaneyek, R., and Stubbs, G. 1989. Visualization of protein-nucleic acid interactions in a virus: Refined structure of intact tobacco mosaic virus at 2.9 A resolution by X-ray fiber diffraction. J. Mol. Biol. 208:307-325.

Namba, K. and Stubbs, G. 1986. Structure of tobacco mosaic virus at 3.6 A resolution: Implications for assembly. Science 231:1401-1406.

Neijdat, A., Cellier, F., Holt, C., Gafny, R., Eggenberger, A., and Beachy, R.N. 1991. Transfer of the movement protein gene between two tobamoviruses: Influence on local lesion development. Virology 180:318-326.

Nelson, R.S., Li, G., Hodgson, R., Beachy, R.N., and Shintaku, M. 1992. Evidence for impeded systemic spread of the masked strain of tobacco mosaic virus. Molec. Plant-Microbe Inter. In Press.

Nishiguchi, M., Kikuchi, S., Kibo, Y., Ohno, T., Meshi, T., and Okada, Y. 1985. Molecular basis of plant viral virulence: The complete nucleotide sequence of an attenuated strain of tobacco mosaic virus. Nucleic Acids Res. 13:5585-5590.

Nishiguchi, M., Motoyoshi, F., and Oshima, N. 1978. Behavior of a temperature sensitive strain of tobacco mosaic virus in tomato leaves and protoplasts. J. Gen. Virol. 39:53-61.

Potschka, M., Koch, M.H.J., Adams, M.L., and Schuster, T.M. 1988. Time-resolved solution X-ray scattering of tobacco mosaic virus coat protein: Kinetics and structure of intermediates. Biochem. 27:8481-8491.

Saito, T., Meshi, T., Takamatsu, N., and Okada, Y. 1987. Coat gene sequence of tobacco mosaic virus encodes host response determinant. Proc. Natl. Acad. Sci. U.S.A. 84:6074-77.

Saito, T., Yamanaka, K., and Okada, Y. 1990. Long-distance movement and viral assembly of tobacco mosaic virus mutants. Virology 176:329-336.

Saito, T., Yamanaka, K., Watanabe, Y., Takamatsu, N., Meshi, T., and Okada, Y. 1989. Mutational analysis of the coat protein gene of tobacco mosaic virus in relation to hypersensitive response in tobacco plants with the N' gene. Virology 173:11-20.

Siegel, A. 1959. Mutual exclusion of strains of tobacco mosaic virus. Virology 8:470-77.

Siegel, A., Zaitlin, M., and Sehgal, O.P. 1962. The isolation of defectived tobacco mosaic virus strains. Proc. Natl. Acad. Sci. U.S.A. 48:1854-1851.

Sulzinski, M.A. and Zaitlin, M. 1982. Tobacco mosaic virus replication in resistant and susceptible plants: In some resistant species virus is confined to a small number of initially infected cells. Virology 121:12-19.

Weber, P.V. 1951. Inheritance of a necrotic-lesion reaction to a mild strain of tobacco mosaic virus. Phytopathology 41:593-609.

Tomato Spotted Wilt Virus: Genome Organization,Transmission, and Symptom Induction

R. Goldbach, R. Kormelink, R. de O. Resende, A. C. de Ávila, F. van Poelwijk, J. van Lent, I. Wijkamp, M. Prins, and D. Peters
Department of Virology
Wageningen Agricultural University
Binnenhaven 11
6709 PD Wageningen
The Netherlands

P. de Haan
S & G Seeds B.V.
Department of Biotechnology
P.O. Box 26
1600 AA Enkhuizen
The Netherlands

Tomato spotted wilt virus (TSWV), type species of the genus Tospovirus (family *Bunyaviridae*) has a genome consisting of one negative-stranded RNA segment (L RNA, 8897 nucleotides) and two ambisense RNA segments (M RNA, 4821 nucleotides; S RNA, 2916 nucleotides). L RNA encodes the putative viral polymerase (331.5 kD), M RNA encodes a common precursor to both glycoproteins (G1 and G2) and a non-structural protein (NSm, 33.6 kD), while S RNA encodes the nucleocapsid (N) protein (28.8 kD) and an additional non-structural protein (NSs). Serological studies and nucleotide sequence comparisons revealed that TSWV is not the sole member of the genus Tospovirus, but that at least three more species exist. By serial mechanical passage in *Nicotiana rustica* two types of TSWV mutants were obtained. One type consists of attenuated mutants, causing (very) mild symptoms and the other of morphological defective mutants, which are devoid of the lipid envelope and glycoproteins. Mutants of the former type were shown to contain defective interfering (DI) RNA species, derived from the L RNA. Using a novel assay system, based on the local lesion host *Petunia X hybrida*, efficient transmission by the thrips *Frankliniella occidentalis* was obtained. A large fraction (80 percent) of those insects which will transmit was found to be capable to transmit the virus before pupation. Various lines of evidence indicate that TSWV multiplies in its insect vector, a.o. the increase of virus titer during larval development, and the active synthesis of NSs in salivary gland tissue.

Introduction

Tomato spotted wilt virus causes worldwide, but mainly in the tropic and subtropic regions, serious diseases in agricultural, horticultural, and ornamental crops. Nearly 400 plant species in more than 50 families are reported to be susceptible (Peters *et al.*, 1990). The virus is biologically transmitted by thrips species (Sakimura, 1962), and the renewed incidence of TSWV in the Northern hemisphere since the eighties has been attributed to the spread of the Western Flower thrips *F. occidentalis* (Pergande) from the West of the U.S. over North America and Europe.

The enveloped TSWV particles have a roughly isometric morphology with a diameter of 70-110 nm. The virion contains four structural proteins; a nucleocapsid (N) protein associated with three genomic RNA segments, two glycoproteins (G1, G2) associated with the viral envelope, and a large (more than 200 kD) protein which has been proposed to be the viral polymerase (Tas *et al.*, 1977; Mohamed, *et al.*, 1973; de Haan, 1991). Its genome consists of three single-stranded RNA molecules denoted small (S) RNA, medium (M) RNA, and large (L) RNA.

Based on its genome organization, TSWV has recently been classified as the sole member of the newly created genus, Tospovirus, within the family *Bunyaviridae* (Francki *et al.*, 1991). This paper will briefly describe the genetic organization and expression of the TSWV genome and present evidence, based on serological and molecular studies, that TSWV is not the sole member of the genus Tospovirus. Furthermore, two types of mutant viruses, obtained upon serial mechanical passages, will be described; namely, attenuated mutants, showing very mild symptoms, and morphological defective mutants, devoid of the envelope and glycoproteins. Finally, some thrips transmission characteristics will be discussed, as determined by a novel transmission assay system; and evidence will be presented showing that TSWV replicates in its insect vector.

Genetic Organization and Expression of the TSWV Genome

The complete sequence determination of the tripartite TSWV genome (de Haan *et al.*, 1989; de Haan *et al.*, 1991; Kormelink *et al.*, 1992) has confirmed earlier suggestions that TSWV represents a bunyavirus able to infect plants. The L RNA (8897 nucleotides long) is of negative polarity, encoding a single large protein of 331.5 kD (Figure 1). This protein represents the putative viral transcriptase as it contains all conserved sequence motifs which are diagnostic for viral polymerases of negative strand RNA viruses (de Haan *et al.*, 1991; Tordo *et al.*, 1992). Highest homology was found with the L protein (transcriptase) of Bunyamwera virus (de Haan *et al.*, 1991). The two other genomic RNA's (S and M) are ambisense which makes TSWV the first member of the *Bunyaviridae* that possesses two ambisense genome segments (de Haan *et al.*, 1989; Kormelink *et al.*, 1992). The M RNA segment (4821 nucleotides) contains two open reading frames (ORF's); one in the viral (v) sense which encodes a non-structural (NSm) protein of

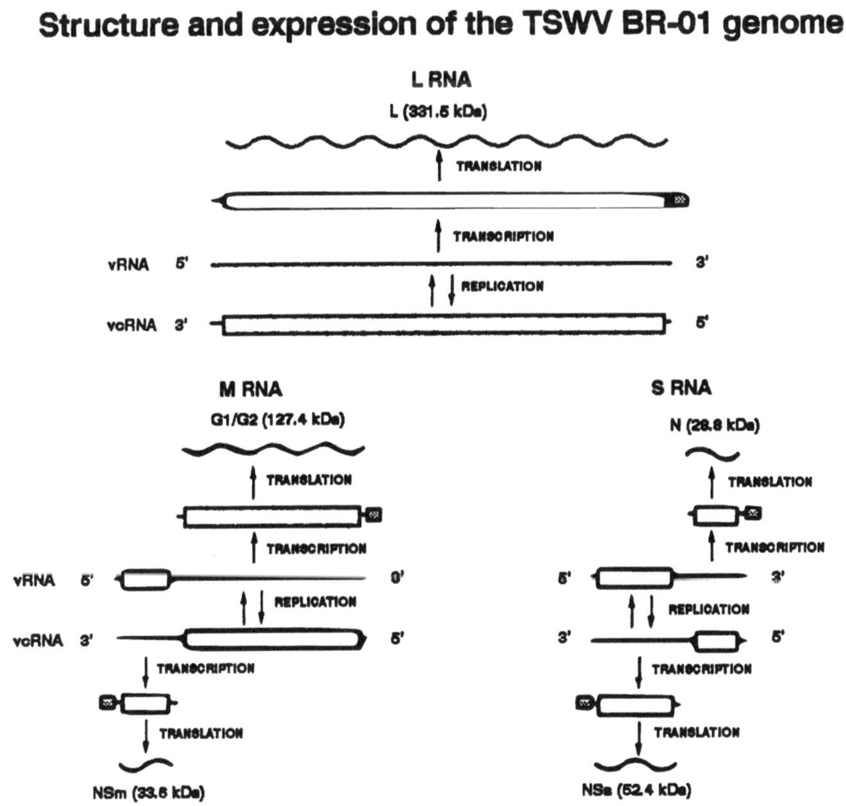

Figure 1. The Genome Structure and Expression Strategy of the Tripartite Genome of TSWV (isolate BR-01). Open reading frames are indicated as open bars; the hatched boxes at the 5'-end of mRNA's represent non-viral leader sequences.

33.6 kD, and another in the viral complementary (vc) sense, which encodes the 127.4 kD precursor to the G1 and G2 envelope glycoproteins (Figure 1). Similarly the smallest (S) RNA (2916 nucleotides) comprises two genes in ambisense arrangement, encoding the N protein (28.8 kD) in the vc sense, and a non-strucutural (NSs) protein of 52.4 kD in the v sense (de Haan *et al.*, 1989; Kormelink *et al.*, 1991) (Figure 1). Both the M and S RNA segments are translationally expressed via the synthesis of subgenomic mRNA's which terminate at stable, A-U rich hairpin structures located in the intergenic regions. Primer extension analysis of purified N and NSs mRNA's have revealed that the TSWV mRNA's, similar to the mRNA's of animal-infecting bunyaviruses and orthomyxoviridae, have non-viral (12-20) nucleotides at their 5'-termini, indicating that transcription is initiated by a process referred to as cap-snatching (Kormelink *et al.*, 1992) (Table 1).

Other Members of the Genus *Tospovirus*

The detailed knowledge of the genetic organization, sequence, and expression of the TSWV genome, as briefly summarized in the previous section, has been obtained by mainly focussing on one particular isolate, the Brazilian isolate BR-01.

In view of the worldwide spread of tospoviruses, able to infect a high number of plant species, one may question whether virus isolates from different parts of the world show enough variation to consider them as belonging to more than one single virus species. Ávila *et al.* (1990; 1992) indeed showed that a selection of 20 isolates originating from different geographical areas and crops can be divided into three distinct serogroups (I-III) using polyclonal and monoclonal antibodies directed to their N protein. Most isolates studied belong to serogroup I, including type isolate BR-01. The serogroup I isolates reacted only weakly with these antibodies to serogroup II viruses and not at all with antibodies raised against serogroup III viruses. This latter serogroup consists of almost completely identical isolates from *Impatiens* plants in the U.S.A. (TSWV-I) (Law and Moyer, 1990; Law *et al.*, 1991) and the Netherlands (NL-07) (de Ávila *et al.*, 1992), differing entirely from the serogroup I and II viruses in a serological way. Therefore, serogroup III isolates may be considered as belonging to a different species, which was denoted *Impatiens* necrotic spot virus (INSV), while serogroup I may represent a species for which the name tomato spotted wilt virus should be reserved. Furthermore, serogroup II splits into two distinct serotypes (de Ávila *et al.*, 1990).

Since the serogrouping of tospoviruses (de Ávila *et al.*, 1990; de Ávila *et al.*, 1992; de Ávila *et al.*, 1992) has been based on analyses using poly- and monoclonal antibodies directed to the N protein, it was decided to choose the nucleotide sequences of the N protein gene as a molecular criterium for the classification of tospoviruses into species, analogous to the use of coat protein sequences for positive strand RNA viruses, like potyviruses and comoviruses (Francki *et al.*, 1991; Shukla and Ward, 1988). To this end, these sequences were determined for a number of serogroup I, II, and III isolates and compared with those of previously published sequence data (Table 2). This analysis indeed confirmed that serogroup I (i.e., TSWV), the two distinct serotypes of serogroup II, and serogroup III (INSV) should be regarded as four distinct species (de Ávila *et al.*, 1992).

The N protein sequences of the two serogroup II isolates were for 76 percent (BR-03) to 78 percent (SA-05) homologous to all serogroup I isolates. The homology between serogroup II and III (INSV) isolates is considerably lower (55-57 percent) and is of the same level as the homology between serogroup I and III N protein sequences (55 percent, Table 2).

Members within serogroup I or serogroup III show 99 percent or more homology to each other. However, within serogroup II, both isolates analyzed (BR-03 and SA-05) significantly diverge, showing only 81 percent homology between each other (de Ávila *et al.*, 1992).

Although this rate of divergence is less than between TSWV and ISNV isolates, we propose to consider them as two distinct species. This proposal is supported by the observation that the N genes within the TSWV and ISNV species, with isolates originating either from Brazil, Hawaii, and Bulgaria or from the U.S.A. and the Netherlands, have an almost 100 percent nucleotide sequence homology (de Ávila *et al.*, 1992). The two novel species may be named tomato chlorotic spot virus (TCSV) and groundnut ringspot virus (GRSV) for the serotype I (BR-03) and II (SA-05) viruses (Table 2).

Factors Influencing Symptom Induction

Except for INSV isolates all other (more than 40) tospovirus isolates tested in our laboratory could not easily be differentiated by the symptoms induced on a number of test plants. All TSWV, TCSV, and GRSV isolates induced necrotic or chlorotic spots and rings, followed by vein necrosis on the inoculated leaves of most solanaceous hosts tested; a mosaic, mottling, and leaf deformation appeared on the top leaves followed by necrosis as the leaves of these hosts aged. There

Table 1. Presence of Cellular Leader Sequences in mRNA's of Multipartite Negative Strand RNA Viruses

Family	Genus	Species	mRNA	Non-Viral leader sequence	3'-terminal base of leader	References
Bunyaviridae	Tospo	TSWV	N, NSs	12 - 20 nts		de Haan et al., 1989; Kormelink et al., 1992
	Phlebo	Rift Valley Fever	G	12 - 14		Collett, 1986
		Punta Toro	N, NSs	11 - 18		Ihara et al., 1985
		Uukuniemi	N, NSs	7 - 25	no consensus	Simons and Pettersson, 1991
		Toscana	N, NSs	? ?		Giorgi et al., 1991
	Bunya	Germiston	N, G	12 - 18	U, C	Bouloy et al., 1990
		Snowshoe hare	N, G, NSs	12 - 17	A	Bishop et al., 1983; Eshita et al., 1985
		La Crosse	N, NSs	10 - 14		Patterson and Kolakofsky, 1984
Arenaviridae		Tacaribe	N	1 - 4		Raju et al., 1990
Orthomyxoviridae	Orthomyxo	Influenza A	M (RNA7) NS (RNA8) H (RNA4)	10 - 18	A, G	Caton and Robertson, 1980; Dhar et al., 1980

Table 2. Proposed Species Within the Genus Tospovirus (*Bunyaviridae*)

Isolate	Origin Country	Origin Host	Sero-group	Sero-type	Species	Amino acid sequence homology *	References
BR-01	Brazil	Tomato	1		Tomato spotted wilt virus (TSWV)	100%	de Haan, 1991; de Haan et al., 1989
BR-03	Brazil	Tomato	11	1	Tomato chlorotic spot (TCSV)	76%	de Ávila et al., 1992
SA-05	South-Africa	Groundnut	11	11	Groundnut ringspot virus (GRSV)	78%	de Ávila et al., 1992
Type 1	U.S.A.	*Impatiens*			*Impatiens* necrotic spot	55%	Law and Moyer, 1990; Law et al., 1991
NL-07	Netherlands		111		(INSV)		de Ávila et al., 1992; de Haan et al., 1992

* relative to isolate BR-01

are, however, differences in severity and length of the incubation period. Thus, severe vein necrosis is mainly induced (for instance on some solanaceous hosts) by the isolates BR-01, E-01, BR-06, and BR-08 (all TSWV), and by isolate BR-03 (TCSV), which all originated from tomato or pepper. Necrosis rarely occurred when these solanaceous hosts were infected with isolate SA-05 (GRSV).

Two factors have been identified which may affect the (severity of) symptom induction by tospovirus isolates; e.g., the rate of NSs production and the generation of defective interfering (DI) RNA's. Using highly specific antibodies, raised against baculovirus vector produced NSs, it was found that NSs is not present in mature virions (and, thus, indeed represents a non-structural protein) but accumulates in the cytoplasm, where it may occur either disperse or clustered into fibrous material or para-crystalline structures, depending on the isolate (Figure 2) (Kormelink *et al.*, 1991; Kitajima *et al.*, 1992). Comparison of a number of isolates provided indications that the accumulation of NSs correlates with the severity of the symptoms. The negative effect of the presence of NSs on the plant's physiology is also supported by the finding that transgenic tissues expressing the NSs gene, regenerated to stunted, yellowing plantlets or did not regenerate at all (unpublished results). However, the primary function of NSs during the viral multiplication process remains to be established. It may not have only a function in the viral multiplication process in plant hosts, but also in the infection of thrips as this protein is actively synthesized in salivary gland tissue of viruliferous thrips vectors (see below).

A second factor which may greatly influence the (severity of) symptom expression by TSWV is the spontaneous generation of defective interfering RNA's. This is observed frequently upon serial mechanical passage of isolates on certain hosts (e.g., *N. rustica*). All DI RNA's characterized so far were found to be derived from the L RNA, by the introduction of internal deletions, which may be as large as 7 kb (Resende *et al.*, 1991; Resende *et al.*, 1992). While some deleted forms were found to be instable, disappearing upon further passages, some of them remained stable and were able to suppress the disease symptoms, also in other host plants. This observation underlines once more that caution must be taken when trying to distinguish and classify tospoviruses on (severity of) symptom expression solely.

A second class of mutants obtained under laboratory conditions are the so-called morphological-defective mutants, which lack the lipid envelope and produce clusters of free nucleocapsids in infected plant tissue (Resende *et al.*, 1991; Kitajima *et al.*, 1992). They are deficient in glycoprotein synthesis and serologically only detectable if antibodies against the N protein are used. Furthermore, they may escape the attention when superficial electron microscopical analysis is performed. A rapid and reliable electron microscopical approach for detecting morphological defective forms has been recently developed (de Ávila *et al.*, 1990; Kitajima *et al.*, 1992). Since the glycoproteins are coded for by the M RNA (see Figure 1) and this genome segment does not have a deviant size in envelope-deficient mutants, it is tempting to assume that the impeded glycoprotein synthesis is caused

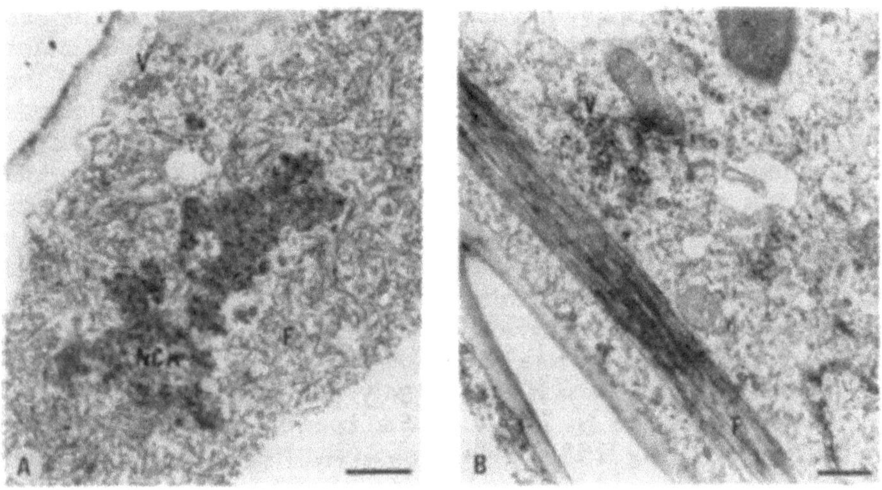

Figure 2. **Cytopathology of *N. rustica* Mesophyll Cells Showing the Accumulation of the NSs Protein into Dispersely Oriented Fibrous Material upon Infection with Isolate DK-01 (panel A) and into Clustered Fibers upon Infection with Isolate SA-05 (panel B). F = fibrous material containing NSs protein, V = virus particles, NCA = nucleocapsid aggregates.**

by mutations, rather than deletions in the G1/G2 precursor ORF. Preliminary sequence analyses indeed confirm this view. A sequenced 500 bases-long region in the G1/G2 gene of an envelope-less line of

TSWV isolate NL-04 was found to accumulate not less than 7 point mutations, one creating a premature stop codon (Resende, unpublished results).

Transmission of Tospoviruses

Tospoviruses are transmitted by thrips (*Thysanoptera; Thripidae*) in a persistent manner. At least eight species have been reported as possible vectors of which *F. occidentalis* (Pergande), *Frankliniella schulzei* (Trybom), *Frankliniella fusca* (Hinds), and *Thrips tabaci* (Lindeman) are most frequently recorded (Sakimura, 1962; Best and Gallus, 1955). Thrips feed by piercing a cell and sucking its content (Hunter and Ullman, 1992). However, the mechanism by which the virus is transmitted is not understood and most parameters describing the virus-vector relationships have not been thoroughly analyzed. It has generally been accepted that only larvae can acquire the virus. The minimal acquisition period is 15 to 30 minutes (Sakimura, 1962). Between acquisition and transmission a period occurs in which the thrips is not infectious. Reported incubation periods of TSWV ranged in *T. tabaci* from 4 to 18 days and in *F. fusca* from 4 to 12 days.

Recently, a novel assay system has been developed allowing accurate studies of tospovirus transmission by thrips (Wijkamp and Peters, 1992). The system is based on the use of the local lesion host *Petunia X hybrida* cv. Blue Magic (Allen and Matteoni, 1991) which develops brown or black, readily scorable lesions. Using this system, efficient transmission by the Western Flower thrips *F. occidentalis* was obtained for two different tospoviruses, TSWV and INSV. Transmission rates of up to 55 and 93 percent were reached for TSWV and INSV respectively, when *Impatiens* sp. plants were used as source plants. Strikingly, L2 larvae were found to be efficient transmitters. When virus was acquired by less-than-1-day-old first stage larvae, 80 percent of these larvae could transmit before pupation (Wijkamp and Peters, 1992). The median latent period (LP50) for this stage was found to be temperature dependent, being approximately 82-84 hours at 27°C, for both TSWV and INSV.

Multiplication of TSWV in its Vector *F. Occidentalis*

Circumstantial evidence exists for several animal-infecting bunyaviruses that they multiply in the insect vector (Beaty and Calisher, 1991), but such evidence has been lacking so far for the tospoviruses. This is probably due to the fact that the vectors are diminutive insects in which accumulation of virus is difficult to assay quantitatively and that cell cultures of thrips have so far not been developed. Conclusive evidence for multiplication of tospoviruses in thrips may, alternatively, be obtained by detection of the non-structural proteins, NSs or NSm, which do not occur in the mature virus particles and whose presence depends on the replication of the viral genome (Kormelink *et al.*, 1991; Kormelink *et al.*, 1992) and by demonstrating an increase of the N protein which may play an essential role in the switch from transcription to replication.

To determine whether TSWV multiplies in its vector *F. occidentalis*, the possible accumulation of the N and NSs proteins was monitored by both ELISA and Western blot analysis. Using cocktail-ELISA and polyclonal antibodies against the N protein, it was found that the titer of this protein dropped within the first few hours after acquisition and increased then after 24 hours to levels significantly above the amount ingested and achieving a plateau before becoming a prepupae (Wijkamp *et al.*, 1992). Lower values for the N protein were consistently observed around 36 hours after acquisition; a phenomenon which coincided with moulting of the larvae from the first instar to the second one.

While the increase of N protein in the thrips larvae already indicated that TSWV multiplies in its vector, further evidence for this was obtained by demonstrating *de novo* synthesis of the NSs protein in thrips larvae. Using ELISA the amount of this protein in thrips immediately after acquisition was found to be zero but started to increase simultaneously with the N protein (Wijkamp *et al.*, 1992).

The production of the N and NSs proteins in thrips was further substantiated by immunoblot analysis. Typical results obtained when monitoring the NSs synthesis during larval development are shown in Figure 3. The NSs protein was consistently found in extracts prepared from all instars, except when the larvae were sampled 2 and 6 hours after acquisition. Somewhat lower amounts of NSs were found in samples prepared from second stage larvae just before pupation, prepupae, and pupae than in samples from larvae 72 hours after acquisition and 1 to 8 day old adults (Figure 3). Accumulation of both N and NSs proteins and of virus particles was further confirmed by *in situ* immunolabeling of the salivary glands and other tissues of adult thrips (Wijkamp *et al.*, 1992). The presence of massive numbers of virus particles in the salivary gland canals demonstrates that the salivary glands are a major site of TSWV replication. The occurrence of virus particles in the salivary vesicles is indicative for the involvement of the Golgi apparatus in the maturation of the virus particles and its transport to the salivary canals.

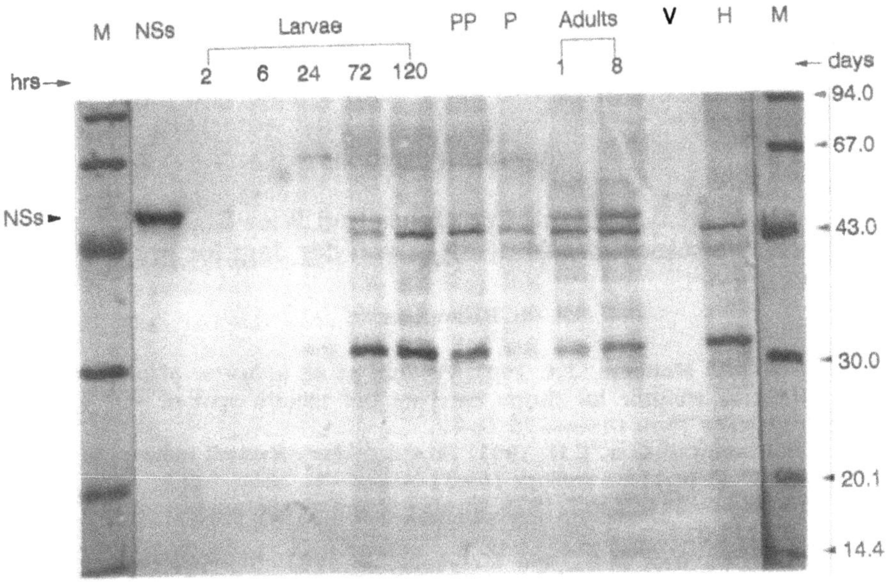

Figure 3. **Accumuation of NSs Protein During Larval Development of *F. occidentalis*, After a 2 Hour Acquisition Access Period (AAP). The immunoblot contains extracts from 15 larvae or adults sampled at subsequent periods after the onset of a 2 hour long AAP, and has been probed with antiserum directed against the NSs protein. M = marker; NSs = purified NSs protein produced from a baculovirus vector (Kormelink *et al.*, 1991); V = purified virus preparation; larvae = sampled after 2, 6, 24, 72, and 120 hours (harvested just before pupation) after the start of AAP; PP = prepupae; P = pupae; Adults, 1 and 8 days old; H = sample of 15 virus-free adults.**

Conclusions

The content of this contribution illustrates that our knowledge on tospoviruses; i.e., their genome organization, diversity, and interactions with both hosts and vectors have greatly increased during the past few years. This again has led to the successful development of engineered forms of resistance, by which host plants even are protected against thrips-mediated infection (Gielen et al., 1991; de Haan et al., 1992). Tospoviruses have not only been recognized as ranking among the most important plant viruses world wide, they also have become a grateful subject for fundamental research. They take a unique position among plant viruses but share many properties with the animal-infecting Bunyaviridae, to which family they belong. Future research will hopefully shed light on the molecular properties which enable tospoviruses to infect both animal and plant cells so successfully.

Acknowledgement

The authors wish to thank Cor Meurs and Wies Smeets for skilled technical resistance and Annemieke van der Jagt for preparing the manuscript.

References

Allen, W.R. and Matteoni, J.A. 1991. Petunia as an indicator plant for use by growers to monitor for thrips carrying the tomato spotted wilt virus in greenhouses. Plant Disease 75:78-82.

Beaty, B.J. and Calisher, C.H. 1991. *Bunyaviridae* - Natural history. Topics in Microbiology and Immunology 169:27-78.

Best, R.J. and Gallus, H.P.C. 1955. Strains of tomato spotted wilt virus. Aust. J. Sci. 15:212-214.

Bishop, D.H.L., Gay, M.E., and Matsuoko, Y. 1983. Nonviral heterogeneous sequences are present at the 5'-ends of one species of snowshoe hare bunyavirus S complementary RNA. Nucleic Acids Research 11:6409-6418.

Bouloy, M., Pardigon, N., Vialat, P., Gerbaud, S., and Girard, M. 1990. Characterization of the 5'- and 3'-ends of viral messenger RNA's isolated from BHK21 cells infected with Germiston virus (bunyavirus). Virology 175:50-58.

Caton, A.J. and Robertson, J.S. 1980. Structure of the host-derived sequences present at the 5'-ends of influenza virus mRNA. Nucleic Acids Research 8:2591-2603.

Collett, M.S. 1986. Messenger RNA of the M segment RNA of Rift Valley fever virus. Virology 151:151-156.

de Ávila, A.C., de Haan, P., Kitajima, E.W., Kormelink, R., Resende, R. de O., Goldbach, R.W., and Peters, D. 1992. Characterization of a distinct isolate of tomato spotted wilt virus (TSWV) from *Impatiens* sp. in the Netherlands. Journal of Phytopathology 134:133-151.

de Ávila, A.C., de Haan, P., Kormelink, R., Resende, R. de O., Goldbach, R.W., and Peters, D. 1992. Classification of Tospoviruses based on phylogeny of nucleoprotein gene sequences. Journal of General Virology In press.

de Ávila, A.C., de Haan, P., Smeets, M.L.L., Resende, R. de O., Kormelink, R., Kitajima, E.W., Goldbach, R.W., and Peters, D. 1992. Distinct levels of relationships between tospovirus isolates. Archives of Virology In Press.

de Ávila, A.C., Huguenot, C., Resende, R. de O., Kitajima, E.W., Goldbach, R.W., and Peters, D. 1990. Serological differentiation of 20 isolates of tomato spotted wilt virus. Journal of General Virology 71:2801-2807.

de Haan, P. 1991. Exploring and exploiting the RNA genome of tomato spotted wilt virus. Ph.D. thesis, pp. 1-135 Wageningen Agricultural University, Wageningen, The Netherlands.

de Haan, P., de Ávila, A.C., Kormelink, R., Westerbroek, A., Gielen, J.J.L., Peters, D., and Goldbach, R.W. 1992. The nucleotide sequence of the S RNA of *Impatiens* necrotic spot virus, a novel tospovirus. FEBS Letters 306:27-32.

de Haan, P., Gielen, J.J.L., Prins, M., Wijkamp, M.G., van Schepen, A., Peters, D., van Grinsven, M.Q.J.M., and Goldbach, R.W. 1992. Characterization of RNA-mediated resistance to tomato spotted wilt virus in transgenic tobacco plants. Bio/Technology In press.

de Haan, P., Kormelink, R., Resende, R. de O., van Poelwijk, F., Peters, D., and Goldbach, R.W. 1991. Tomato spotted wilt virus L RNA encodes a putative RNA polymerase. Journal of General Virology 72:2207-2216.

de Haan, P., Wagemakers, L., Peters, D., and Goldbach, R. 1989. The S RNA segment of tomato spotted wilt virus has an ambisense character. Journal of General Virology 71:1001-1007

Dhar, R., Chanock, R.M., and Lai, C.-J. 1980. Non-viral oligonucleotides at the 5'-terminus of cytoplasmic influenza viral mRNA deduced from cloned complete genomic sequences. Cell 21:495-500.

Eshita, Y., Ericson, B., Romanowski, V., and Bishop, D.H.L. 1985. Analyses of the mRNA transcription processes of snowshoe hare bunyavirus S and M RNA species. Journal of Virology 55:681-689.

Francki, R.I.B., Fauquet, C.M., Knudson, D.D., and Brown, F. 1991. Fifth report of the International Committee on Taxonomy of Viruses. Archives of Virology Supplement 2:1-450.

Gielen, J.J.L., de Haan, P., Kool, A.J., Peters, D., van Grinsven, M.Q.J.M., and Goldbach, R.W. 1991. Engineered resistance to tomato spotted wilt virus, a negative-strand RNA virus. Bio/Technology 9:1363-1367.

Giorgi, C., di Bonito, P., and Gro, C. 1991. Characterization of the 5'- and 3'-ends of the two subgenomic mRNA's transcribed from the S segment of Toscana virus (Phlebovirus). In: *Abstracts from the Negative-Strand Virus Meeting*, Charleston, South Carolina, U.S.A.

Hunter, W.B. and Ullman, D.E. 1992. Anatomy and ultrastructure of the piercing-sucking mouthparts and paraglossal sensilla of *Frankliniella occidentalis* (Pergande) (Thysanoptera: Thripidae). Journal of Insect Morphology and Embryology 21:17-35.

Ihara, T., Matsuura, Y., and Bishop, D.H.L. 1985. Analyses of the mRNA transcription processes of Punta Toro phlebovirus (*Bunyaviridae*). Virology 147:317-325.

Kitajima, E.W., de Ávila, A.C., Resende, R. de O., Goldbach, R.W., and Peters, D. 1992. Comparative cytological and immunogold labelling studies on different isolates of tomato spotted wilt virus. Journal of Submicroscopy and Cytological Pathology 24:1-14.

Kitajima, E.W., Resende, R. de O., de Ávila, A.C., Goldbach, R.W., and Peters, D. 1992. Immuno-electron microscopical detection of tomato spotted wilt virus and its nucleocapsids in crude plant extracts. Journal of Virological Methods 38:313-322.

Kormelink, R., de Haan, P., Meurs, C., Peters, D., and Goldbach, R.W. 1992. The nucleotide sequence of the M RNA segment of tomato spotted wilt virus: A bunyavirus with two ambisense RNA segments. Journal of General Virology In Press.

Kormelink, R., de Haan, P., Peters, D., and Goldbach, R.W. 1992. Viral RNA synthesis in tomato spotted wilt virus-infected *Nicotiana rustica* plants. Journal of General Virology 73:687-693.

Kormelink, R., Kitajima, E.W., de Haan, P., Zuidema, D., Peters, D., and Goldbach, R.W. 1991. The nonstructural protein (NSs) encoded by the ambisense S RNA segment of tomato spotted wilt virus is associated with fibrous structures in infected plant cells. Virology 181:459-468.

Kormelink, R., van Poelwijk, F., Peters, D., and Goldbach, R.W. 1992. Non-viral heterogeneous sequences at the 5'-ends of tomato spotted wilt virus (TSWV) mRNA's. Journal of General Virology 73:2125-2128.

Law, M.D. and Moyer, J.W. 1990. A tomato spotted wilt virus-like virus with a serologically distinct N protein. Journal of General Virology 71:933-938.

Law, M.D., Speck, J., and Moyer, J.W. 1991. Nucleotide sequence of the 3'-non-coding region and N gene of the S RNA of a serologically distinct tospovirus. Journal of General Virology 72:2597-2601.

Mohamed, N.A., Randles, J.W., and Francki, R.I.B. 1973. Protein composition of tomato spotted wilt virus. Virology 56:12-21.

Patterson, J.L. and Kolakofsky, D. 1984. Characterization of La Crosse virus small-genome transcripts. Journal of Virology 49:680-685.

Peters, D., de Ávila, A.C., Kitajima, E.W., Resende, R. de O., de Haan, P., and Goldbach, R.W. 1990. An overview of tomato spotted wilt virus. In: *Virus-Thrips-Plant Interactions of Tomato Spotted Wilt Virus.* H.-T. Hsu. and R.H. Lawson (eds.), pp. 1-14. Proceedings of a USDA Workshop, Beltsville, Maryland, U.S.A., April 18-19.

Raju, R, Raju, L., Hacker, D., Garcin, D., Compans, R., and Kolakofsky, D. 1990. Non-templated bases at the 5'-ends of Tacaribe virus mRNA's. Virology 174:53-59.

Resende, R. de O., de Haan, P., de Ávila, A.C., Kitajima, E.W., Kormelink, R., Goldbach, R.W., and Peters, D. 1991. Generation of envelope and defective interfering RNA mutants of tomato spotted wilt virus by mechanical passage. Journal of General Virology 72:2375-2383.

Resende, R. de O., de Haan, P., van de Vossen, E., de Ávila, A.C., Goldbach, R.W., and Peters, D. 1992. Defective interfering (DI) L RNA segments of tomato spotted wilt virus (TSWV) retain both virus genome termini and have extensive internal deletions. Journal of General Virology 73:2503-2506.

Sakimura, K. 1962. The present status of thrips-borne viruses. In: *Biological Transmission of Disease Agents.* K. Maramorosch (ed.). Acad. Press, New York.

Shukla, D.D. and Ward, C.W. 1988. Amino acid sequence homology of coat proteins as a basis for identification and classification of the potyvirus group. Journal of General Virology 69:2703-2710.

Simons, J.F. and Pettersson, R.F. 1991. Host-derived ends and overlapping complementary 3'-ends of the two mRNA's transcribed from the ambisense S segment of Uukuniemi virus. Journal of General Virology 65:4741-4748.

Tas, P.W.L., Boerjan, M.L., and Peters, D. 1977. The structural proteins of tomato spotted wilt virus. Journal of General Virology 36:267-279.

Tordo, N., de Haan, P., Goldbach, R.W., and Poch, O. 1992. Evolution of negative-stranded RNA genomes. Seminars in Virology 3:340-357.

Wijkamp, M.G. and Peters, D. 1993. Determination of the median latent period of two tospoviruses in *Frankliniella occidentalis* using anovel leaf disc assay. Phytopathology 83:986-991.

Wijkamp, M.G., van Lent, J., Kormelink, R., Goldbach, R.W., and Peters, D. 1993. Multiplication of tomato spotted wilt virus in its insect vector *Frankliniella occidentalis*. Journal of General Virology 74:341-349.

Genetic and Environmental Factors That Determine the Ability of Cauliflower Mosaic Virus to Systemically Infect Solanaceous Hosts

J. E. Schoelz, S. G. Qiu, and W. M. Wintermantel
Department of Plant Pathology
University of Missouri
Columbia, Missouri 65211

Cauliflower mosaic virus infects a wide range of crucifers, and several strains such as D4 and W260 are also able to systemically infect solanaceous species. We have shown that gene VI of CaMV strain D4 determines systemic infection of *Nicotiana bigelovii*. However, the viral genomic regions required for W260 to systemically infect *N. bigelovii* differ considerably from the D4 requirements. Three separate regions of W260, containing primarily genes I, IV, and VI, influenced the ability of W260 to systemically infect *N. bigelovii*. The genetic evidence suggests that gene VI may directly interact with host components to influence host specificity. Genes I and IV may affect host specificity by their interaction with gene VI, rather than through their interaction with host proteins. Recently, we expressed the D4 gene VI product in transgenic *N. bigelovii* and demonstrated that some CaMV chimeric viruses are complemented for long-distance transport while at least one CaMV strain is capable of recombination with CaMV sequences present in the transgenic plants. The factors that determine complementation or recombination may have important implications for mechanisms within *N. bigelovii* that determine resistance to CaMV infections.

Introduction

One of the most important methods for controlling plant virus diseases has been to breed plants that are resistant to virus infections. It has been relatively easy to introduce virus resistance genes into crop plants from wild species by using conventional plant breeding techniques because resistance has frequently been shown to be determined by a single, dominant gene. The only constraint on breeding virus resistant plants has been that in order for the progeny to be fertile, the plant which is the source of resistance must be closely related to the plant which is susceptible to viral infections. Occasionally, plant viruses have evolved to overcome host resistance. With the advent of recombinant DNA techniques, it has been possible to identify the viral gene, or even the precise mutation, that is responsible for overcoming host resistance.

One plant virus that has been very amenable to molecular studies is cauliflower mosaic virus. Its genome is composed of circular, double-stranded DNA (Shepherd, 1981) that is easily cloned in infectious form (Howell *et al.*, 1980; Lebeurier *et al.*, 1980). Several strains have been completely sequenced, revealing the presence of six to eight open reading frames (ORF's) (Frank *et al.*, 1980; Gardner *et al.*, 1981; Balazs *et al.*, 1982). Gene products have been identified *in vivo* for the six largest ORF's. The product of ORF I has an apparent molecular weight of 41-45 kDa, is present in the viral inclusion body, (Martinez-Izquierdo *et al.*, 1987; Young *et al.*, 1987) and is involved in cell-to-cell movement (Hull and Covey, 1985; Hull *et al* ., 1986; Albrecht *et al.*, 1988; Linstead *et al.*, 1988; Citovsky *et al.*, 1991). The product of ORF II is an 18 kDa protein that is found in inclusion bodies and is required for aphid transmission (Woolston *et al.*, 1983; Armour *et al.*, 1983; Daubert *et al.*, 1984). The product of ORF III, an 11-15 kDa protein, has been reported to be associated with virions and to be a DNA binding protein (Xiong *et al.*, 1984; Giband *et al.*, 1986; Mesnard *et al.*, 1990). ORF IV encodes the 42 kDa capsid protein (Frank *et al.*, 1980; Daubert *et al.*, 1982). The product of ORF V is a reverse transcriptase (Toh *et al.*, 1983; Laquel *et al.*, 1986; Takatsuji *et al.*, 1986). ORF VI encodes a 62 kDa inclusion body matrix protein (Covey and Hull, 1981; Xiong *et al.*, 1982). Recent evidence suggests that the gene VI product facilitates translation of genes on the 35 S RNA (Gowda *et al.*, 1989; Bonneville *et al.*, 1989).

CaMV strains infect a wide range of crucifers and several strains also infect solanaceous species. We have been interested in identifying which genes of CaMV determine the ability of the virus to infect solanaceous species. In the past, we have found that gene VI of one CaMV strain is responsible for allowing this virus to systemically infect several solanaceous species (Daubert *et al.*, 1982; Schoelz *et al.*, 1986). We have recently completed the analysis of a second CaMV strain that systemically infects a more restricted range of solanaceous species (Qiu and Schoelz, 1992) and have started to identify environmental conditions that influence systemic infections of both solanaceous and cruciferous hosts. Lastly, we present evidence that the gene VI product of CaMV may act as a dominant inhibitor of systemic infection.

Experimental

CaMV Strains and Chimeric Viruses: The wild-type CaMV strains CM1841, W260, and D4 have previously been cloned in infectious form into bacterial plasmid vectors (Howarth *et al.*, 1981; Schoelz *et al.*, 1986; Schoelz and Shepherd, 1988). The wild-type CaMV strains California, NY8153, and Cabbage B-Davis were obtained from the American Type Culture Collection and cloned at their unique *Sal I* site into the plasmid vector pUC18. The infectious clones of the chimeric viruses H25, H31, and H86 have been constructed previously (Qiu and Schoelz, 1992; Schoelz and Shepherd, 1988; Wintermantel and Schoelz, 1992). To eliminate sequence variation within the viral populations, the infectious clones of the CaMV strains and chimeric viruses were inoculated to turnips as described (Howell *et al.*, 1980; Lebeurier *et al.*, 1980). The infected turnip leaves were subsequently lyophilized and stored at 4 C to serve as inoculum for future experiments.

Recombinant DNA Techniques: Methods for DNA ligation, plasmid transformation, and plasmid purification have been described (Maniatis *et al.*, 1982). All plasmids were propagated in *Escherichia coli* strain JM101 (Messing, 1979). Viral DNA's were purified from infected turnip leaves according to the procedure of Gardner and Shepherd (1980). Restriction enzymes and T4 DNA ligase were purchased from Promega Corp. (Madison, Wisconsin) or New England Biolabs (Beverly, Massachusettes). To determine CaMV DNA sequences, viral DNA segments were subcloned into the plasmid vector pUC18 and both DNA strands were sequenced following the dideoxy chain termination method of Sanger *et al.* (1975). Oligonucleotide primers were synthesized by the DNA core lab of the University of Missouri-Columbia. Enzymes and chemicals for DNA sequencing were purchased in the Sequenase Version 2.0 kit (United States Biochemical Corp., Cleveland, Ohio).

Analysis of Host Specificity: Virus inoculum was prepared as described previously (Schoelz *et al.*, 1986). Solanaceous plants were inoculated 4-5 weeks and crucifers 2-3 weeks after seeds were sown. The inoculated solanaceous plants were maintained either in growth chambers set for a 10 hour day at 18-20 C or in the greenhouse during the months of October through April. Symptoms were recorded daily, and virus concentration quantitated by enzyme-linked immunosorbent assay (ELISA) (Clark and Adams, 1977) as described by Anderson *et al.* (1991).

Results and Discussion

A Comparison of the Ability of Solanaceous and Cruciferous Hosts to Support CaMV Infections: The host range of CaMV was originally reported to be limited to crucifers (Tompkins, 1937; Walker *et al.*, 1945). Through the years, there have been several reports that CaMV strains could infect solanaceous species (Schoelz *et al.*, 1986; Hills and Campbell, 1968; Lung and Pirone, 1972; Gracia and Shepherd, 1985); but the general consensus was that very few strains were able to systemically infect plants outside of the cruciferae family (Shepherd, 1981). We have recently started to test greater numbers of CaMV strains on a wider range of solanaceous hosts; and surprisingly, we have found that a majority of the CaMV strains could systemically infect at least three solanaceous species.

The viruses could be divided into three groups, based on their interaction with six solanaceous hosts (Table 1). The first group consisted of CaMV strains CM1841 and California, two viruses that were unable to systemically infect any solanaceous host under even the most permissive environmental conditions. CM1841 was able to replicate in the inoculated leaves of several solanaceous species, as indicated by the presence of chlorotic or necrotic local lesions, but systemic infection was blocked. The second group, which consisted of CaMV strains NY8153, W260, and Cabbage B Davis, were able to systemically infect three solanaceous species, although the viruses were not able to replicate to the same extent in each host. Upon inoculation to *Datura stramonium*, all three viruses induced necrotic lesions and systemic necrosis. A characteristic feature of the systemic necrosis symptom in *D. stramonium* was that viral titer in systemically infected leaves remained very low. NY8153, Cabbage B-Davis, and W260 accumulated to higher levels and induced more severe symptoms in *N. bigelovii* and *Nicotiana clevelandii* than in *D. stramonium*. In contrast to *D. stramonium*, *N. clevelandii* and *N. bigelovii* initially developed a systemic mosaic after inoculation with W260, NY8153, and Cabbage B-Davis, with systemic necrosis developing later except in the case of the interaction between W260 and *N. bigelovii*. The third group, which consisted of the D4 CaMV strain, was able to infect the greatest number of solanaceous species. D4 induced chlorotic primary lesions and a systemic mosaic on all six species.

Although many CaMV strains are capable of infecting at least a few solanaceous species, the most common host for research involving CaMV has been turnips (cv Just Right). 'Just Right' turnips

Table 1. Systemic Response of Solanaceous Hosts to Infection by CaMV Isolates.

Host	CM1841	California	NY8153	Cabbage B-Davis	W260	D4
D. stramonium	None	None	Systemic Necrosis	Systemic Necrosis	Systemic Necrosis	Mosaic
N. bigelovii	None	None	Mosaic/ Necrosis	Mosaic/ Necrosis	Mosaic	Mosaic
N. clevelandii	None	None	Mosaic/ Necrosis	Mosaic/ Necrosis	Mosaic/ Necrosis	Mosaic
N. edwardsonii	None	nt[a]	nt	None	None	Mosaic
N. glutinosa	None	nt	nt	nt	None	Mosaic
N. benthamiana	None	nt	nt	nt	None	Mosaic

Selected CaMV Isolates: CM1841, California, NY8153, Cabbage B-Davis, W260, D4.

[a]nt = Not tested.

have been used almost exclusively because CaMV can infect this host under a broader range of environmental conditions and can accumulate to higher levels in infected leaves than in other hosts. Covey and coworkers (Covey *et al.*, 1990; Saunders *et al.*, 1990) have compared the ability of several crucifers to support CaMV infections and have found that crucifers could be divided into three groups. The *Brassica* species that developed the most severe infections included turnip (*B. rapa* subspecies *rapifera*), chinese cabbage (*B. rapa* subscpecies *chinensis*) and pak choy (*B. rapa* subspecies *pekinensis*). Infected plants were extremely chlorotic and stunted. The 19S and 35S RNA transcripts were readily detected in turnip leaves while the unecapsidated viral DNA consisted primarily of reverse transcription products and relatively little supercoiled DNA.

In contrast to the highly susceptible hosts, *Brassica* species in the second group, which included kohlrabi (*B. oleracea* subspecies *gongylodes*) remained essentially symptomless. Infected plants contained extremely low levels of viral RNA and unencapsidated viral DNA consisted primarily of supercoiled DNA (Covey *et al.*, 1990). The third group, consisting of *Brassica* species such as spring rape (*B. napus* subspecies *olerifera*) had a response to CaMV infection that was intermediate between turnip and kohlrabi. Infected plants developed a mild mosaic, viral RNA levels were higher than kohlrabi but less than turnip, and the DNA products of reverse transcription were more prominent relative to CaMV infections in kohlrabi. A comparison of the reactions of turnip, spring rape, and kohlrabi to CaMV infections suggested that transcription is regulated, in part, by the host and that the level of transcription has an important role in determining the degree of susceptibility of crucifers to CaMV infections (Covey *et al.*, 1990). Sanfacon and Wieczorek (1992) have noted that the CaMV 35S RNA contains a 180 bp terminal redundancy and have suggested that the decreased synthesis of the 35S RNA in resistant *Brassica* species such as kohlrabi might also be due to a "less efficient type bypass of the poly(A) signal" during transcription.

We have been interested in comparing the ability of solanaceous and cruciferous species to support CaMV infections. This knowledge could be very useful for standardizing the interactions of CaMV with all of its hosts. In our tests, we have included turnip, cauliflower, and cabbage as representatives of crucifer species that support high, medium, and low levels of CaMV replication, respectively. In the test presented in Table 2, we inoculated plants with two strains of CaMV and compared the responses of the crucifer species to four solanaceous species. The two CaMV strains were chosen because they systemically infect a range of solanaceous species in addition to crucifers, and the environmental conditions were optimized for infection of solanaceous species (Schoelz *et al.*, 1986).

A comparison of CaMV levels in turnip, cauliflower, and cabbage confirmed that these hosts support different levels of CaMV replication. Turnips supported the highest level of replication of CaMV strains D4 and W260, cauliflower supported an intermediate level of replication, and cabbage supported the lowest level of replication. W260 did not induce any symptoms in cabbage and virus was barely detected by ELISA, while D4 was not detected in this host. One striking feature that has not been reported previously was that cauliflower was differentially susceptible to CaMV strains. The D4 virus did not induce any symptoms in cauliflower, and this virus was present at a very low level in cauliflower leaves. In contrast, W260 induced a strong mosaic in cauliflower and was readily detected by ELISA.

As with crucifers, the solanaceous species differed in their ability to support CaMV infections. Based on the virus levels measured in solanaceous species, the solanaceous hosts could be divided into four classes (Table 2). The CaMV levels measured in *N. bigelovii* and *N. clevelandii* indicated that these hosts were capable of supporting the same level of replication of CaMV as cauliflower. The amount of D4 virus in *Nicotiana glutinosa* was very low, resembling the level of W260 in cabbage. The difference between W260 infections of cabbage and D4 infections of *N. glutinosa* was that D4 systemic symptoms in *N. glutinosa* were clearly visible as a chlorotic mottle while the W260 infection of cabbage was symptomless. The concentration of D4 in *Nicotiana edwardsonii* was higher than in *N. glutinosa* but lower than in *N. bigelovii* and *N. clevelandii*.

The results in Table 2 suggest that two resistance mechanisms may operate against CaMV. First, the host may regulate the capacity for replication, probably by controlling the level of transcription (Covey

Table 2. Virus Concentration in Solanaceous and Cruciferous Hosts as Determined by ELISA[a].

		Virus Concentration (μg/g Leaf Tissue)	
Host	***Class***	***W260***	***D4***
Turnip	I	25.88 ± 6.71	32.88 ±17.06
Cauliflower	II	10.18 ± 4.58	0.71 ± 0.92
N. clevelandii	II	4.69 ± 2.95	12.96 ± 3.74
N. bigelovii	II	5.63 ± 2.35	10.18 ± 3.13
N. edwardsonii	III	0.00 ± 0.00	3.24 ± 2.00
Cabbage	IV	0.10 ± 0.14	0.00 ± 0.00
N. glutinosa	IV	0.00 ± 0.00	0.46 ± 0.41

[a] Each virus was inoculated to eight plants. For the ELISA, a single sample was taken from a mature leaf approximately one-third of the distance from the top. The ELISA value presented is the average of eight samples.

et al., 1990; Sanfacon and Wieczorek, 1992). Second, individual viruses may differ in their ability to infect a given host. The genetic studies described below demonstrate that this resistance mechanism differs substantially from a mechanism that regulates transcription.

Identification of CaMV Regions that Influence the Ability of CaMV to Systemically Infect Solanaceous Hosts: We have identified the regions of CaMV strains W260 and D4 that influence their ability to systemically infect solanaceous hosts by constructing chimeric viruses between these two viruses and CM1841. The results of those studies are summarized in Figure 1. All the chimeric viruses systemically infected turnips, the common host for all three CaMV strains; but they differed in their ability to infect the solanaceous hosts. We constructed seven chimeric viruses between D4 and CM1841, demonstrating that D4 sequences within the 5'- half of the gene VI coding region determine systemic infection of *N. bigelovii*, *D. stramonium*, and *N. edwardsonii* (Daubert *et al.*, 1984; Schoelz, *et al.*,

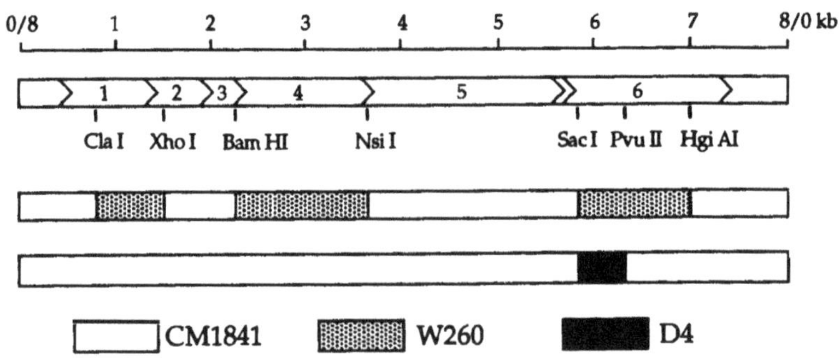

Figure 1. A Comparison of Regions of CaMV Strains W260 and D4 That Influence Systemic Infection of Solanaceous Hosts. The composite virus genomes are presented linearly to facilitate comparisons with the origin of the nucleotide sequence at each end of the diagram. The genetic map at the top shows the position of the six major cauliflower mosaic virus open reading frames and important restriction enzyme sites.

1986; Schoelz and Shepherd, 1988). We constructed 11 chimeric viruses between CaMV strains CM1841 and W260 in order to demonstrate that three separate regions of W260 influence the ability of W260 to systemically infect *N. bigelovii* and induce systemic necrosis in *D. stramonium*. These regions consist of the 3'-half of gene I and the first 263 bp of gene II, all of gene IV and the first 146 bp of gene V, and the gene VI coding sequence (Qiu and Schoelz, 1992). Any model that describes the interaction of CaMV with its solanaceous hosts must resolve the differences between the D4 and W260 maps presented in Figure 1.

Significantly, the regions that contain the 19S and 35S promoters did not play a role in the ability of D4 and W260 to systemically infect solanaceous species. The 35S promoter has been shown to map within nucleotides 7092 to 7435 (Odell *et al.*, 1985; Ow *et al.*, 1987; Odell *et al.*, 1988; Benfey *et al.*, 1989), a region that does not overlap with any of the W260 or D4 regions that influence systemic infection of solanaceous species (Figure 1). It is also probable that none of the W260 or D4 sequences that we have identified overlap with the 19S promoter. Therefore, it is unlikely that a resistance mechanism that functions by inhibiting transcription could be responsible for the differential abilities of D4, W260, and CM1841 to systemically infect solanaceous species.

One important observation is that gene VI has contributed to the host specificity of every chimeric virus that has been constructed (Daubert *et al.*, 1984; Schoelz *et al.*, 1986; Schoelz and Shepherd, 1988; Daubert and Routh, 1990). In contrast, genes I (cell-to-cell movement protein) and IV (coat protein) affect host specificity only in constructs between CM1841 and W260. The absolute involvement of gene VI in host specificity indicates that its contribution might differ from that of genes I and IV. Gene VI may interact with host proteins (Sanger *et al.*, 1991) and that interaction might be explained in one of two ways. The first explanation is related to the function of the CaMV gene VI product, which is to facilitate translation of genes on the 35S RNA (Gowda *et al.*, 1989; Bonneville *et al.*, 1989; Scholthof *et al.*, 1992). The CM1841 gene VI product may be unable to properly interact with unidentified host components in *N. bigelovii*, possibly host ribosomes. A second explanation may be that the CM1841 gene VI product elicits a plant defense response that limits further spread of the virus. The D4 and W260 gene VI products may not elicit a defense response, allowing the virus to move unimpaired throughout the plant.

Results obtained with chimeric viruses constructed between D4 and CM1841 suggest that genes I and IV may affect host specificity by their interaction with gene VI, rather than through their interaction with host proteins. A chimeric virus in which gene VI is derived from D4 and genes I-V from CM1841 systemically infects *N. bigelovii* (Schoelz *et al.*, 1986). If CM1841 genes I and IV were inherently defective in *N. bigelovii*, then this virus should have been limited to the inoculated leaf; and we would have generated exactly the same maps when identifying the regions of D4 and W260 that determine systemic infection of *N. bigelovii.* The gene VI product may interact with nucleotide sequences which occur within genes I and IV that are necessary for proper translation of the 35S RNA. The W260 gene VI product may be able to interact properly with its own W260 sequences, but unable to properly recognize CM1841 genes I and IV. The D4 gene VI product, in contrast, might be able to interact properly with either CM1841, W260, or D4 sequences in *N. bigelovii.*

An alternative explanation might be that the gene VI protein interacts with the gene I and IV proteins in some manner that affects long distance transport. The W260 gene VI protein might be unable to function properly with the CM1841 gene I and IV products. Oligonucleotide mutagenesis will be essential for identifying the nucleotide substitutions responsible for the involvement of regions containing genes I and IV and whether the involvement of these regions is at the protein or nucleic acid level.

Variations in Light Intensity Influence the Involvement of CaMV Regions in Systemic Infection of N. bigelovii: In the process of evaluating the CM1841/W260 chimeric viruses for their ability to systemically infect *N. bigelovii,* we found that some of the chimeric viruses had a leaky phenotype. For example, the chimeric virus H25 (Figure 2) originally was scored as unable to systemically infect *N. bigelovii.* In subsequent tests, approximately 10 percent of the plants

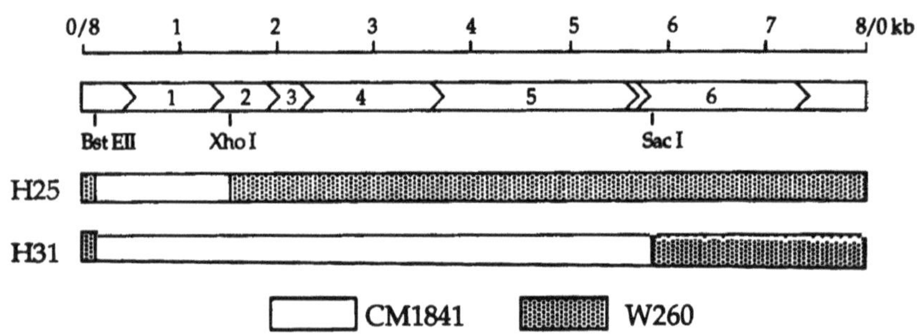

Figure 2. The Genetic Composition of Chimeric Viruses H25 and H31, Two Viruses Constructed Between the Wild-Type Strains CM1841 and W260. The virus genomes are presented linearly to facilitate comparisons with the origin of the nucleotide sequence at each end of the diagram. The genetic map at the top shows the position of the six major cauliflower mosaic virus open reading frames and important restriction enzyme sites.

inoculated with H25 developed systemic symptoms; and in one test conducted under winter greenhouse conditions, 100 percent of the H25-inoculated plants became systemically infected. In contrast to H25, the results obtained with most other chimeric viruses were invariant. For example, none of the *N. bigelovii* plants inoculated with H31 (Figure 2) have developed systemic symptoms in the 10+ times this virus has been tested.

The inconsistent results obtained with H25 suggested that either a mutant H25 virus was selected upon inoculation to *N. bigelovii* or that environmental conditions affected the ability of H25 to systemically infect *N. bigelovii*. To investigate whether or not a mutation was responsible for the ability of H25 to move systemically, we recovered a representative H25 isolate from systemically infected leaves, designated it H25S, and cloned the full length viral DNA into pBR322 as its unique *Sal I* site. We determined the nucleotide sequence of the 1516 bp *Bst EII* - *Xho I* DNA segment of H25S and compared it to the CM1841 nucleotide sequence. The nucleotide sequence of the *Bst EII* - *Xho I* DNA segment of H25S agreed exactly with the CM1841 sequence, eliminating the possibility that a mutation within this sequence influenced the ability of H25S to systemically infect *N. bigelovii* (Qiu and Schoelz, 1992).

To investigate whether environmental conditions could influence the ability of H25 to systemically infect *N. bigelovii*, we inoculated *N. bigelovii* with H25, H25S, CM1841, and W260 and grew the plants under three different light intensity levels. We chose to vary light intensity after reviewing all of the tests that included H25 and concluding that variations in light intensity might be more likely to be responsible for the leaky phenotype than fluctuations in temperature. The conditions most permissive for viral infection, in which light intensity was set at 170 $\mu E\ M^{-2}\ S^{-1}$, day length at 9.5 hours and temperature at 18°C, duplicated as closely as possible the winter greenhouse conditions. In other growth chamber settings, the temperature remained at 18°C, the day length was increased to 10 hours, and the light intensity was increased to either 320 $\mu E\ M^{-2}\ S^{-1}$ or 550 $\mu E\ M^{-2} S^{-1}$. *N. bigelovii* seedlings were preincubated in these environments for 10-12 days prior to virus inoculation.

The chimeric viruses H25 and H25S exhibited a sensitivity to light intensity (Table 3). H25 systemically infected *N. bigelovii* when plants were grown at the lowest level of light intensity, but systemic movement was restricted at medium and high light sensitivity levels. H25S-inoculated plants reacted similarly to light intensity as plants inoculated with H25, further evidence that H25S did not carry a mutation that enhanced its ability to move long distances in *N. bigelovii* relative to H25. In contrast, variations in light intensity did not significantly influence the ability of W260 and CM1841 to infect *N. bigelovii* (Table 3). W260 systemically infected all plants in each of the treatments while CM1841 failed to systemically infect *N. bigelovii* under even the most permissive conditions.

Table 3. Development of Systemic Infection of Wild-Type and Chimeric Viruses in *N. bigelovii* Under Three Levels of Light Intensity.

	Light Intensity Level		
Virus	170 $\mu E\ M^{-2}\ S^{-1}$	320 $\mu E\ M^{-2}\ S^{-1}$	550 $\mu E\ M^{-2}\ S^{-1}$
W260	8/8[a]	8/8	8/8
CM1841	0/8	0/8	0/8
H25	8/8	1/8	0/8
H25S	8/8	0/8	0/8

[a] Number of plants that developed systemic symptoms/number of plants inoculated.

The observation that variations in light intensity can influence the ability of the chimeric virus H25 to systemically infect solanaceous hosts explains the leaky phenotype of H25. Further research may provide an important insight into how environmental conditions affect CaMV infections. The region of H25 involved in systemic infection of *N. bigelovii* that is sensitive to light intensity has now been narrowed down to the 3'-half of gene I and the first 263 bp of gene II. Within this DNA segment, there are 19 nucleotide differences between W260 and CM1841, and five of the differences result in amino acid substitutions (Qiu and Schoelz, 1992). The principal question that we need to resolve now is whether the sequences responsible for alterations in host specifity act at the nucleotide level or amino acid level. Such information will be useful for understanding which part of the virus infection cycle has been disrupted. If alterations in CaMV proteins within this region are responsible for the inability of H25 to systemically infect *N. bigelovii* at high light intensity, then it is likely that those proteins play a role in long distance transport of the virus. If, however, silent mutations are implicated, then it would suggest a disruption in expression of genes on the 35S RNA.

There is very little information on how light intensity affects CaMV infections. Previous studies have focused primarily on how temperature influences the ability of CaMV to replicate. Yamaoka *et al.* (1982) reported that the optimum temperature for multiplication of CaMV in turnips was 20°C and that multiplication ceased when plants were maintained at 33°C and 16 hours of daylight. In contrast, Schoelz *et al.* (1986) found that the optimum temperature for CaMV strain D4 to infect *D. stramonium* and *N. bigelovii* was 18°C and the optimum daylength was 10 hours, while Daubert and Routh (1990) reported that D4 failed to systemically infect *D. stramonium* when plants were maintained at 24°C with 14 hours of light at 160 µE $M^{-2}S^{-1}$ light intensity. Our work should provide important information about how a second environmental variable, light intensity, affects CaMV infections.

The CM1841 Gene VI Product May Act as a Dominant Inhibitor of Systemic Infection: The genetic studies involving D4, W260, and CM1841 demonstrated that the involvement of gene VI in host specificity differs from that of genes I and IV because the gene VI product must interact with a host component to influence host specificity. In the case of the CM1841 gene VI product, it might be unable to function properly in *N. bigelovii* or it might elicit a plant defense response. To distinguish between these two possibilities, we created transgenic plants that expressed the D4 gene VI product (Schoelz *et al.*, 1991). We hypothesized that if the CM1841 gene VI product was defective, then transgenic *N. bigelovii* that express the D4 gene VI product would complement CM1841 for systemic infection. However, if the CM1841 gene VI product elicits a plant defense response, then we would predict that the transgenic *N. bigelovii* which express the D4 gene VI product might be unable to complement CM1841 for systemic infection.

To demonstrate that the D4 gene VI product expressed by the transgenic plants was biologically active, we first showed that the transgenic plants could complement chimeric virus H31 (Figure 2) (Schoelz *et al.*, 1991), a virus composed of CaMV strains W260 and CM1841. Previous studies showed that H31 could not systemically infect nontransformed *N. bigelovii* (Qiu and Schoelz, 1992; Schoelz and Shepherd, 1988), possibly because of a defect in the interaction of the W260 gene VI with CM1841 genes I - V. Transgenic *N. bigelovii* inoculated with H31 developed systemic symptoms 3 - 4 weeks after inoculation, while the nontransformed controls inoculated with H31 remained symptomless. To show that no changes had occurred within the H31 genome after passage through transgenic *N. bigelovii*, we compared the restriction enzyme map of H31 recovered from systemically infected leaves of transgenic plants to the H31 inoculum and found that they were identical. More importantly, we demonstrated that H31 isolated from transgenic plants had not acquired the ability to systemically infect nontransformed *N. bigelovii*. Lastly, a sequence analysis of the 5'-half of gene VI of the H31 recovered from transgenic plants matched the predicted H31 sequence, further evidence that no genetic changes had occurred in the H31 genome that affected host specificity. This study demonstrated that the D4 gene VI product expressed by the transgenic plants was able to complement a CaMV isolate for systemic infection. It also provided further evidence that the involvement of CM1841 genes I and IV in precluding systemic infection was due to a defect in the interaction of W260 gene VI with CM1841 genes I and IV.

By demonstrating that the gene VI product produced by the transgenic plants was functional, we could now investigate whether or not the wild-type virus CM1841 would be complemented by the transgenic plants. Complementation would indicate that the CM1841 gene VI product is unable to function properly in *N. bigelovii* while a lack of complementation would suggest that CM1841 elicits a plant defense response.

We compared the abilities of CM1841, H31, and W260 to systemically infect transgenic *N. bigelovii* that express the D4 gene VI product with their ability to infect nontransformed controls (Table 4). The nontransformed *N. bigelovii* inoculated with CM1841 did not develop systemic symptoms and no virus could be detected in the upper, noninoculated leaves by ELISA. Transgenic *N. bigelovii* inoculated with CM1841 developed systemic symptoms, but the appearance of symptoms was delayed 10 days relative to chimeric virus H31. In contrast to H31, the CM1841 virus recovered from the systemically infected leaves (designated CM1841R) had a gene VI restriction enzyme map that corresponded to gene VI of D4, the source of gene VI in the transgenic plants. Other restriction enzyme sites characteristic of CM1841 that mapped to genes I-V had been retained in CM1841R, suggesting that changes had only occurred within gene VI of CM1841R. In attempts to recover CM1841 from systemically infected leaves of eleven transgenic plants, we have recovered recombinant viruses resembling CM1841R in all eleven plants.

We cloned a representative CM1841R viral DNA into pUC18, determined the nucleotide sequence of the gene VI coding region, and compared its sequence to the CM1841 inoculum and the D4 sequence used for construction of the transgenic *N. bigelovii*. The CM1841 and the D4 sequence present in the transgenic plants differ at 42 nucleotides within the gene VI (Figure 3). The CM1841R sequence agreed with the D4 sequence at 41 of the 42 sites. The one difference between CM1841R and D4 within the *Sac I* - *Hgi AI* DNA segment occurred at the extreme 3'-end of the DNA segment. The results obtained with CM1841 suggested that CM1841 was not complemented by the transgenic *N. bigelovii* and that wild-type CM1841 remained localized within the inoculated leaf. The nucleotide sequence of CM1841R showed that it resulted from a recombination event between CM1841 and the D4 transgene in the inoculated leaves. CM1841R was responsible for the systemic symptoms in CM1841-inoculated transgenic *N. bigelovii*.

CM1841R differed phenotypically from the wild-type CM1841 in two ways. CM1841R induced symptomless infections in turnips, while wild-type CM1841 induced a distinct chlorotic mosaic. Second, CM1841R could systemically infect nontransformed *N. bigelovii*, while CM1841 could not. Both of these phenotypic traits could be predicted from the gene VI coding sequence of CM1841R because of the genetic studies that we have conducted in the past. The chimeric virus H7 (Daubert *et al.*, 1984), a virus constructed *in vitro* from D4 and CM1841 that has a genome composition similar to CM1841R, induced symptomless infections in turnips (Daubert *et al.*, 1984) and systemically infected *N. bigelovii* (Schoelz *et al.*, 1986).

Table 4. Development of Systemic Symptoms in Transgenic and Nontransformed *N. bigelovii*.

Virus	Nontransformed *N. bigelovii*	Transgenic *N. bigelovii* that Express the D4 Gene VI
W260	10/10[a]	10/10
H31	0/10	10/10
CM1841	0/10	8/10

[a] Number of plants that developed systemic symptoms/number of plants inoculated.

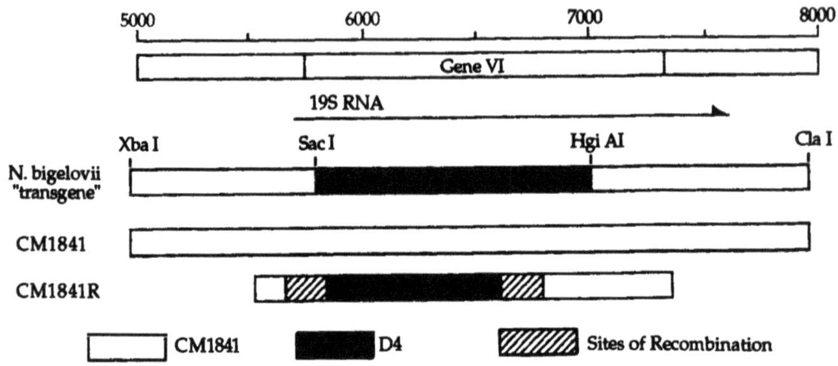

Figure 3. A Comparison of the Nucleotide Composition of Gene VI and Flanking Regions Present in the Transgenic *N. bigelovii* to the Gene VI Nucleotide Composition and Flanking Regions of CM1841 and CM1841R.

Gal *et al.* (1992) have also shown that CaMV is capable of recombination with transgenic plants that contain CaMV gene VI coding sequences. In their study, they agroinoculated a CaMV construct that lacked gene VI onto transgenic *B. napus* that contained the gene VI coding sequence. CaMV was detected in upper, noninoculated leaves of the transgenic *B. napus* by dot blot analysis. The viral DNA recovered from the upper leaves consisted of an infectious recombinant virus. The authors could not find any evidence for complementation, possibly because the expression of the gene VI product was very low. Alternatively, complementation might not have been possible because of the removal of sequences within gene VI that may be required *in cis* for transactivation (Scholthof *et al.*, 1992).

Two lines of evidence suggest that gene VI of CM1841 acts as a dominant inhibitor of systemic infection in *N. bigelovii*. First, a previous study with chimeric viruses constructed between CM1841 and D4 showed that gene VI of CM1841 was responsible for the inability of chimeric viruses to systemically infect *N. bigelovii* (Schoelz *et al.*, 1986). Second, a comparison of the genome compositions of H31 (Figure 2) and CM1841 indicates that gene VI of CM1841 is responsible for inhibition of CM1841 in *N. bigelovii*, leading to the selective advantage of CM1841R, the recombinant virus, over CM1841. The observation that H31 was complemented for systemic infection suggests that viruses that are unable to systemically infect a given host may be divided into two groups. One group consists of viruses such as H31 that are unable to function properly in certain hosts while the second group is composed of viruses like CM1841 that may contain sequences that actively inhibit systemic infection.

In future research, we will be interested in identifying what host proteins interact with the CaMV gene VI product to influence host specificity. Bonneville *et al.* (1989) suggested that gene VI might interact with host ribosomes to influence translation. It is possible that the same host protein might also play a role in determining resistance or susceptibility of solanaceous species to CaMV infection.

Acknowledgements

The authors thank M. Shababi for contributing the infectious clones of NY8153, California, and Cabbage B. This research was supported by a grant from the Food for the 21st Century Program of the University of Missouri, Columbia, and by USDA Grant 89-37263-4812.

References

Albrecht, H., Geldreich, A., Menissier de Murcia, J., Kirchherr, D., Mesnard, J.-M, and Lebeurier, G. 1988. Cauliflower mosaic virus gene I product detected in a cell-wall-enriched fraction. Virology 163:503-508.

Anderson, E.J., Qiu, S.G., and Schoelz, J.E. 1991. Genetic analysis of disease severity and virus concentration determinants in cauliflower mosaic virus. Virology 181:647-655.

Armour, S.L., Melcher, U., Pirone, T.P., Lyttle, D.J., and Essenberg, R.C. 1983. Helper component for aphid transmission encoded by region II of cauliflower mosaic virus DNA. Virology 129:25-30.

Balazs, E., Guilley, H., Jonard, G., and Richards, K. 1982. Nucleotide sequence of DNA from an altered-virulence isolate D/H of cauliflower mosaic virus. Gene 19:239-249.

Benfey, P.N., Ren, L., and Chua, N.-H. 1989. The CaMV 35S enhancer contains at least two domains which can confer different developmental and tissue-specific expression patterns. EMBO J. 8:2195-2202.

Bonneville, J.M., Sanfacon, H., Futterer, J., and Hohn, T. 1989. Posttranscriptional trans-activation in cauliflower mosaic virus. Cell 59:1135-1143.

Citovsky, V., Knorr, D., and Zambryski, P. 1991. Gene I, a potential cell-to-cell movement locus of cauliflower mosaic virus, encodes an RNA-binding protein. Proc. Natl. Acad. Sci. U.S.A. 88:2476-2480.

Clark, M.F. and Adams, A.N. 1977. Characteristics of the microplate method of enzyme-linked immunosorbant assay for the detection of plant viruses. J. Gen. Virol. 34:475-483.

Covey, S.N. and Hull, R. 1981. Transcription of cauliflower mosaic virus DNA: Detection of transcripts, properties, and location of the gene encoding the virus inclusion body protein. Virology 111:463-474.

Covey, S.N., Turner, D.S., Lucy, A.P., and Saunders, K. 1990. Host regulation of the cauliflower mosaic virus multiplication cycle. Proc. Natl. Acad. Sci. U.S.A. 87:1633-1637.

Daubert, S., Schoelz, J., Debao, L., and Shepherd, R.J. 1984. Expression of disease symptoms in cauliflower mosaic virus genomic hybrids. J. Mol. Appl. Genet. 2:537-546.

Daubert, S., Richins, R., Shepherd, R., and Gardner, R. 1982. Mapping of the coat protein gene of CaMV by its expression in a prokaryotic system. Virology 122:444-447.

Daubert, S. and Routh, G. 1990. Point mutations in cauliflower mosaic virus gene VI confer host-specific symptom changes. Mol. Plant-Microbe Interact. 3:341-345.

Frank, A., Guilley, H., Jonard, G., Richards, K., and Hirth, L. 1980. Nucleotide sequences of cauliflower mosaic virus DNA. Cell 21:285-294.

Gal, S., Pisan, B., Hohn, T., Grimsley, N., and Hohn, B. 1992. Agroinfection of transgenic plants leads to viable cauliflower mosaic virus by intermolecular recombination. Virology 187:525-533.

Gardner, R.C., Howarth, A., Hahn, P., Brown-Leudi, M., Shepherd, R.J., and Messing, J. 1981. The complete nucleotide sequence for an infectious clone of cauliflower mosaic virus by M13mp7 shotgun sequencing. Nucleic Acids Res. 9:2871-2888.

Gardner, R. and Shepherd, R. 1980. A procedure for rapid isolation and analysis of cauliflower mosaic virus DNA. Virology 106:159-161.

Giband, M., Mesnard, J.M., and Lebeurier, G. 1986. The gene III product (P15) of cauliflower mosaic virus is a DNA-binding protein while an immunologically related P11 polypeptide is associated with virions. EMBO J. 5:2433-2438.

Gowda, S., Wu, F.C., Scholthof, H.B., and Shepherd, R.J. 1989. Gene VI of figwort mosaic virus (caulimovirus group) functions in posttranscriptional expression of genes on the full-length RNA transcript. Proc. Natl. Acad. Sci. U.S.A. 86:9203-9207.

Gracia, O. and Shepherd, R.J. 1985. Cauliflower mosaic virus in the nucleus of *Nicotiana*. Virology 146:141-145.

Hills, B.J. and Campbell, R.N. 1968. Morphology of broccoli necrotic yellows virus. J. Ultrastructure Res. 24:134-144.

Howarth, A.J., Gardner, R.C., Messing, J., and Shepherd, R.J. 1981. Nucleotide sequence of naturally occurring deletion mutants of cauliflower mosaic virus. Virology 112:678-685.

Howell, S.H., Walker, L.L., and Dudley, R.K. 1980. Cloned cauliflower mosaic virus DNA infects turnips (*Brassica rapa*). Science 208:1255-1267.

Hull, R. and Covey, S.N. 1985. Cauliflower mosaic virus: Pathways of infection. BioEssays 3:160-163.

Hull, R., Sadler, J., and Longstaff, M. 1986. The sequence of carnation etched ring virus DNA: Comparison with cauliflower mosaic virus and retroviruses. EMBO J. 5:3083-3090.

Laquel, P., Ziegler, V., and Hirth, L. 1986. The 80K polypeptide associated with the replication complexes of cauliflower mosaic virus is recognized by antibodies to gene V translation product. J. Gen. Virol. 67:197-201.

Lebeurier, G.L., Hirth, L., Hohn, B., and Hohn, T. 1980. Infectivities of native and cloned DNA of cauliflower mosaic virus. Gene 12:139-146.

Linstead, P.J., Hills, G.J., Plaskitt, K.A., Wilson, I.G., Harker, C.L., and Maule, A.J. 1988. The subcellular location of the gene I product of cauliflower mosaic virus is consistent with a function associated with virus spread. J. Gen. Virol. 69:1809-1818.

Lung, M.C.Y. and Pirone, T.P. 1972. *Datura stramonium*, a local lesion host for certain isolates of cauliflower mosaic virus. Phytopathology 62:1473-1474.

Maniatis, T., Fritsch, E.F., and Sambrook, J. 1982. Molecular Cloning: A Laboratory Manual Cold Spring Harbor Laboratory, Cold Spring Harbor, New York. 545 pp.

Martinez-Izquierdo, J.A., Futterer, J., and Hohn, T. 1987. Protein encoded by ORF I of cauliflower mosaic virus is part of the viral inclusion body. Virology 160:527-530.

Mesnard, J.-M., Kirchherr, D., Wurch, T., and Lebeurier, G. 1990. The cauliflower mosaic virus gene III product is a non-sequence-specific DNA binding protein. Virology 174:622-624.

Messing, J. 1979. A multipurpose cloning system based on single-stranded DNA bacteriophage M13. Recomb. DNA Tech. Bull. 2:43-48.

Odell, J.T., Knowlton, S., Lin, W., and Mauvais, C.J. 1988. Properties of an isolated transcription stimulating sequences derived from the cauliflower mosaic virus 35S promoter. Plant Mol. Biol. 10:263-272.

Odell, J.T., Nagy, F., Chua, N.-H. 1985. Identification of sequences required for the activity of the cauliflower mosaic virus 35S promoter. Nature 313:810-812.

Ow, D.W., Jacobs, J.D., and Howell, S.H. 1987. Functional regions of the cauliflower mosaic virus 35S RNA promoter determined by use of the firefly luciferase gene as a reporter of promoter activity. Proc. Natl. Acad. Sci. U.S.A. 84:4870-4874.

Qiu, S.G. and Schoelz, J.E. 1992. Three regions of cauliflower mosaic virus strain W260 are involved in systemic infection of solanaceous hosts. Virology 190:773-782.

Sanfacon, H. and Wieczorek, A. 1992. Analysis of cauliflower mosaic virus RNA's in *Brassica* species showing a range of susceptibility to infection. Virology 190:30-39.

Sanger, F., Nicklen, S., and Coulson, A.R. 1975. DNA sequencing with chain-terminating inhibitors. Proc. Natl. Acad. Sci. U.S.A. 74:5463-5467.

Sanger, M., Daubert, S., and Goodman, R.M. 1991. The regions of sequence variation in caulimovirus gene VI. Virology 182:830-834.

Saunders, K., Lucy, A.P., and Covey, S.N. 1990. Characterization of cDNA clones of host RNA's isolated from cauliflower mosaic virus-infected turnip leaves. Physiol. and Mol. Plant Path. 35:339-346.

Schoelz, J.E., Goldberg, K.-B., and Kiernan, J. 1991. Expression of cauliflower mosaic virus (CaMV) gene VI in transgenic *Nicotiana bigelovii* complements a strain of CaMV defective in long-distance movement in nontransformed *N. bigelovii*. Mol. Plant-Microbe Interact. 4:350-355.

Schoelz, J.E. and Shepherd, R.J. 1988. Host range control of cauliflower mosaic virus. Virology 162:30-37.

Schoelz, J.E., Shepherd, R.J., and Daubert, S.D. 1986. Region VI of cauliflower mosaic virus encodes a host range determinant. Mol. Cell Biol. 6:2632-2637.

Schoelz, J.E., Shepherd, R.J., and Richins, R.D. 1986. Properties of an unusual strain of cauliflower mosaic virus. Phytopathology 76:451-454.

Scholthof, H.B., Gowda, S., Wu, F.C., and Shepherd, R.J. 1992. The full-length transcript of a caulimovirus is a polycistronic mRNA whose genes are *trans* activated by the product of gene VI. J. Virol. 66:3131-3139.

Scholthof, H.B., Wu, F.C., Gowda, S., and Shepherd, R.J. 1992. Regulation of caulimovirus gene expression and the involvement of *cis*-acting elements on both viral transcripts. Virology 190:403-412.

Shepherd, R.J. 1981. Cauliflower mosaic virus. CMI/AAB Descriptions of Plant Viruses 243:6 pp.

Takatsuji, H., Hirochika, H., Fukushi, T., and Ikeda, J.-E. 1986. Expression of cauliflower mosaic virus reverse transcriptase in yeast. Nature 319:240-243.

Toh, H., Hayashida, H., and Miyata, T. 1983. Sequence homology between retrovial reverse transcriptase and putative polymerases of hepatitis B virus and cauliflower mosaic virus. Nature 305:827-829.

Tompkins, C.M. 1937. A transmissible mosaic disease of cauliflower. J. Agric. Res. 55:33-46.

Walker, J.C., LeBeau, F.J., and Pound, G.S. 1945. Viruses associated with cabbage mosaic. J. Agric. Res. 70:379-404.

Wintermantel, W.M., Anderson, E.J., and Schoelz, J.E. 1993. Identification of domains within gene VI of cauliflower mosaic virus that influence systemic infection of *Nicotiana bigelovii* in a light-dependent manner. Virology 196:789-798.

Woolston, C.J., Covey, S.N., Penswick, J.R., and Davies, J.W. 1983. Aphid transmission and polypeptide are specified by a defined region of the cauliflower mosaic virus genome. Gene 23:15-23.

Xiong, C., Lebeuier, G., and Hirth, L. 1984. Detection *in vivo* of a new gene product (gene III) of cauliflower mosaic virus. Proc. Natl. Acad. Sci. U.S.A. 81:6608-6612.

Xiong, C., Muller, S., Lebeurier, G., and Hirth, L. 1982. Identification by immunoprecipitation of cauliflower mosaic virus *in vitro* major translation product with a specific serum against viroplasm protein. EMBO J. 1:971-976.

Yamaoka, N., Morita, T., Furusawa, I., and Yamamoto, M. 1982. Effect of temperature on the multiplication of cauliflower mosaic virus. J. Gen. Virol. 61:283-287.

Young, M.J., Daubert, S.D., and Shepherd, R.J. 1987. Gene I products of cauliflower mosaic virus detected in extracts of infected tissue. Virology 158:444-446.

A "Gene for Gene" Interaction and Potato Virus X

D. C. Baulcombe, B. Köhm, M. G. Goulden, S. Santa-Cruz, and A. Forsyth
The Sainsbury Laboratory
Colney Lane, Norwich NR4 7UH
United Kingdom

S. Chapman
The University of Florida
Lake Alfred, Florida 33850

T. Kavanagh
Department of Genetics, Trinity College
Lincoln Place Gate
Dublin 2, Eire

In many instances, resistance to infection by plant viruses is conferred by single dominant genes. The resistance conferred by these genes may be either induced or constitutive. To determine which of these mechanisms underlies the resistance to PVX conferred by various resistance genes in potato, we have carried out an analysis of different strains of potato virus X. These strains differ in the ability to overcome the resistance conferred by Rx, Nx, or Nb resistance genes. The analysis has shown first that the interaction between the virus and the product of the resistance gene can be affected by different viral genes. We have also used this type of analysis to confirm that these resistance genes confer an induced resistance in which the recognition phase is quite distinct from the response phase. The recognition phase may be highly strain specific; whereas the response phase may have broad spectrum effectiveness. The identification of virus determinants is the first step in understanding of resistance interactions at the molecular level.

Introduction

It is useful to consider plant resistance to viruses within the context of resistance to bacteria and fungi as, in some of these examples, the genetics or biochemistry have been well studied and provide a useful conceptual framework (Keen, 1990). In broad terms, it appears that the resistance depends on a highly specific interaction between the product of a resistance gene in the host and a determinant in the pathogen. There are apparently two major types of resistance (Figures 1a and 1b): there may be an induced resistance, in which the pathogen is constrained as a result of processes induced in the infected plant, or constitutive resistance, in which the product of a resistance gene may directly inactivate a factor from the pathogen which is necessary for pathogenicity. These alternatives are summarized diagrammatically in Figures 1a and 1b in which the product of the resistance gene is labelled R, and the viral component interacting with R is labelled Y. The component X in Figure 1b is either from the plant or the pathogen and needs to interact with Y for pathogenesis to take place.

Induced resistance is often characterized by necrotic lesions at the site of inoculation and arrest of the pathogen at, or near to, the lesion. The difference between induced and the constitutive resistance is that, in the induced resistance, the mechanism by which the pathogen is restricted to the site of inoculation is secondary and distinct from the process by which the plant recognizes the pathogen: hypersensitive necrosis and the associated defense reactions are induced in the plant as a consequence of the recognition event (Figure 1a). With constitutive resistance the primary interaction arrests growth of the pathogen (Figure 1b) by prevention of the interaction between X and Y.

At an experimental level, it is possible to determine which of these mechanisms applies to a particular resistance interaction by analysis of the component in the pathogen (Y in Figure 1) which is recognized by the resistance gene product. Manipulation of the gene for this component of the pathogen may have different effects, depending on the type of resistance. In this article, we discuss three different ways in which analysis of the Y component (Figure 1) can shed light on the type of the reaction. For example, with constitutive resistance mechanisms (Figure 1b), the pathogen component in the interaction is necessary for pathogenesis. If a mutant pathogen does not produce that factor, it will not infect (Figure 1b). In the instance of the induced resistance, the pathogen component recognized by the resistance gene product may not be essential for pathogenesis but is required for initiation of the defense responses (Figure 1a). Consequently, disruption of the gene for Y may allow a pathogen to overcome a resistance gene giving an induced resistance. Thus, the alternative types of resistance can be distinguished if mutation of the gene for Y causes a transition from resistance sensitivity to resistance-breaking.

A second test of these two types of resistance mechanism can be carried out if there is allelic variation in the pathogen, at the locus encoding Y. In this situation, different isolates of the pathogen would

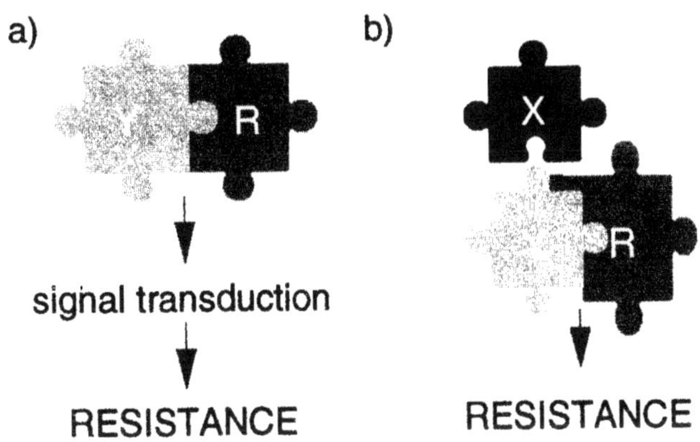

Figure 1. **Two Types of Resistance Mechanism. A: Induced resistance. Avirulence determinant in the pathogen (Y) interacts with the product of a resistance gene (R) and triggers a sequence of events resulting in induction of resistance. B: Direct resistance. In the susceptible plant, a product of the pathogen (Y) interacts with a component (X) either from the plant or from the pathogen to bring about pathogenesis. In the resistant plant, the interaction is prevented due to the interaction of the resistance gene product (R) with Y.**

either be sensitive to the effects of the resistance gene or able to break the resistance. If the resistance is induced, the Y function of the resistance-sensitive strain will be genetically dominant over that from the resistance-breaking isolate: resistance will be fully induced by Y from the resistance-sensitive strain irrespective of whether the Y from the resistance-breaking strain is also present. Conversely, if the resistance is constitutive, as depicted in Figure 1b, the interaction of R with Y of the resistance-sensitive strain will have no effect on Y from the resistance-breaking strain. Consequently, in cells expressing both types of Y, there will be full pathogenicity, mediated by the resistance-breaking Y.

For several fungal and bacterial pathogens, the Y function in induced resistance is referred to as the avirulence determinant and the test of dominant avirulence is a basic test of this type of induced resistance. There is now increasing molecular and biochemical evidence for induced resistance mechanisms in plants; but the original paradigm was developed as the "gene for gene" hypothesis, based on deletion and dominance tests of pathogen components in the resistance gene interaction (Keen, 1990).

A third test of the type of resistance interaction also based on the analysis of the pathogen involves treatment of resistant plants with the purified Y factor. If the resistance is of the induced type, the Y factor of the resistance sensitive strains will elicit some, or all of the defence reactions, associated with the resistance response. If the resistance is constitutive, application of Y will not induce resistance.

Although there are several well characterized examples of resistance to virus infection in plants (Fraser, 1990), only one system has been characterized in sufficient detail for it to be evident whether it is a direct or indirect resistance mechanism, although many of those interactions associated with HR will almost certainly be the induced type. This single example involves the induced resistance in tobacco to tobacco mosaic virus (TMV) mediated by *N'*. Extensive studies (Saito *et al.*, 1987; Culver and Dawson, 1989) have identified amino acid residues in the viral coat protein involved in the interaction with the product of *N'*; and, in an analysis of transgenic plants carrying *N'* and expressing coat protein of the resistance-sensitive isolate of TMV, it was shown that the CP is an elicitor of the hypersensitive necrosis (Culver and Dawson, 1991). In effect, this transgenic analysis was an example of the third test for constitutive or induced resistance involving application of the virus-encoded specificity determinant to the resistant plant. The results were conclusively in support of the induced mechanism (Culver and Dawson, 1991).

An example of direct resistance to virus infection involves cowpea mosaic virus and the Arlington cowpea. This cultivar of cowpea contains a single dominant gene conferring resistance to cowpea mosaic virus. There is very little known about the genetical characteristics of the specificity determinants in the viral genome; but, from a biochemical analysis, it seems likely that this is an example of direct, constitutive resistance: the genome of cowpea mosaic virus is expressed into two polyproteins which are normally processed into the biologically active constituent proteins by a virus encoded protease acting in *cis*. The resistance gene in Arlington cowpea apparently encodes an inhibitor of this protease (Ponz *et al.*, 1988).

Aanalysis of Resistance Gene Interactions with Potato and Potato Virus X

To extend understanding of virus resistance mechanisms in plants we have initiated an analysis of the interaction between potato and potato virus X (PVX). The attractions of this as an experimental system are several fold:

1) There are three classes of PVX resistance loci (Cockerham, 1970) (*Nx, Nb, Rx*) which determine either hypersensitive resistance (HR) to PVX (*Nx, Nb*) or an extreme resistance (ER). The ER is controlled by *Rx* and not normally associated with production of necrotic local lesions but prevents virus accumulation in the inoculated leaf and in inoculated protoplasts (Adams *et al.*, 1986; Saladrigas *et al.*, 1990). The potato/PVX interaction thus proves the opportunity to analyze different types of resistance mechanism with the same virus.

2) There is genetic variation between virus isolates such that, for each resistance locus, there are isolates which are either resistance-sensitive or resistance-breaking. This variation forms the basis of the classification of strains of PVX (Cockerham, 1970) (Table 1).

3) The potato genome is almost entirely co-linear with the genome of tomato (Bonierbale *et al.*, 1988) such that detailed molecular analysis of the plant-encoded resistance genes could exploit the extensive analysis of the tomato genome.

Resistance Gene Specificity Determinants in PVX: As discussed above, the initial determination of whether a resistance mechanism is of the constitutive or the induced type can be made by analysis of the resistance specificity determinants in the viral genome. In effect, it is necessary first to identify Y (Figure 1). To this end, in the analysis of resistance to PVX, cDNA clones were produced of resistance-sensitive and resistance-breaking strains from which infectious RNA could be produced *in vitro*. The subsequent steps would be to construct and analyze hybrid genomes incorporating elements from the different cDNA's.

Table 1. Resistance Interactions of PVX Strains on Potato Cultivars

plant genotype	Strain of PVX				
	group 1	group 2	group 3	group 4	PVX_{HB}
nx nb rx	S	S	S	S	S
nx Nb rx	HR	HR	S	S	S
Nx nb rx	HR	S	HR	S	S
Nx Nb rx	HR	HR	HR	S	S
nx nb Rx	ER	ER	ER	ER	S

S=susceptible interaction, HR=hypersensitive response, ER=extreme resistance.

The first strain for which the necessary cDNA clones were constructed was PVX_{UK3}. This strain gives the HR on plants carrying *Nx* and the ER when inoculated to *Rx* potato. Analysis of the specific infectivity of the *in vitro* transcripts from this cDNA showed that there was an influence of non-viral sequences at the 5'- and 3'-ends. The most infectious transcripts were produced from a derivative with 80 nucleotides at the 3'-end transcribed into a polyadenylate tail in the RNA (Chapman, 1991) and with the promoter for RNA synthesis fixed directly to the 5'-terminus of the PVX cDNA. However, this homopolymer tail was not stable when the cDNA was amplified in a plasmid vector; and the experiments to be described were all carried out with a derivative of the cDNA with a 25 nucleotide homopolymer tail and which gave rise to transcripts that were 5 percent as infectious as viral RNA isolated from virus particles (Chapman, 1991). This reduced infectivity was substantially higher than with a previously reported cDNA of PVX (Hemenway *et al.*, 1990) and was not a problem in the analysis of interactions with potato, as the titre of the inoculum was increased prior to inoculation to potato by passage through tobacco.

Control experiments carried out with transcript-derived inoculum confirmed that the PVX_{UK3} cDNA was typical of the progenitor virus: the symptoms on tobacco were systemic mild mosaic; on *Chenopodium amaranticolor* there were necrotic local lesions; on PVX susceptible cultivars of potato there was systemic infection accompanied in some instances with a mild mosaic and HR or ER responses on potato cultivars with *Nx* or *Rx*, respectively.

The first series of hybrids were constructed by substitution of PVX_{HB} cDNA into the homologous region of PVX_{UK3}. PVX_{HB} overcomes (Moreira and Jones, 1980) the effects of *Nb*, *Nx*, and *Rx* (Table 1). Surprisingly, for reasons discussed below, the hybrids with only the coat protein gene of PVX_{HB} were resistance-breaking; they were able to overcome both *Nx*- and *Rx*-mediated resistance (Kavanagh *et al.*, 1992). These observations indicated that the viral specificity determinant(s) implicated in the interaction with both *Nx* and *Rx* were within, or encoded by, the coat protein gene.

Following up this initial observation was less straightforward than the initial step, due to the paucity of shared restriction enzyme sites in the two cDNA's which complicated the construction of hybrid genomes. For this reason, the experiments were split into separate analysis of *Nx* and *Rx* specificity determinants in the genome of PVX.

The Rx Specificity Determinants in PVX: The analysis of the *Rx* determinants was based on PVX_{HB} as before, but with PVX_{cp4} as the resistance-sensitive strain. Within the coat protein gene, PVX_{cp4} is more than 95 percent identical at the nucleotide level to PVX_{HB} (Goulden and Santa-Cruz, unpublished). It is similar to PVX_{HB} in that it overcomes the *Nx* and *Nb* genes but differs from PVX_{HB} in that it does not overcome the *Rx*-mediated resistance (Jones, 1985). The high degree of nucleotide similarity facilitated construction of a series of intra-genic hybrids based on the coat protein gene, which were tested by inoculation to potato cultivar Cara (*Rx*), and to various susceptible cultivars.

Each of the tested hybrid genomes was either sensitive to *Rx* or was clearly able to overcome the resistance gene (Goulden, unpublished); and, from the various constructions, the *Rx* specificity determinant in the viral genome was located in the central part of the coat protein gene (Figure 2). Within this region there were two coding sequence differences between PVX_{HB} and PVX_{cp4} (Goulden, unpublished) which, as discussed below, are thought to affect the structure of the viral determinant involved in the interaction with *Rx* (Y, Figure 1).

The Nx Specificity Determinants in PVX: Following from the analysis of PVX_{UK3}/PVX_{HB} hybrids, a series of PVX_{UK3}/PVX_{cp4} hybrids was constructed, with substitutions of parts of the PVX_{cp4} coat protein gene into the PVX_{UK3} genome. However, the analysis of these genomes was complicated by apparent incompatibility of the components of the

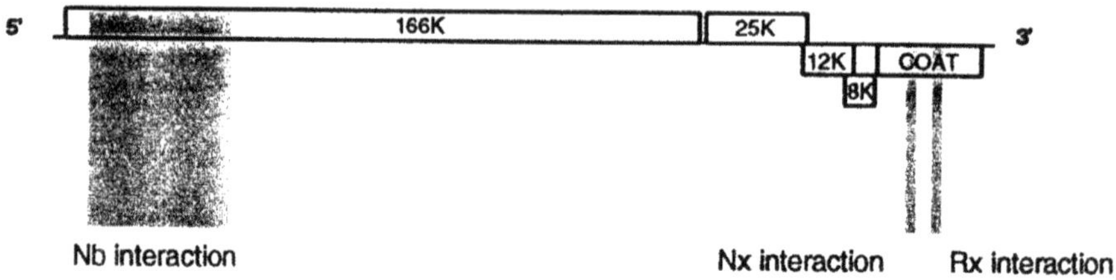

Figure 2. Resistance Specificity Determinants in the PVX Genome. The diagram shows a scale diagram of the PVX genome with the M_r of the gene products (K=kDa) indicated, except for the coat protein gene identified as "COAT." The location of resistance specificity determinants, as located by molcular analysis, is indicated by the shading.

intragenic hybrids resulting, in some instances, in viral genomes which were only poorly infectious, if at all. Nevertheless, it was possible to make the limited conclusion, from the hybrid and genomes which did infect, that the *Nx* specificity determinant is in, or encoded by, the 5'-half of the coat protein gene (Santa-Cruz, unpublished data).

More precise information about the *Nx* specificity determinants was obtained by analysis of mutant strains of PVX which were selected, for transition from resistance-sensitivity to resistance-breaking, by passage of inocula on plants carrying *Nx*. These mutants were first described by Jones (1982) using strain PVX_{DX} which, on passage in the *Nx* potato, gave rise to resistance-breaking strain PVX_{DX4}. Substitution of the coat protein gene, as cDNA, from these strains into the transcription clone of PVX_{UK3}, produced two hybrid genomes designated PVX_{KD} and PVX_{KD4}. The strain PVX_{KD} with PVX_{DX} coat protein gene was sensitive to *Nx*; whereas the PVX_{KD4} strain with the PVX_{DX4} coat protein gene was resistance-breaking. This result confirmed the localization of the *Nx* specificity determinant in the coat protein gene. From sequence analysis, it was further concluded that amino acid, or codon 78, is involved. This position is likely to be part of a more extensive domain, as further mutants had coding sequence changes close to, but not precisely at, codon 78 (Santa-Cruz, unpublished data). One mutant derivative of PVX_{KD}, selected for *Nx* breaking, as described above, had a change at codon 76; whereas a derivative of PVX_{KD4}, selected as producing necrotic lesions on *Nx* plants, had a change at codon 62 (Figure 3a). In both instances the coat protein gene of the mutant viruses was substituted into the transcription clone of PVX_{UK3}; and it was demonstrated by inoculation to potato that these mutations were responsible for the transitions to, or from, resistance breaking (Santa-Cruz, unpublished data).

The available data do not allow the firm conclusion that the coat protein, rather than the coat protein gene, is the viral determinant (Y, Figure 1) in the interaction with *Nx*. However, it is the most likely interpretation, as all of the mutations affecting the interaction are coding sequence changes. A protein-mediated effect would certainly be consistent with the available information about the structure of the PVX coat protein (Baratova *et al.*, 1992): the region around amino acids 62-78 implicated in the *Nx* interaction is one of the minor exposed domains in the PVX virion, as assessed by tritium planigraphy (Baratova *et al.*, 1992). There is, therefore, the potential for direct interaction of this region of the coat protein with the product of *Nx* (Figure 3). However, the structural information would also be consistent with the suggestion (Culver and Dawson, 1989; 1991), as for *N'* interactions of TMV, that the amino acid residues identified as affecting the resistance-breaking phenotype have a secondary effect on the structure of the coat protein or inter-subunit interactions in the virus particle (see also Dawson, this volume).

Other Resistance Gene Specificity Determinants: Not all resistance gene interactions involve virus coat proteins or coat protein genes. An analysis of hybrid PVX genomes incorporating elements of strains PVX_{cp2} and PVX_{cp4} indicates that interactions with *Nb* also involve sequences outside the coat protein gene. The PVX_{cp2} strain is from strain group 2 (Table 1) and gives an HR response on potato carrying the *Nb* gene (Jones, 1985). The PVX_{cp4} strain was derived from PVX_{cp2}, as described above for the transition from PVX_{DX} to PVX_{DX4}, but with selection by passage on potato carrying the *Nb* gene. The analysis of the PVX_{cp2}/PVX_{cp4} hybrids has so far implicated the region in the replicase gene encoded between nucleotides 291 and 1254 (Forsyth, unpublished data), although there is no information as to whether the effect is mediated by RNA or protein (Figure 2).

With TMV, deletion of the viral coat protein gene had no major effect on the ability to induce local lesions (Takamatsu *et al.*, 1987; Dawson *et al.*, 1986) on plants carrying *N* (distinct from *N'*) although the specificity determinant for *N* is not yet known. There is also evidence that TMV resistance genes in tomato, *Tm1* and *Tm2*, interact respectively with the viral replicase component (Meshi *et al.*, 1988) and the 30 kDa movement protein (Meshi *et al.*, 1989).

The identification of resistance gene specificity determinants of viral genomes is the first stage in the detailed characterization of how the resistance gene product recognizes the virus and brings about resistance. With the PVX systems described above, it will be possible to determine whether the interactions involve the virus encoded protein or viral RNA by introduction of silent mutations in the regions implicated in the resistance gene interactions. It will also be possible, having identified the gene controlling resistance-breaking, to determine whether, for the hypersensitive resistance genes, resistance-breaking is due to absence of an avirulence function or whether, in the resistance-breaking strains, there is the capability to suppress the primary interaction of the avirulence function; i.e., whether the resistance-breaking determinant is a virulence function. The analysis of the *N'* TMV resistance gene of tobacco resolved this issue by transgenic expression of the viral coat protein of different isolates of TMV in plants carrying *N'*: the plants expressing the coat protein of the resistance sensitive strains displayed a necrotic phenotype indicating that this protein was implicated in the elicitation of the HR response (Culver and Dawson, 1991). In the terms used to describe "gene for gene" interactions, the coat protein of certain strains of TMV is the avirulence determinant with respect to *N'* and is probably the viral component undergoing the primary interaction with the product of the resistance gene. Similar experiments to be carried out with the PVX coat protein gene and also the replicase component will clarify whether these are the avirulence determinants with respect to the *Nx* and *Nb* genes.

Further Analysis of the Rx-Mediated Resistance to PVX: Having identified the PVX coat protein as the probable Y factor (Figure 1) in the interaction with *Rx*, it was possible to determine, by deletion and dominance tests, whether *Rx* mediates induced or constitutive resistance. The identification of the coat protein determinant associated with the *Rx* breaking was surprising as it was known that the extreme resistance phenotype was expressed in protoplasts of resistant plants inoculated with PVX (Adams *et al.*, 1986; Saladrigas *et al.*, 1990). From this previous result it was assumed that *Rx* inhibited virus replication rather like *Tm1* from tomato, which confers resistance to TMV and which is influenced by viral sequences in the gene for the replicase (Tavantzis, 1990). The model proposed for the

a)

	group3	group4	residue
DX>DX4	Q	P	78
KD>KD76	M	I	76
KD4/62<KD4	V	I	62

b)

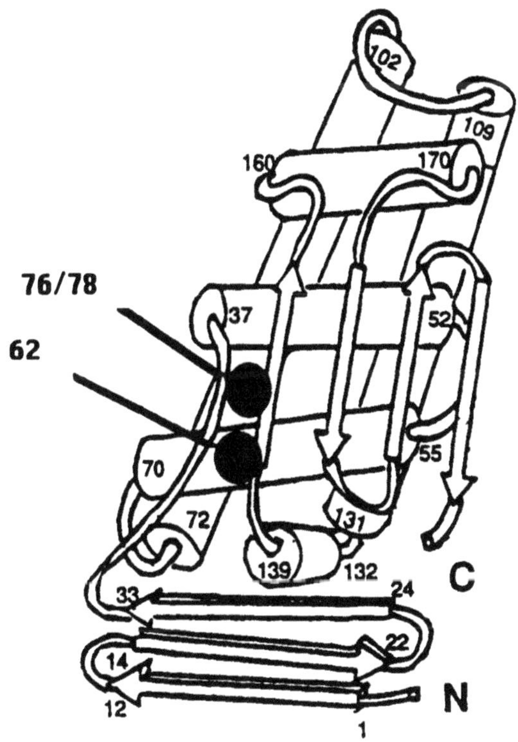

Figure 3. The *Nx* Specificity Determinants and the Structure of the PXV Coat Protein. A: Amino acid changes associated with the indicated transitions between PVX strain groups 3 and 4. B: The structure of the PVX coat protein gene taken from the model of Baratova *et al.* (1992) showing the location of the positions identified in A.

mode of action of *Tm1* involves a modified interaction between the TMV replicase subunit and the product of *Tm1*, resulting directly in inhibition of replicase activity (Meshi *et al.*, 1988). Recombination analysis of PVX_{UK3} and PVX_{HB}, pointing to the coat protein gene, clearly ruled out that idea for *Rx* (Kavanagh *et al.*, 1992); but one interpretation of recent data on the analysis of coat protein function in PVX suggested to us an alternative way in which *Rx* might bring about direct resistance against PVX. This new data was obtained from deletion analysis of the coat protein gene in PVX inoculated either to plants or protoplasts (Chapman *et al.*, 1992) and was interpreted in terms of a two phase infection cycle (Figure 4). The forms of PVX with coat protein mutations nearly all failed to accumulate on the inoculated leaves and in protoplasts failed to accumulate high levels of genomic RNA although the levels of subgenomic mRNA and (-)-ve strand RNA were not affected. The first phase of the infection cycle was, therefore, interpreted as involving production of the subgenomic mRNA's, the (-)ve strand RNA, and low level production of (+)ve strand RNA but without efficient movement between cells and independently of expression of the coat protein. The second phase of the infection in this model would be coat protein dependent, would lead to accumulation of high levels of all viral RNA's, and would be associated with efficient spread of the infection (Figure 4). As it was known that *Rx* prevented efficient accumulation of PVX genomic RNA, both in protoplasts and on the inoculated leaves (Adams *et al.*, 1986; Saladrigas *et al.*, 1990; Meshi *et al.*, 1989), it was possible that the product of *Rx* would interact with the coat protein and inhibit the transition between the first and second phases of the infection cycle as a variation on the constitutive resistance (Figure 1b) model.

A test of the involvement of *Rx* in this transition was a simple northern analysis of the RNA from protoplasts of *Rx* or *rx* potato after inoculation with PVX, in order to determine whether the inhibitory effect of Rx was specific to the (+)strand genomic RNA rather than the (-) strand and sub genomic RNA's. In fact, in *Rx* protoplasts, there was reduced accumulation of all of the PVX RNA's (Kōhm, unpublished data) which implies a block in the infection cycle at, or before, the first stage in the model given above (Figure 4, Kōhm, unpublished data) and makes out, therefore, that there is constitutive resistance through inactivation of the coat protein.

The combined data from the analysis of resistance-breaking determinants in PVX_{HB} and of PVX RNA accumulation in *Rx* protoplasts was, therefore, difficult to reconcile with a direct inhibition of the infection cycle by the *Rx* product interacting with either the replicase component or with the coat protein. We have, therefore, carried out a series of experiments designed to test whether *Rx* might mediate an induced type of resistance and whether the features identified through an analysis of resistance-breaking strains are the primary specificity determinants of the interaction with the *Rx* product. These tests were designed as virological parallels to the deletion and dominance tests described in the **Introduction** as being relevant to the analysis of the resistance specificity determinants in bacterial and fungal pathogens.

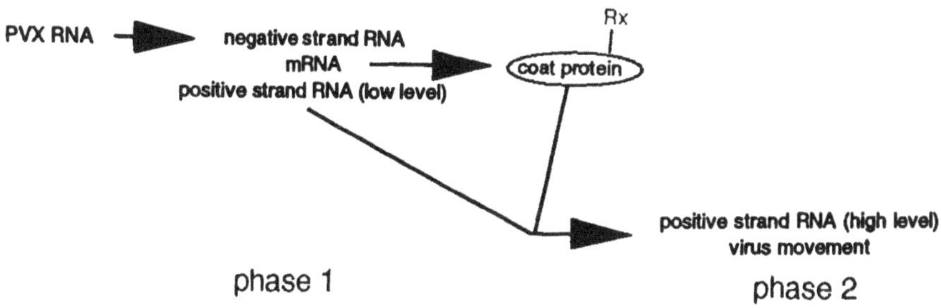

Figure 4. A Two Phase Model for the Infection Cycle of PVX. The model is based on mutation analysis of the coat protein gene (Chapman *et al.*, 1989) indicating the lack of requirement for the coat protein in phase 1 and a definite requirement for the coat protein for phase 2.

The deletion tests were carried out on the coat protein gene of PVX_{UK3} by infection of protoplasts. The indication that *Rx* initiates induced resistance and that the coat protein has the role of an avirulence determinant, in the sense of a "gene for gene" type of interaction, would be the loss of avirulence in the deletion mutant.

The results from this deletion analysis are summarized in Figure 5. Each of the coat protein mutant forms of the PVX_{UK3} genome shows an effect on genomic RNA accumulation in susceptible protoplasts due to the absence of a functional coat protein. The mutant form Δ5 with a deletion in the 3'-part of the coat protein gene had the same sensitivity to *Rx* as did the wild-type PVX_{UK3}. However, a form with a deletion in the 5'-part of the gene (Δ2) or a frame-shift mutation close to the initiation codon (Δ1) had both lost the ability to be resisted by *Rx*, indicating that the avirulence determinant is protein, rather than RNA, and encoded within the region of Δ2. It is noteworthy that amino acids identified by recombination analysis of PVX_{HB} and other strains of PVX had identified the same region of the gene (Goulden, unpublished data). We, therefore, conclude that the resistance-breaking phenotype of PVX_{HB} is due to the absence of a functional avirulence determinant in the coat protein of this strain and that avirulence to *Rx* in other strains of PVX is affected by the identity of amino acids in the region of Δ2.

In principle, it should be possible to carry out the dominance test of the PVX coat protein with a construct of PVX which is diploid and heterozygous for the coat protein genes from *Rx* sensitive and *Rx* breaking strains of PVX. However, there are likely to be complications associated with the use of such a construct due to the difference in expression levels from duplicated copies of the coat protein gene and the genetical stability of the modified virus (Chapman *et al.*, 1992). An alternative approach to the assay of dominance of the viral avirulence determinants involved co-inoculation of different strains of PVX into the same cell by electroporation of viral RNA at high concentration into potato protoplasts.

In these experiments, the two strains of PVX were PVX_{UK3} and PVX_{KH2}, a resistance-breaking derivative of PVX_{UK3} in which the coat protein gene of PVX_{HB} was substituted into the PVX_{UK3} genome (Kavanagh *et al.*, 1992). There was no effect of the strains on each other when inoculated to protoplasts of potato with the *rx* genotype, but in protoplasts with the *Rx* genotype there was a marked reduction in the accumulation of the PVX_{KH2} RNA when inoculated together with the PVX_{UK3} RNA (Kõhm, unpublished data). This was precisely as predicted in the model for induced-resistance which predicts that the simultaneous presence of the avirulence determinant and the resistance gene product would initiate the antiviral defense reactions. Equally conclusively, the data are not in line with the direct resistance model, which predicts either that the Y function of the resistance-breaking strain would complement the deficiency in the avirulent strain caused by the action of the resistance gene product or that there would be no effect of the avirulent strain on the resistance-breaking isolate.

Rx and PVX as a "Gene for Gene" Interaction: These several lines of evidence indicate that the interaction of PVX with the *Rx* gene product brings about induced resistance and are consistent with the "gene for gene" model proposed for resistance to fungal and bacterial pathogens of plants: the deletion and co-inoculation tests both show that a determinant in the viral coat protein elicited an antiviral mechanism in the infected cells. Neither of these tests, nor the analysis of viral RNA accumulation in protoplasts with the *Rx* genotype, was consistent with the alternative model which invokes a direct inactivating interaction of the resistance gene product and an essential (for the virus) viral gene product.

The induction of the *Rx*-mediated resistance reaction in protoplasts was apparently independent of the elicitation of a hypersensitive necrosis. However, this finding may not necessarily indicate that *Rx*-mediated resistance is fundamentally distinct from the more widely studied induced resistance interactions. There is evidence to indicate that, under certain environmental conditions or following graft inoculation, there is some limited PVX accumulation and necrosis in plants carrying *Rx* (von Delhey, 1974; Bagnall, 1961; Benson and Hooker, 1960). The potential in *Rx*-mediated resistance for the elicitation of HR may be limited under normal conditions by resistance mechanisms independent of HR and which keep virus accumulation in the inoculated cell below the level necessary for elicitation of HR.

The identification of viral determinants in the resistance interaction is, of course, only the first step in the understanding of resistance interactions. At the moment there is absolutely nothing known about the identity of the plant component involved in the primary recognition process and very little about the likely steps in the transduction pathway linking the recognition process with the ultimate resistance mechanism. With regard to the transduction pathways there is some indication from work with fungal elicitors that protein kinases may be involved in the intra-cellular events (Felix *et al.*, 1991) and from the analysis of systemic acquired resistance that salicylic acid and other small molecules may mediate inter-cellular

a)

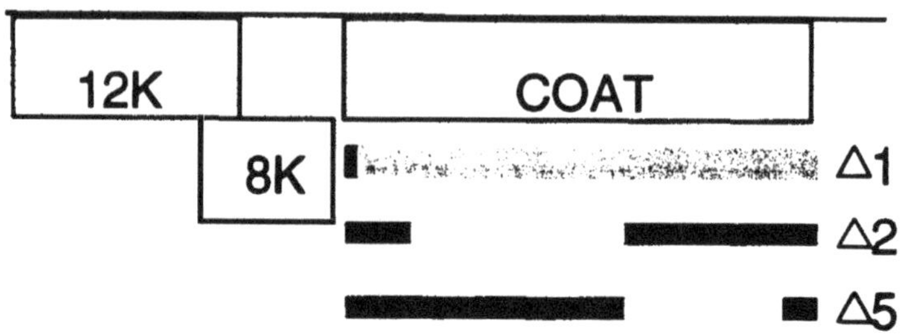

b)

	(+)-ve RNA		mRNA	
	Rx	*rx*	*Rx*	*rx*
Δ1	+	+	+++	+++
Δ2	+	+	+++	+++
Δ5	-	+	-	+++
WT	-	+++	-	+++

Figure 5. Deletion Analysis of *Rx* Specificity Determinants in the Coat Protein Gene of PVX_{UK3}. A: A scale diagram of the coat protein gene of PVX indicating the mutations used for analysis of *Rx* determinants. Portions of the mutant coat protein gene in frame with the wild-type (WT) gene are shown in black. The shaded region indicates where the coat protein has been put out of frame by the mutation. Δ1, Δ2 and Δ5 are different mutants (Tavantzis, 1990). B: The phenotype of PVX_{UK3} and mutant derivatives in protoplasts of *Rx* or *rx* potato indicating the levels of accumulation of various viral RNA's. The relative levels of viral RNA accumulation are in the ratio 100:10:<5 (indicated as +++:+:-).

signalling (Enyedi *et al.*, 1992). The potential to investigate a resistance in protoplasts, as with the *Rx*-mediated phenomenon, will allow relatively easy testing of the role of various biochemical processes in the resistance interaction using exogenously applied inhibitors. The *Rx* system may have some features which will facilitate the isolation of the product of the resistance gene and its gene. For example, if genome map-based methods of gene isolation are to be used, it may well be possible to build on the identification of the position of *Rx* on the genetic map of potato (Ritter *et al.*, 1991) by exploiting the significant progress made to date in the characterization of the tomato genome which is largely co-linear with the potato genome (Bonierbale *et al.*, 1988). There may also be the possibility of identification of the *Rx* product through its potential to interact with the viral coat protein.

References

Adams, S.E., Jones, R.C.A., and Coutts, R.H.A. 1986. Expression of potato virus X resistance gene Rx in potato leaf protoplasts. J. Gen. Virol. 67:2341-2345.

Bagnall, R.H. 1961. Recovery of virus X from leaves of graft-inoculated immune potato plants. Phytopath. 51:338-340.

Baratova, L.A., Grebenshchikov, N.I., Dobrov, E.N., Gedrovich, A.V., Kashirin, I.A., Shishkov, A.V., Efimov, A.V., Järvekülg, L., Radavsky, Y.L., and Saarma, M. 1992. The conformation of coat proteins of potato virus X and their packaging arrangements in the virus particles studied by tritium planigraphy, immunological analysis and model building. Virology 188:175-180.

Benson, A.P. and Hooker, W.J. 1960. Isolation of virus X from "immune" varieties of potato, *Solanum tuberosum*. Phytopath. 50:231-234.

Bonierbale, M.W., Plaisted, R.L., and Tanksley, S.D. 1988. RFLP maps based on a common set of clones reveal modes of chromosomal evolution in potato and tomato. Genetics 120:1095-1103.

Chapman, S., Hills, G., Watts, J., and Baulcombe, D. 1992. Mutational analysis of the coat protein gene of potato virus X: Effects on virion morphology and viral pathogenicity. Virology 194:223-230.

Chapman, S., Kavanagh, T., and Baulcombe, D. 1992. Potato virus X as a vector for gene expression in plants. The Plant J. 2:549-557.

Chapman, S.N. 1991. A molecular analysis of potato virus X. Ph.D. Thesis Cambridge University, United Kingdom.

Cockerham, G. 1970. Genetic studies on resistance to potato viruses X and Y. Heredity 25:309-348.

Culver, J.N. and Dawson, W.O. 1989. Tobacco mosaic virus coat protein: An elicitor of the hypersensitive reaction but not required for the development of mosaic symptoms in *Nicotiana sylvestris*. Virology 173:755-758.

Culver, J.N. and Dawson, W.O. 1991. Tobacco mosaic virus elicitor coat protein genes produce a hypersensitive phenotype in transgenic *Nicotiana sylvestris* plants. Mol. Plant-Microb. Interact. 4:458-463.

Dawson, W.O., Bubrick, P., and Grantham, G.L. 1986. Modifications of the tobacco mosaic virus coat protein gene affecting replication, movement, and symptomatology. Mol. Plant Pathol. 78:783-789.

Enyedi, A.J., Yalpani, N., Silverman, P., and Raskin, I. 1992. Signal molecules in systemic plant resistance to pathogens and pests. Cell 70:879-886.

Felix, G., Grosskopf, D.G., Regenass, M., and Boller, T. 1991. Rapid changes of protein phosphorylation are involved in transduction of the elicitor signal in plant cells. Proc. Natl. Acad. Sci. U.S.A. 88:8831-8834.

Fraser, R.S.S. 1990. The genetics of resistance to plant viruses. Annu. Rev. Phytopathol. 28:179-200.

Hemenway, C., Weiss, J., O'Connell, K., and Tumer, N.E. 1990. Characterization of infectious transcripts from a potato virus X cDNA clone. Virology 175:365-371.

Jones, R.A.C. 1982. Breakdown of potato virus X resistance gene Nx: Selection of a group four strain from strain group three. Plant Pathol. 31:325-331.

Jones, R.A.C. 1985. Further studies on resistance-breaking strains of potato virus X. Plant Pathol. 34:182-189.

Kavanagh, T., Goulden, M., Santa-Cruz, S., Chapman, S., Barker, I., and Baulcombe, D. 1992. Molecular analysis of a resistance-breaking strain of potato virus X. Virology 189:609-617.

Keen, N.T. 1990. Gene-for-gene complementarity in plant-pathogen interactions. Annu. Rev. Genet. 24:447-463.

Meshi, T., Motoyoshi, F., Adachi, A., Watanabe, Y., Takamatsu, N., and Okada, Y. 1988. Two concomitant base substitutions in the putative replicase genes of tobacco mosaic virus confer the ability to overcome the effects of a tomato resistance gene, *Tm-1*. The EMBO J. 7:1575-1581.

Meshi, T., Motoyoshi, F., Maeda, T., Yoshiwoka, S., Watanabe, H., and Okada, Y. 1989. Mutations in the tobacco mosaic virus 30 kD protein gene overcome *Tm-2* resistance in tomato. The Plant Cell 1:515-522.

Moreira, A. and Jones, R.A.C. 1980. Properties of a resistance-breaking strain of potato virus X. Ann. Appl. Biol. 95:93-103.

Ponz, F., Glascock, C.B., and Bruening, G. 1988. An inhibitor of polyprotein processing with the characteristics of a natural virus resistance factor. Mol. Plant Microb. Interact. 1:25-31.

Ritter, E., Debener, T., Barone, A., Salamini, F., and Gebhardt, C. 1991. RFLP mapping on potato chromosomes of two genes controlling extreme resistance to potato virus X (PVX). Mol. Gen. Genet. 227:81-85.

Saito, T., Meshi, T., Takamatsu, N., and Okada, Y. 1987. Coat protein gene sequence of tobacco mosaic virus encodes a host response determinant. Proc. Natl. Acad. Sci. U.S.A. 84:6074-6077.

Saladrigas, M.V., Ceriani, M.F., Tozzini, A.C., Arese, A.I., and Hopp, H.E. 1990. Potato gene X confers inoculum dependent resistance to potato virus X replication in protoplasts. Plant Cell Physiol. 31:90-97.

Takamatsu, N., Ishikawa, M., Meshi, T., and Okada, Y. 1987. Expression of bacterial chloramphenicol acetyltransferase gene in tobacco plants mediated by TMV-RNA. The EMBO J. 6:307-311.

Tavantzis, S.M. 1990. Outlook on the molecular biology of virus resistance in potato. In: *The Molecular Biology of the Potato*. M.E. Vayda and W.D. Park (eds.), pp. 113-135. CAB International.

von Delhey, R. 1974. Zur natur der extremen virusresistenz bei der kartoffel. Phytopath Z. 80:97-119.

Note: Some of the experiments described in this paper have been described in the following papers:

Goulden, M.G., Köhm, B.A., Santa Cruz, S., Kavanagh, T.A., and Baulcombe, D.C. 1993. A feature of the coat protein of potato virus X affects both induced virus resistance in potato and viral fitness. Virology 197:293-302.

Goulden, M.G. and Baulcombe, D.C. 1993. Functionally homologous host components recognize potato virus X in *Gomphrena globosa* and potato. Plant Cell 5:921-930.

Köhm, B.A., Goulden, M.G., Gilbert, J.E., Kavanagh, T.A., and Baulcombe, D.C. 1993. A potato virus X resistance gene mediates an induced, nonspecific resistance in protoplasts. Plant Cell 5:913-920.

Santa Cruz, S. and Baulcombe, D.C. 1993. Molecular analysis of potato virus X isolates in relation to the potato hypersensitivity gene Nx. Mol. Plant-Microbe Interact. 6:707-712.

PLRV Gene Expression in Potato

E. Tacke, B. Kull, D. Prüfer, S. Reinold, J. Schmitz, F. Salamini, and W. Rohde
Max-Planck-Institut für Züchtungsforschung
50829 Köln
Germany

The RNA genome of potato leafroll luteovirus (PLRV) codes for six large open reading frames (ORF's) divided into two gene clusters by an intergenic region of 201 nucleotides. The 3'-cluster is expressed via a subgenomic RNA (sgRNA1) and encodes the capsid protein CP (ORF3), a single-stranded nucleic acid-binding 17 kDa protein (pr17; ORF4) and a putative aphid transmission factor of 56 kDa (pr56; ORF5). Among these proteins, pr17 has attracted most of our attention, as it shows properties of a transport protein (homodimer formation, nucleic acid binding) and might, therefore, play an important role during PLRV multiplication. High-level expression of pr17 wild-type or mutant proteins in transgenic plants was considered to interfere with viral replication according to the concept of pathogen-derived resistance. Consequently, transgenic potato lines were established that express mutant pr17 and tested for induced resistance to PLRV. Grafting experiments showed that lines expressing modified pr17 are resistant to virus infection. These results are interpreted as the possible function of a pr17 "dominant suppressor" gene in that modified pr17 preexisting in the transgenic plant and wild-type pr17 of the challenging virus cannot form a functional multimeric complex necessary for virus spread and/or replication. These results provide another example of genetically engineered virus resistance based on the expression of a nonstructural protein.

Introduction

Studies in the expression of viral genes and the physiological role of the corresponding gene products for viral replication, the induction of pathogenicity mechanisms, interaction with host resistance genes or with transmissibility by an insect vector demand for a variety of experimental strategies. These include:

(1) The synthesis of infectious full-length cDNA copies of the viral genome in conjunction with site-specific mutations;

(2) The transient and stable *in vivo* expression of single genes, gene clusters, or regulatory regions; or

(3) The *in vitro* expression of viral proteins and analysis of their biochemical properties.

With potato leafroll virus (PLRV) we are using these three strategies for a functional analysis of various viral gene products. Like other members of the luteovirus group, PLRV is transmitted by aphids in a persistent manner and replicates specifically in the phloem tissue (Francki *et al.*, 1985). This specificity was originally observed by electronmicroscopical detection of virus particles (Francki *et al.*, 1985), the end product of the replication cycle, and could be verified by *in situ* hybridization (Reinold, 1989) (Figure 1).

In thin sections from PLRV-infected plants of *Solanum tuberosum* or the experimental host *Physalis floridana* ^{35}S-labeled antisense RNA detected viral (+) sense RNA only in the phloem tissue (Figure 1, Lower panels). The PLRV (-) sense RNA strand, the template for the synthesis of subgenomic RNA as well as progeny genomic RNA, is present in amounts not detectable by *in situ* hybridization with a (+) riboprobe (Figure 1, Middle panels). At the macroscopic level, PLRV infections result in a reduced growth rate and the yellowing and rolling of leaves (Figure 5, Panel E).

The virus particles contain a single-stranded (+) sense genomic RNA of about 5.8 kb linked to a small, virus-coded protein (VPg) (Harrison, 1984; Mayo *et al.*, 1989). Sequence analysis of genomic RNA's for four PLRV isolates (Mayo *et al.*, 1989; van der Wilk *et al.*, 1989; Keese *et al.*, 1990; Kull, 1992) has revealed six large open reading frames (ORF1, ORF2a, ORF2b, ORF3, ORF4, ORF5) which are organized into two gene clusters separated by an intergenic region (Figure 2). Genes located in the 3'-cluster are the three large open reading frames ORF3, ORF4, and ORF5 translated from subgenomic sgRNA1 (Tacke *et al.*, 1990) as depicted in Figure 2. The 5'-gene cluster contains an open reading frame (ORF1) of unknown function followed by ORF2a and ORF2b, two large ORF's possibly representing the catalytically active part of the viral replicase complex.

Among the translational strategies used for the synthesis of PLRV proteins are internal initiation, ribosomal frameshifting, and UAG amber stop codon suppression. ORF2a, for example, is synthesized *in planta* from genomic RNA by internal translation initiation *in planta* in amounts comparable to those of ORF1 (unpublished data), while ORF2b is produced at a frequency of only 1 percent as an ORF2a/2b transframe protein by -1 ribosomal frameshifting. A similar expression level is found for ORF5 synthesized from sgRNA1

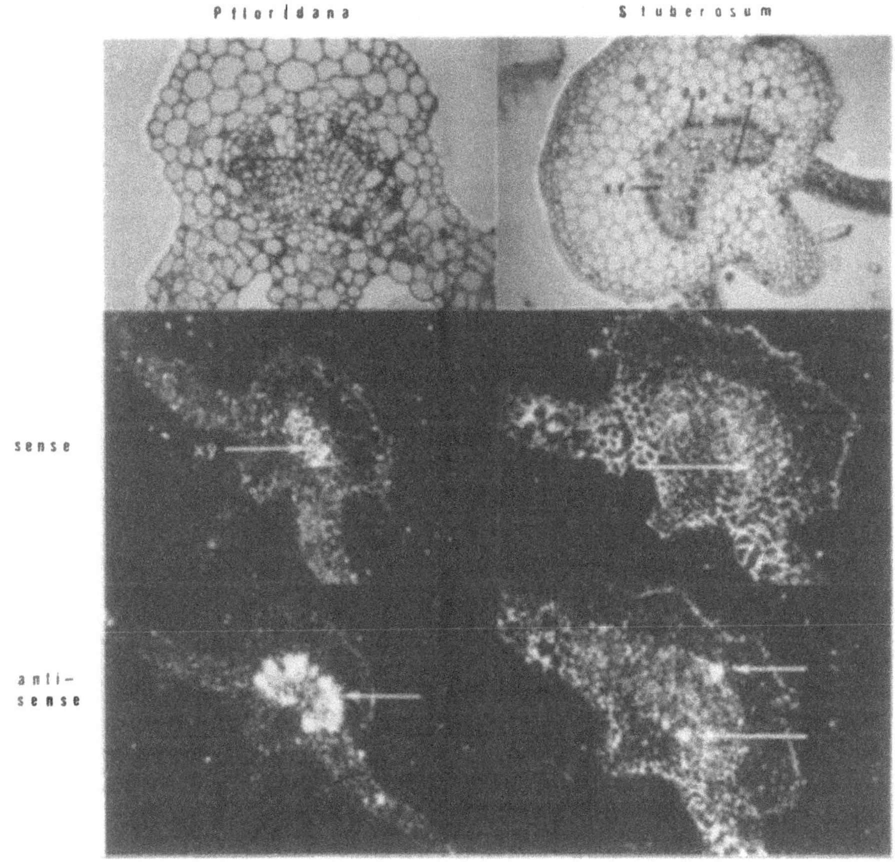

Figure 1. ***In situ*** **Hybridization of PLRV-Infected *S. tuberosum* and *P. floridana*. Upper panels: Cross-sections were taken from methacrylate-embedded mature foliage leaves of PLRV-infected *P. floridana* and *S. tuberosum* plants and stained with toluidine blue to visualize tissue structures. Inner (ip), outer phloem (op), and xylem tissue (xy) are indicated. Middle panels: Hybridization of paraplast-embedded thin sections with sense riboprobe. Only background hybridization signals are visible. Autofluorescence of xylem tissue under UV is indicated by an arrow. Lower panels: Hybridization with an antisense PLRV riboprobe detects PLRV-specific RNA only in the inner and outer phloem as indicated by arrows.**

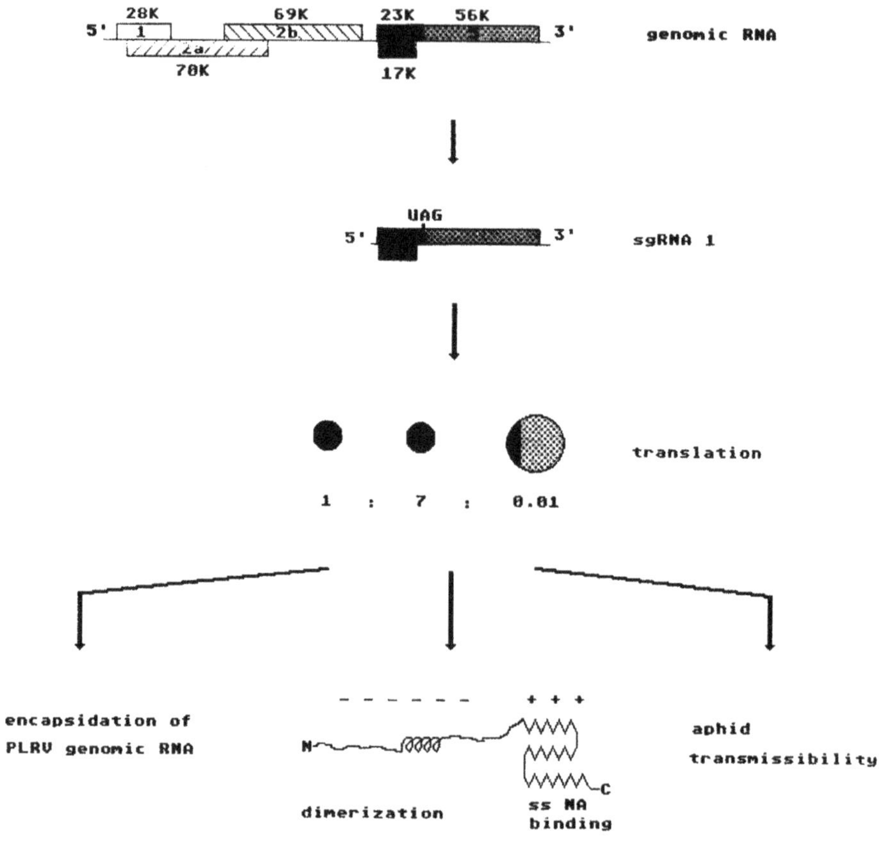

Figure 2. Schematic Representation of the PLRV Genome Organization and the Expression of the 3'-Gene Cluster. The 3'-located gene cluster consists of the 23 kDa coat protein CP gene (ORF3) which is separated by an UAG stop codon from a gene coding for a 56K protein (ORF5) expressed by suppression of the amber stop codon. The gene coding for the 17K nucleic acid-binding protein (ORF4) is nested within the CP gene. Relative translation efficiencies of these genes from sgRNA1 as determined by *in vitro* and *in vivo* experiments are indicated together with the known (23K) or presumed (56K) physiological functions of their gene products. For the 17K protein the identified biochemical functions are indicated.

as an inframe protein with CP (ORF3) by UAG suppression (Tacke *et al.*, 1990). In contrast to the relative rate of ORF1/ORF2a synthesis, translation efficiencies for the pr17 gene (ORF4, Figure 2) which is nested within the CP gene and translated by internal initiation, exceed that of the CP gene by a factor of seven (Tacke *et al.*, 1990); and the 17K protein accumulates to high levels in the PLRV-infected host plant (Tacke *et al.*, in preparation).

In vitro studies on the biochemical properties of the 17K protein had shown its single-stranded nucleic acid-binding activity (Tacke *et al.*, 1991) which resides in the basic C-proximal part of the pr17 molecule. The acidic aminoterminal region contains an amphipathic helix which confers to the protein the capacity to form homodimers and higher oligomers (Tacke *et al.*, in preparation). Such covalently linked dimers resistant to monomerization by treatment with mercapto reagents were also isolated *in vivo* from transgenic plants expressing pr17 (Tacke *et al.*, in preparation). In addition, pr17 produced *in planta* had the ability to bind nucleic acids thereby confirming the *in vitro* experiments. These properties of pr17 (homodimer formation, nucleic acid binding) are characteristic for several families of transcriptional activators. The high level of pr17 expression in the infected host was considered unusual for a solely regulatory protein, and it was suggested (Tacke *et al.*, in preparation) that pr17 might participate as a transport protein in PLRV multiplication.

According to the concept of pathogen-derived resistance (Sanford and Johnston, 1985), expression of an essential gene of the pathogen might interfere with its replication. For several viruses, this concept was successfully applied by the expression in transgenics of wild-type or deletion mutant replicases (Braun and Hemenway, 1992; MacFarlane and Davies, 1992; Golemboski *et al.*, 1990). As a second approach genetic cross protection by the expression of capsid protein genes in transgenic plants (Beachy *et al.*, 1990; Hanley-Bowdoin and Hemenway, in press) has been known since several years to induce virus resistance. Following these concepts we established transgenic potato lines that express the PLRV CP or 17K protein genes. Here we report on several aspects of PLRV gene expression in the potato.

Experimental

Construction of Vectors for the Expression of PLRV CP and 17K Genes in Transgenic Potato: For high-level expression of the PLRV capsid protein, an oligodeoxynucleotide representing the tobacco mosaic virus (TMV) omega sequence with a synthetic *Xho*I restriction site at the 5'-end and a *Sac*I restriction site at the 3'-end was annealed to the complementary oligodeoxynucleotide according to the protocol (Altered sites™ *in vitro* mutagenesis system; Promega). After restriction with the appropriate enzymes the resulting DNA fragment was ligated into a pUC-based vector carrying a 35S CaMV turbo promoter and terminator (Tacke *et al.*, in preparation).

The PLRV CP gene was amplified from cDNA clone pCPL1 (Tacke *et al.*, 1989) by the polymerase chain reaction (PCR) using primers 36 nucleotides in length and homologous to the 5'-(sense primer) or complementary to the 3'-end of the gene (antisense primer). The sense primer was used to optimize the sequences surrounding the CP start codon for translation initiation (ACAATGGC) (Lütcke *et al.*, 1987) and to delete by point mutation the translational start codon(s) of the 17K gene without altering the amino acid sequence of the CP gene (AUG --> ACG). Amplified DNA fragments were cloned into the vector described above between the TMV omega sequence and the 35S CaMV terminator.

The gene for the 17K protein was amplified as described elsewhere (Tacke *et al.*, in preparation) and cloned into the vector pRT103 (Töpfer *et al.*, 1987) restricted by *Xho*I and treated with Klenow polymerase to produce flush ends. Construction of a plant expression cassette for pr17 modified by six c-terminal histidine codons is described elsewhere (Tacke *et al.*, in preparation). For the expression of a pr17/GUS fusion protein, the PLRV 17K gene was amplified by PCR with a 36mer primer homologous to the 5'-end of the gene and containing a 5'-located *Xho*I restriction site. A PCR primer complementary to the 3'-end of the gene was used to add a *Bcl*I restriction site. The resulting PCR DNA fragment was restricted with the appropriate enzymes and ligated into the *Bcl*I site of the vector pRT104GUS (Töpfer *et al.*, 1988) which resulted in translational fusion of pr17 to the N-terminal region of the GUS gene.

Plant Transformation: Ligation of plant expression cassettes into the binary vector pBIN19 (Bevan, 1984) and subsequent transformation of *S. tuberosum* var. *Desiree* followed protocols described elsewhere (Tacke *et al.*, in preparation).

Northern Blot Analysis: Extraction of total plant RNA and Northern blot analysis was described previously (Tacke *et al.*, 1990). The ^{32}P-labelled 2.2 kb *Eco*RI insert of plasmid pCPL1 (Tacke *et al.*, 1989) was used for detection of CP mRNA's and the 3.6 kb *Bam*HI insert of plasmid pTA250.2 (Shirzadegan *et al.*, 1991) was used for the detection of ribosomal RNA.

Measurement of GUS Activity: Activity of the GUS enzyme was determined by a fluorimetric assay (Jefferson, 1987) as mean value of six independent measurements. GUS activity was normalized with respect to background activity which had a mean value of 22 pmol MU formed/mg protein/min.

In situ Hybridization: Leaf sections of healthy or PLRV-infected *S. tuberosum* and *P. floridana* plants were processed for *in situ* hybridization by fixation and paraplast embedding according to Meyerowitz (1987). Cross-sections of 8 to 10 um thickness were prepared with a microtome type 1165/rotocut (Reichert-Jung) and treated with pronase before hybridization with ^{35}S-labeled sense and antisense riboprobes (Schmelzer *et al.*, 1988).

Results and Discussion

Construction of a PLRV CP Gene Expression Cassette: A binary plant vector cassette for the expression of the PLRV CP gene was constructed using the CaMV 35S promoter and terminator as regulatory sequences. As multimerization of the CaMV 35S enhancer leads to a linear increase in transcriptional efficiency (Fang *et al.*, 1989), a turbo CaMV promoter with four copies of the enhancer (Tacke *et al.*, in preparation) was used. Also the TMV omega sequence which increases translational efficiency by 2 to 20fold (Wilson *et al.*, 1990) was included in the construct. Additionally, the CP translational start codon was optimized for translation initiation according to published sequences (Lütcke *et al.*, 1987); and the AUG start codon(s) of the 17K gene were eliminated by mutation to ACG, thus leaving the amino acid sequence of the coat protein unchanged (Figure 3, Construct 1).

Construction of PLRV 17K Gene Expression Cassettes: For the induction of virus resistance according to the concept of pathogen-derived resistance, binary vectors containing the wild-type 17K gene or 17K derivatives were constructed (Figure 3, Constructs 2-4). Derivatives of pr17 ($pr17_{mut}$) were expected to retain their ability for protein/protein interactions due to an undisturbed amphipathic helix and, thus, could possibly form pr17/$pr17_{mut}$ heterodimers with wild-type pr17 upon PLRV infection of transgenic potato, thereby converting pr17 into a complex unable to exert its physiological role during virus replication. These $pr17_{mut}$ genes would thus act as dominant suppressor genes during viral replication. As only the intact

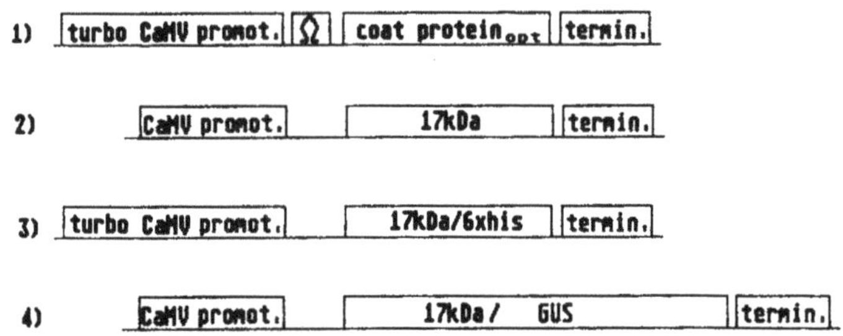

Figure 3. Construction of Plant Expression Cassettes. The various open reading frames (ORF's) are transcriptionally controlled either by the 35S CaMV promoter (CaMV promot.) or a modified version with four copies of the enhancer region (turbo CaMV promot.) and the 35S CaMV terminator (termin.). Construct 1: For the enhancement of translation the TMV omega sequence (Ω) was inserted into the expression cassette. Sequences surrounding the CP translational start codon gene were modified for optimal translation initiation in plants (coat protein$_{opt}$). Construct 2: The 17K gene without the CP translational start codons (17kDa). Construct 3: The 17K gene modified at its 3'-terminus by the addition of six histidine codons (17kDa/6xhis). Construct 4: The 17K gene translationally fused at its 3'-end to the 5'-end of the GUS gene (17kDa/GUS).

pr17 protein without internal deletions is able to dimerize (Tacke *et al.*, in preparation), derivatives of pr17 were engineered at the carboxy-terminus of the protein. For this, six histidine codons were fused to the 3'-end resulting in the 17K/his gene (Figure 3, Construct 3). The corresponding protein would be able to chelate divalent metal ions, thus possibly interfering with the nucleic acid-binding activity of the pr17 carboxy-terminal half. In addition, this pr17 derivative could be isolated from transgenic lines by affinity chromatography on Ni^{2+}/chelate columns (Tacke *et al.*, in preparation). In another construct, the GUS gene was translationally fused to the carboxy-terminus of pr17 (Figure 3, Construct 4) resulting in a chimeric pr17/GUS protein of approximately 87 kDa. Expression of this gene in transgenic plants was expected to be monitored by the GUS activity of the chimeric protein, as was previously shown for the PLRV CP/GUS protein (unpublished results).

Analysis of Transgenic Potato Plants Expressing the PLRV CP Gene: Transformation of the tetraploid potato variety *Desiree* followed established protocols (Tacke *et al.*, in preparation). After transfer of individual kanamycin-resistant plants to the greenhouse, Northern blot analyses were performed on total RNA's. Electrophoretically separated RNA was transferred to a nylon membrane and hybridized with a radiolabelled PLRV CP probe. Integrity of the RNA was examined by hybridization with a rRNA probe.

Eight out of thirteen independent transformants showed the presence of variable amounts of PLRV CP RNA (Figure 4). Phenotypic differences of these plants as compared to the pBIN-transformed control plants were not visible. The eight plants positive for CP gene expression were tested for the presence of coat protein by Western blot analysis. Although CP gene expression was driven by the CaMV turbo promoter and translation enhanced by the presence of the TMV omega sequence on the transcript and although the CP AUG start codon was optimized for translation initiation, no coat protein could be detected (data not shown). Thus, resistance to PLRV was first evaluated with the 17K transgenic plants (see below).

Analysis of Plants Transformed With the PLRV 17K Gene: Potato leaf disc transformation was done with the binary vector pBIN19 harbouring expression cassettes for the PLRV wild-type 17K gene, the 17K/his or the 17K/GUS genes; and 10 plants each were regenerated and further analyzed. Plants expressing 17K/GUS were assayed for gene expression by monitoring GUS activity. Only a single plant (P13-8) showed GUS activity, although at a very low level (280 pmol MU formed/mg protein/min). This low GUS activity can be explained from comparisons with transgenic potato lines expressing the acidic N-terminal half (97 amino acid residues) or the basic C-proximal part (93 amino acid residues) of the 17K protein as GUS fusion proteins. Mean values (four individual transformants each) for GUS activity ranged from 9843 (N-terminus) to 5370 (C-terminus) pMol MU formed/mg protein/min (unpublished data). Thus, fusion of the full-length 17K protein (156 amino acid residues) to GUS results in a dramatic decline of GUS enzyme activity of the fusion protein; and similar data on the influence of N-terminal GUS extensions on enzyme activity have been obtained for ORF1/GUS fusions (unpublished data). In contrast, the full-length CP/GUS protein shows an only marginal reduction of GUS activity when compared to CP deletion variants fused to the GUS gene (Tacke *et al.*, 1990; unpublished data).

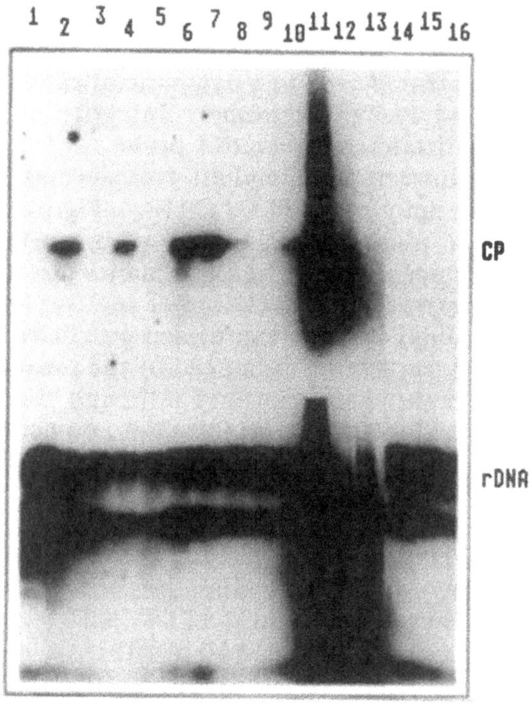

Figure 4. Northern Blot Analysis of Total RNA From Plants Transformed With the PLRV CP Gene. Ten ug of total, glyoxylated RNA's from 13 *S. tuberosum* plants transformed with the PLRV CP gene (Construct 1, Figure 3) were separated on 1.2 percent agarose gels (lanes 1-13). As a control RNA of control plants transformed with the binary vector pBIN19 alone was loaded (lanes 14-16). The gel was blotted on duplicate filters and hybridized with ^{32}P-labelled DNA representing the PLRV CP gene (CP) or ribosomal sequences (rRNA).

Furthermore, the P13-8 plant showed phenotypic abnormalities like reduced growth rate, leaf rolling, pale green coloration, and formation of stem-borne tubers (Figure 5, Panel B). Progenies originating from this plant by cuttings (C1 generation) developed a close to normal growth pattern, when compared to pBIN-transformed control plants; and no areal tubers were formed. Leaves, however, showed a typical curling and a dark green/pale green variegation pattern. This chimera phenotype was explained by the somatic inactivation of the 17K/GUS gene in the potato LII layer. In several angiosperms the shoot apex is organized in layers. The shoot apex of potato consists of three layers; LI and LII, constituting the tunica; and LIII, the corpus. All three layers participate to the formation of the leaf primordium. The LIII layer, however, contributes only to the part of the leaf included between the epidermis (derived from LI) and the subepidermis (derived from LII). The LIII derived leaf tissues, moreover, do not extend along all leaf lamina: they occupy only a central fraction of the leaf. If the transgenic tissues react to the expression of the foreign gene with typical symptoms, then somatic inactivation of the 17K/GUS gene in LII would result in the variegated phenotype (Figure 5 C) which has a normal growth in the dark green leaf areas derived from LII and a reduced growth in the pale green central regions of LIII origin. This assumption was substantiated by the finding that in leaves of the C1 plants GUS activity was shown only in the central pale green regions (Figure 5 D). Northern analyses and *in situ* hybridization experiments are under way to further support this finding. We conclude that the low percentage of transformed, kanamycin-resistant plants expressing GUS activity is a result of very early inactivation of the 17K/GUS gene leading to transformed plants without GUS expression. Due to the uniqueness and characteristics of the single plant expressing the 17K/GUS gene, this gene was not considered to be a good choice for the induction of resistance to PLRV infection.

Potato plants transgenic for the 17K/his gene (Figure 3, Construct 3) under the transcriptional control of the CaMV turbo promoter were monitored for gene expression by Western blot analysis with polyclonal antisera raised against bacterially expressed pr17 (Tacke *et al.*, in preparation). One plant (P17-1) was found to express 17K/his at high levels (Tacke *et al.*, in preparation). A second independent transformant (P17-2) accumulated 7K/his to low levels. P17-2 had a normal phenotype as compared to control plants transformed with pBIN19 or with the GUS gene alone (Figure 5, Panel A). P17-1 showed at first reduction in growth rate, but 10 weeks after transfer to the greenhouse it could no longer be distinguished phenotypically from control plants.

Resistance tests were initiated with these as well as appropriate control plants by grafting shoots from PLRV-infected potato plants onto cuttings of P17-1 and P17-2 as well as on pBIN-transformed plants. Western blot analysis of transgenic tissues using PLRV CP antisera indicated that P17-2 and P17-1 showed resistance to PLRV, and further experiments are under way to substantiate these results. The high-level expression of the nonstructural 17K protein in PLRV-infected plants might argue for multiple functions of this protein during the viral replication cycle. Due to its capactiy for homodimerization and binding to single-stranded nucleic acids (Tacke

Figure 5. **Phenotypes of Transformed and PLRV-Infected Plants. Panel A: Transformed plants 6 weeks after transfer to the greenhouse. From left to right: plants (P17-2 and P17-1) transformed with Construct 3 (Figure 3), a plant transformed with pBIN19 without an expression cassette as control, and a plant with the GUS expression cassette of pRT104GUS as an additional control for the 17K/GUS transformed plant (Panel B). Panel B: Plant 13-8 transgenic for Construct 4 (Figure 3) and with gene activity in the GUS assay. Panel C: First generation of plants (C1) regenerated from cuttings made from the transformant P13-8. Panel D: Leaves from the C1 plants shown in Panel C. The inner pale green area showed GUS activity, whereas in the outer dark green regions only background activity could be detected. Panel E: Phenotypes of non-infected (left) and PLRV-infected (right) *P. floridana* plants.**

et al., 1990; Tacke *et al.*, in preparation), pr17 might protect viral RNA in the plant cell before virus morphogenesis takes place. Additionally, it may have transport functions during the multiplication of viral RNA's. The capsid proteins of tobacco streak and alfalfa mosaic virus are examples of viral proteins performing dual functions. They not only package viral RNA into progeny particles, but they are also necessary for infection by forming specific protein/RNA interactions and activating replication complexes (Jaspars, 1985; van Vloten-Doting, 1975).

In conclusion, the data reported here suggest that expression in transgenic potato of a nonstructural protein like the PLRV 17K protein that exhibits characteristics of a transport protein may result in induced virus resistance. Induced resistance mediated by pr17 would substantiate the usefulness of the concept of pathogen-derived resistance.

Acknowledgements

The technical assistance of A. Kaufmann and B. Förster is gratefully acknowledged. This work was supported by grants from Agrobios Metapontum and Deutsche Forschungsgemeinschaft to W. Rohde.

References

Beachy, R.N., Loesch-Fries, S., and Tumer, N.E. 1990. Coat protein-mediated resistance against virus infection. Ann. Rev. Phytopath. 28:451-474.

Bevan, M. 1984. Binary *Agrobacterium* vectors for plant transformation. Nucleic Acids Res. 12:8711-8721.

Braun, C.J. and Hemenway, C.L. 1992. Expression of amino-terminal portions or full-length viral replicase genes in transgenic plants confers resistance to potato virus X infection. Plant Cell 4:735-744.

Fang, R.-X., Nagy, F., Sivasubramaniam, S., and Chua, N.-H. 1989. Multiple *cis* regulatory elements for maximal expression of the cauliflower mosaic virus 35S promoter in transgenic plants. Plant Cell 1:141-150.

Francki, R.I.B., Milne, R.G., and Hatta, T. 1985. *Atlas of Plant Viruses*. Vol. 1, pp. 137-152. CRC Press, Boca Raton, Florida.

Golemboski, D.B., Lomonossof, G.P., and Zaitlin, M. 1990. Plants transformed with a tobacco mosaic virus nonstructural gene sequence are resistant to the virus. Proc. Natl. Acad. Sci. U.S.A. 87:6311-6315.

Hanley-Bowdoin, L. and Hemenway, C.L. 1992. Expression of plant viral genes in transgenic plants. In: *Genetic Engineering with Plant Viruses*. T.M.A. Wilson and J.W. Davies (eds.), pp. 251-295. CRC Press, Boca Raton, Florida, Ann Arbor, London, Tokyo.

Harrison, B.D. 1984. CMI/AAB Descriptions of Plant Viruses. No. 291.

Jaspars, B.M.J. 1985. *Molecular Plant Virology.* J.W. Davies (ed.), pp. 155-221. CRC Press, Boca Raton, Florida.
Jefferson, R.A. 1987. Assaying chimeric genes in plants: The GUS gene fusion system. Plant Mol. Biol. Rep. 5:387-405.
Keese, P., Martin, R.R., Kawchuk, L.M., Waterhouse, L.M., and Gerlach, W.L. 1990. Nucleotide sequences of an Australian and a Canadian isolate of potato leafroll luteovirus and their relationships with two European isolates. J. Gen Virol. 71:719-724.
Kull, B. 1992. Die *in vitro* Synthese infektiöser RNA des Potato Leafroll Virus. Diploma thesis. University of Cologne.
Lütcke, H.A., Chow, K.C., Mickel, F.S., Moss, K.A., Kern, H.F., and Scheele, G.A. 1987. Selection of AUG initiation codons differs in plants and animals. EMBO J. 6:43-48.
MacFarlane, S. and Davies, J.W. 1992. Plants transformed with a region of the 201K replicase gene from pea early browning virus RNA1 are resistant to virus infection. Proc. Natl. Acad. Sci. U.S.A. 89:5829-5833.
Mayo, M.A., Barker, H., Robinson, D.J., Tamada, T., and Harrison, B.D. 1982. Evidence that potato leafroll virus RNA is positive-stranded, is linked to a small protein, and does not contain polyadenylate. J. Gen. Virol. 59:163-167.
Mayo, M.A., Robinson, D.J., Jolly, C.A., and Hyman, L. 1989. Nucleotide sequence of potato leafroll luteovirus. J. Gen. Virol. 70:1037-1051.
Meyerowitz, E.M. 1987. *In situ* hybridization to RNA in plant tissue. Plant Mol. Biol. Rep. 5:242-250.
Reinold, S. 1989. Diploma thesis. University of Cologne.
Sanford, J.C. and Johnston, S.A. 1985. The concept of pathogen derived resistance: Deriving resistance genes from the parasite's own genome. J. Theor. Biol. 113:395-405.
Schmelzer, E., Jahnen, W., and Hahlbrock, K. 1988. *In situ* localization of light-induced chalcone synthase mRNA, chalcone synthase, and flavonoid end products in epidermal cells of parsley leaves. Proc. Natl. Acad. Sci. U.S.A. 85:2989-2993.
Shirzadegan, M., Christie, P., and Seemann, J.R. 1991. An efficient method for isolation of RNA from tissue cultured plant cells. Nucleic Acids Res. 19:6055.
Tacke, E., Prüfer, D., Salamini, F., and Rohde, W. 1990. Characterization of a potato leafroll luteovirus subgenomic RNA: Differential expression by internal translation initiation and UAG suppression. J. Gen. Virol. 71:2265-2272.
Tacke, E., Prüfer, D., Schmitz, J., and Rohde, W. 1991. The potato leafroll luteovirus 17K protein is a single-stranded nucleic acid-binding protein. J. Gen. Virol. 72:2035-2038.
Tacke, E., Sarkar, S., Salamini, F., and Rohde, W. 1989. Cloning of the gene for the capsid protein of potato leafroll virus. Archives Virol. 105:153-163.
Tacke, E., Schmitz, J., Prüfer, D., and Rohde, W. 1993. Mutational analysis of the nucleic acid-binding 17 kDa phosphoprotein of potato leafroll luteovirus identifies an amphipathic α-helix as the domain for protein/protein interactions. Virology 197:274-282.
Töpfer, R., Matzeit, V., Gronenborn, B., Schell, J., and Steinbiss, H.-H. 1987. A set of plant expression vectors for transcriptional and translational fusions. Nucleic Acids Res. 15:5890.

Töpfer, R., Pröls, M., Schell, J., and Steinbiss, H.-H. 1988. Transient gene expression in tobacco protoplasts: II. Comparison of the reporter gene systems for CAT, NPTII, and GUS. Plant Cell Rep.7:225-228.

van der Wilk, P., Huisman, M.J., Cornelissen, B.J.C., Huttinga, H., and Goldbach, R. 1989. Nucleotide sequence and organization of potato leafroll virus genomic RNA. FEBS Letters 245:51-56.

van-Vloten-Doting, L. 1975. Coat protein is required for invectivity of tobacco streak virus: Biological equivalence of the coat proteins of tobacco streak and alfalfa mosaic virus. Virology 65:215-225.

Wilson, T.M.A., Saunders, K., Dowson-Day, M.J., Sleat, D.E., Trachsel, H., and Mundry, K.W. 1990. Effects of the 5'-leader sequence of tobacco mosaic virus RNA, or derivatives thereof, on foreign mRNA and native vrial gene expression. *Post-Transcriptional Control of Gene Expression.* J.E.G. McCarthy and M.F. Tuite (eds.), pp. 261-275. Springer-Verlag, Berlin, Heidelberg.

Resistance to Bean Yellow Mosaic Virus (BYMV) and Other Potyviruses in Transgenic Plants Expressing BYMV Antisense RNA, Coat Protein, or Chimeric Coat Proteins

J. Hammond and K. K. Kamo
Florist and Nursery Crops Laboratory
Agricultural Research Service, USDA
Beltsville Agricultural Research Center
Beltsville, Maryland 20705

Nicotiana benthamiana expressing BYMV antisense (AS) RNA from the 3'-half of the coat protein (CP) gene, with the complete 3'-non-coding sequence and a short poly-A tract, were resistant to BYMV but not to pepper mottle virus (PeMV) or turnip mosaic virus (TuMV). Either of two BYMV-CP constructs also conferred resistance to BYMV, but not PeMV or TuMV. Inoculation with BYMV RNA did not overcome AS resistance. Two types of resistance were observed for both AS and CP constructs. Some transformants resisted infection, while others became infected but resisted replication and/or transport of BYMV; both types of resistance were combined in other transformants. One AS transformant was apparently immune to BYMV. Chimeric CP constructs, with N-terminal domains from BYMV and C-terminal sequence from PeMV or zucchini yellow mosaic virus, each conferred resistance to BYMV; the BYMV/PeMV chimera also conferred partial resistance to potato virus Y and TuMV.

Introduction

Coat protein (CP)-mediated resistance has been demonstrated to be effective against viral challenge for several of the major groups of RNA plant viruses (reviewed by Beachy) (Beachy *et al.*, 1990). The mechanisms of CP-mediated protection are somewhat different for the different groups and are, therefore, not universal. The differences are not correlated with virus architecture; RNA of the rod-shaped tobamovirus TMV (Nelson *et al.*, 1987; Register and Beachy, 1988), tobacco rattle tobravirus (Angenent *et al.*, 1990) (TRV), and of the quasi-isometric alfalfa mosaic virus (Van Dun *et al.*, 1987) (AMV) is able to overcome the protection conferred against virions while the CP's of the rod-shaped potexvirus potato virus X (Hemenway *et al.*, 1988) (PVX) and carlavirus potato virus S (Mackenzie and Tremaine, 1990) (PVS) were found to protect against RNA inoculation as well. There are at least two separate mechanisms for CP-mediated resistance to TMV, as systemic spread of TMV in CP-expressing plants is reduced even when the initial resistance to infection is overcome by RNA inoculation (Wisniewski *et al.*, 1990). Similar distinctions between resistance to initial infection and symptom amelioration have been reported for plants expressing the CP's of the potyviruses papaya ringspot virus (Ling *et al.*, 1991) (PRV), plum pox virus (Regner *et al.*, 1992; Lindbo and Dougherty, 1992b) (PPV), tobacco etch virus (Lindbo and Dougherty, 1992a) (TEV), and bean yellow mosaic virus (Hammond and Kamo, 1993; Hammond and Kamo, 1994) (BYMV).

Antisense (AS) RNA has been demonstrated to be an effective means of regulating gene expression (e.g., Oeller *et al.*, 1991), but there have been mixed results when AS RNA has been tested for control of plant viruses. A part of this difference is probably because viruses replicate; and, therefore, a more complete repression is required for control. AS RNA of cucumber mosaic virus (Cuozzo *et al.*, 1988) (CMV), PVX (Hemenway *et al.*, 1988), and TMV (Powell *et al.*, 1989) was found less effective than CP-mediated protection, while Rezaian *et al.* (1988) found that only a single transformant of one of three CMV AS constructs conferred any protection at all. In contrast, AS RNA of the potyviruses TEV (Lindbo and Dougherty, 1992a; 1992b) and BYMV (Hammond and Kamo, 1993) were found able to confer protection equivalent or superior to that from CP expression. Dual mechanisms of resistance to BYMV (to initial infection and to systemic accumulation or movement) were also observed in plants expressing AS RNA from BYMV (Hammond and Kamo, 1993).

We are examining the mechanisms of resistance to BYMV and other potyviruses conferred by AS RNA and variants of the CP gene (Hammond and Kamo, 1993; Hammond and Kamo, 1994). By use of deleted and chimeric constructs, we expect to be able to separate the two types of resistance observed and to determine the types of interactions involved. *N. benthamiana* was chosen as the model system because it is susceptible to BYMV and a large number of other potyviruses (Quacquarelli and Avgelis, 1975; Christie and Crawford, 1978), thus allowing us to challenge with both the homologous virus and heterologous viruses that naturally infect a spectrum of crop types. Additional differences in the virus specificity of the protection from different constructs will offer further insights into the mechanisms of resistance.

Experimental

Gene Constructs: The CP gene of BYMV isolate GDD was previously cloned as pBY9 (Hammond and Hammond, 1989). For the AS RNA, the BYMV insert of pBY6 was used; this has about two-thirds of the CP gene, the complete 3'-non-coding sequence, and 19 residues of the poly(A) tract (Hammond and Hammond, 1989). For CP expression, the coding sequence was fused to the 3'-non-coding sequence by ligation of clones pBY6 and pBY9 at an internal Dde 1 site, creating pBY96; which contains the CP gene, complete 3'-non-coding sequence, and 19 residues from the poly(A) tract. Two derivatives of pBY96 were produced to provide leader sequences and translational initiation codons in a context functional in plants (Lutcke *et al.*, 1987), as potyvirus CP's are normally cleaved by a viral protease from a polyprotein precursor (Dougherty and Carrington, 1988). Construct pBY.CP has a portion of the 5'-leader sequence, initiation codon, and 13 additional codons from the nopaline synthase gene fused in-frame to a Sau3A site in the second codon of the BYMV CP gene; the CP product of pBY.CP, and plants expressing it, are designated CP-1. A second construct, pBY.oCP, has a short synthetic leader and initiation codon derived from a pair of partially complementary oligonucleotides, fused to the same Sau3A site; the CP product of pBY.oCP, and plants expressing it, are termed CP-2. Carboxy-terminal deletion mutants of the BYMV CP gene were derived by Bal 31 digestion of pBY96; an oligonucleotide with termination codons in all three reading frames was then introduced to replace the deleted viral termination codon and minimize non-viral coding sequence (Hammond *et al.*, 1990). Each of the deleted CP constructs was then placed behind the leader sequence of pBY.oCP. Chimeric CP genes were made by in-frame fusions of the N-terminal coding sequence of pBY.oCP with the C-terminal sequence (including the non-coding region) of pepper mottle virus (Dougherty *et al.*, 1985a) (PeMV; from W.G. Dougherty) or zucchini yellow mosaic virus (Grumet and Fang, 1990) (ZYMV; from R. Grumet) at common restriction sites; these CP constructs were designated BY/Pe and BY/ZY, and plants expressing them BY/Pe and BY/ZY respectively.

Plant Transformation and Assay: Each of the gene constructs was transferred to the binary vector pGA643, between the cauliflower mosaic virus 35S promoter and *Agrobacterium* gene 7 transcription terminator (An *et al.*, 1988). The binary vector was then transformed into *Agrobacterium tumefaciens* strain 5922 (LBA4404 containing Ti helper plasmid PC2760) and used to transform cotyledons or primary leaf disks of *N. benthamiana*, essentially as described (Horsch *et al.*, 1985). Selection of transformed shoots was in the presence of 300ug/ml kanamycin, 100ug/ml cefotaxime, and 1mg/ml carbenicillin. Shoots were transferred to rooting medium containing kanamycin and carbenicillin, and rooted cuttings transferred to Promix; rooted plants were grown either in a growth chamber or greenhouse. Leaf tissue samples were taken during tissue culture, and again from rooted plants, for detection of the transgene and its expression. CP plants (CP-1, CP-2, BY/Pe, and BY/ZY) were analyzed by PCR, essentially as described (McGarvey and Kaper, 1991) except that PCR products were analyzed on acrylamide gels; and by ELISA with monoclonal antibodies (MAb's) PTY 1 and PTY 43 (Jordan and Hammond, 1991). AS plants were analyzed by PCR. In each case, the upstream PCR primer was within the CaMV 35S promoter (JH 006, 5' CCTTCGCAAGACCCTTCC 3'); for each of the CP constructs a degenerate reverse primer (JH 008, 5' T[AC]A[CGA][TA]GCTGC[TCG]GC[CT]TTCAT 3', where degeneracies are indicated inside[]) was designed to account for differences in the CP sequences of BYMV, PeMV, and ZYMV. The AS reverse primer (JH 010, 5' GCAAAGCCAACATTCCGC 3') is BYMV-specific. The PCR products of CP-1 and CP-2 are easily distinguished by size, owing to the additional leader and coding sequence in CP-1 (data not shown). The appropriate pGA643 plasmid derivative was used as a positive control for PCR product size, and a DNA extract of healthy non-transgenic plants as a negative control for all amplifications.

The original transformants (R0) were allowed to self, and the R1 progeny analyzed by PCR and/or ELISA, as appropriate; in some instances seed were also germinated in the presence of 300ug/ml kanamycin. Putative homozygous R1 seedlings were selfed to obtain R2 seed, and R2 populations tested for homozygosity by PCR and/or ELISA and/or kanamycin resistance. R2 seed were also collected from some AS(-) R1 segregants for use as additional negative controls. All challenge inoculations were made to putative homozygous R2 populations.

Virus and RNA Purification, Inoculation, and Assay: Turnip mosaic virus (TuMV) isolate Linc-2 (Hammond and Chastagner, 1988), potato virus Y isolate PVY-3, PeMV isolate NC 165, BYMV-GDD, and BYMV isolate Ideal A were purified, and RNA of BYMV-GDD purified as described (Hammond and Lawson, 1988); RNA was diluted in sterile water to 50ug/ml, and 50ul inoculated to a single bentonite-dusted leaf per plant with a sterile glass rod. Purified virus was diluted to 100, 20, or 2ug/ml, and either a pinch of Celite added to the virus solution, or leaves were dusted with Celite or bentonite; 50ul of the appropriate diluted virus preparation was used to inoculate each of three expanded leaves of plants at the 8-10 leaf stage. Four to eight plants of each population were inoculated for each treatment, including non-transgenic controls. Non-inoculated plants of each transformant and non-transgenic plants served as negative controls and for controls for the expression of the transgene CP.

Plants were observed daily for symptom development for 28 days after inoculation and were then tested by ELISA essentially as described (Jordan and Hammond, 1991) with MAb's PTY 1, PTY 10, PTY 43 (Jordan and Hammond, 1991), and TBV 27C2H2 (Hsu *et al.*, 1988) as appropriate to confirm the diagnosis of symptoms and the identity of the causal virus (Table 1). BYMV symptom severity was expressed as the average score of the inoculated plants on the following scale: 0, no symptoms; 1, vein-clearing; 2, interveinal chlorosis (IC), with or without recovery from symptoms in upper leaves; 3, mild mosaic (mm), mm + IC, or mosaic + IC; 4, mm/mosaic; 5, strong mosaic. Plants were scored on the predominant or emerging symptom type. Additional ELISA tests were performed on leaves from lower, middle, and upper positions on the main stem of selected plants. Extracts of some upper leaves were bioassayed by inoculation to non-transgenic *N. benthamiana*. In some experiments, BYMV was also detected by dot-blot hybridization with a riboprobe transcribed from clone pBYr5 (Hammond and Hammond, 1989) subcloned into pGEM7Zf(+). Clone pBYr5 does not cross-hybridize to the CP gene or 3'-non-coding region, being derived from an upstream portion of the genome (J. Hammond, unpublished).

Table 1. Reactions of the Monoclonal Antibodies with the Potyvirus Isolates Used.

Monoclonal Antibody	Virus				
	BYMV-GDD	BYMV Ideal A	PeMV NC 165	PVY-3	TuMV Linc
PTY 1	+++	+++	+++	+++	+++
PTY 43	+++	-	-	-	-
PTY 10	-	-	+++	++	-
TBV 27C2H2	-	-	-	-	+++

The amount of CP produced in transgenic plants was compared by ELISA to the level in an active BYMV infection; for some transformants the level of expression was analyzed by leaf position to determine whether expression levels differed with leaf age or position. The CP from the transgene was specifically identified by use of MAb PTY 43, which reacts only with the CP of BYMV-GDD and not the isolate BYMV-Ideal A which was used to challenge CP-expressing plants; MAb PTY 1 reacts with both isolates of BYMV, as well as PeMV, ZYMV, PVY, and TuMV (Jordan and Hammond, 1991) (Table 1). The epitope with which MAb PTY 43 reacts is in the N-terminal domain of BYMV-GDD (Jordan and Hammond, 1991; Jordan, 1992) and is included in all of the CP constructs.

Results & Discussion

Transformation, Gene Detection, and Inheritance: At least 50 individual putative transformants of each gene construct were selected on the basis of kanamycin resistance; after further selection by PCR and/or ELISA some regenerants were discarded as apparent chimeras or false positives. Plants which were of normal appearance and set adequate numbers of seed to allow determination of R1 segregation ratios were further examined. Most R1 populations segregated in an approximately 3:1 ratio of expressors:non-expressors (by PCR and/or ELISA, and germination in the presence of kanamycin), suggesting a single insertion of the T-DNA carrying the BYMV gene. Others segregated approximately 15:1 in the R1 generation and, thus, apparently had two independent insertions of the T-DNA. A few R1 populations did not segregate in a simple Mendelian fashion and were not examined further. R2 populations in which each of 24 seedlings tested positive for CP or AS were selected for virus challenge.

In general, CP expression was readily detectable by ELISA in seedlings; the range of CP accumulation in individual transformants was estimated to be from 0.1 percent to 10 percent of the accumulation in an active virus infection. The variation between comparable leaves of R2 plants of a single transformant was within a factor of three. When different leaf positions were analyzed in plants 6 to 8 weeks after transplanting, in most cases the upper leaves were found to have higher CP levels than lower leaves (Table 2), and overall levels of CP were lower than in younger plants. No CP could be detected in some transformants at this time (Table 2), although readily detectable in younger seedlings.

Challenge Inoculation of AS Plants: The R2 progeny of four separate transformants were examined (AS 30, AS 42, AS 112, and AS 140); an AS(-) segregant R2 population of one transformant (AS 30) was included as an additional control. In initial experiments, several populations of each transformant were challenged with BYMV; results from each population were similar, so a single population was used for all subsequent tests.

Table 2. CP Expression Relative to Leaf Position, 6-8 Weeks After Transplanting.

Plant	Leaf A[a]	Leaf B	Leaf C	Leaf D	CP level[b]
AS 140	0.03[c]	0.02	0.03	0.03	-
CP-1 30	0.03	0.03	0.04	0.05	nld[d]
CP-1 33	0.32	0.21	0.08	0.05	<1%
CP-1 79	0.39	0.23	0.14	0.09	<1%
CP-2 37	0.22	0.19	0.14	0.11	<1%
CP-2 44	0.91	0.74	0.52	0.23	1-2%
BY/Pe 9	0.25	0.30	0.23	0.28	<1%
BY/Pe 25	0.05	0.03	0.03	0.04	nld
BY/Pe 30	1.18	1.48	1.44	1.11	2-5%
BY/Pe 46	0.67	0.85	0.39	0.27	1-2%
BY/ZY 7	0.04	0.04	0.05	0.05	nld
BY/ZY 34	0.04	0.03	0.04	0.04	nld
BY/ZY 43	0.42	0.27	0.14	0.11	<1%

[a] Leaf A was the uppermost expanded leaf, with leaf B,C, and D each respectively two nodes further down the main stem.

[b] Approximate level of CP expression at time of testing compared to an active BYMV infection; higher levels were detected in younger plants of some transformants.

[c] Value are absorbance at 405 nm.

[d] nld = CP no longer detectable at significant level, although readily detectable in young seedlings.

One transformant, AS 140, was not infected by any concentration of purified virus tested, up to 100ug/ml. No symptoms were observed in the inoculated plants, nor was any virus detectable by ELISA (Figure 1), bioassay or dot-blot hybridization from the challenged plants. This transformant, thus, has apparent immunity to infection by BYMV; and we believe this is the first case of an AS RNA conferring immunity to a virus and certainly any plant virus.

For the other three transformants, all of the inoculated plants became infected, with symptoms initially indistinguishable from the non-transgenic control plants and the non-expressing R2 population of AS 30 (Figure 1). For one of these three transformants, AS 112, the only distinction was a slight delay in symptom expression and severity; however, virus titer was not significantly different from the controls. For AS 30, symptom development paralleled the controls until about 2 weeks after inoculation, when subsequently developing leaves showed a novel symptom type never observed in non-transgenic plants. The newly formed upper leaves of these plants did not develop the typical mottle/mosaic pattern; but instead chlorosis was confined to the area between the major veins; and the tissue either side of the veins was as green as the uninoculated controls. The interveinal chlorosis became progressively more restricted until at 4 weeks after inoculation the upper leaves had a few chlorotic flecks on an otherwise normal leaf. Virus titer in these leaves was significantly less than leaves from equivalent positions on the control plants. For AS 42 plants are similar, but more complete recovery from symptom expression occurred; plants did not develop the full severity of systemic symptoms, and upper leaves recovered completely from symptom expression. Virus titer in leaves displaying interveinal chlorosis was markedly reduced; and no virus was detectable by ELISA (Figure 1), dot-blot hybridization, or bioassay from the symptomless upper leaves. The pattern of recovery observed, and the novel symptom type that always preceded full recovery from symptoms, are strongly suggestive of the pattern of expression of the CaMV 35S promoter (Jefferson *et al.*, 1987). The first signs of recovery were green vein-banding, resulting in restriction of chlorosis to the lamina between the major veins; as recovery progressed, the chlorosis became progressively localized within the lamina, and the virus titer was reduced. The initial phase of recovery may be reduced viral replication around the vascular tissue, resulting in reduced pools of virus to be transported to the developing leaves; alternatively, there may be a direct inhibition of transport into the vascular system. In either case, a reduction in the amount of virus available for transport would result in a reduced quantity of virus reaching the developing leaves. Movement out of the vascular bundle may also be inhibited.

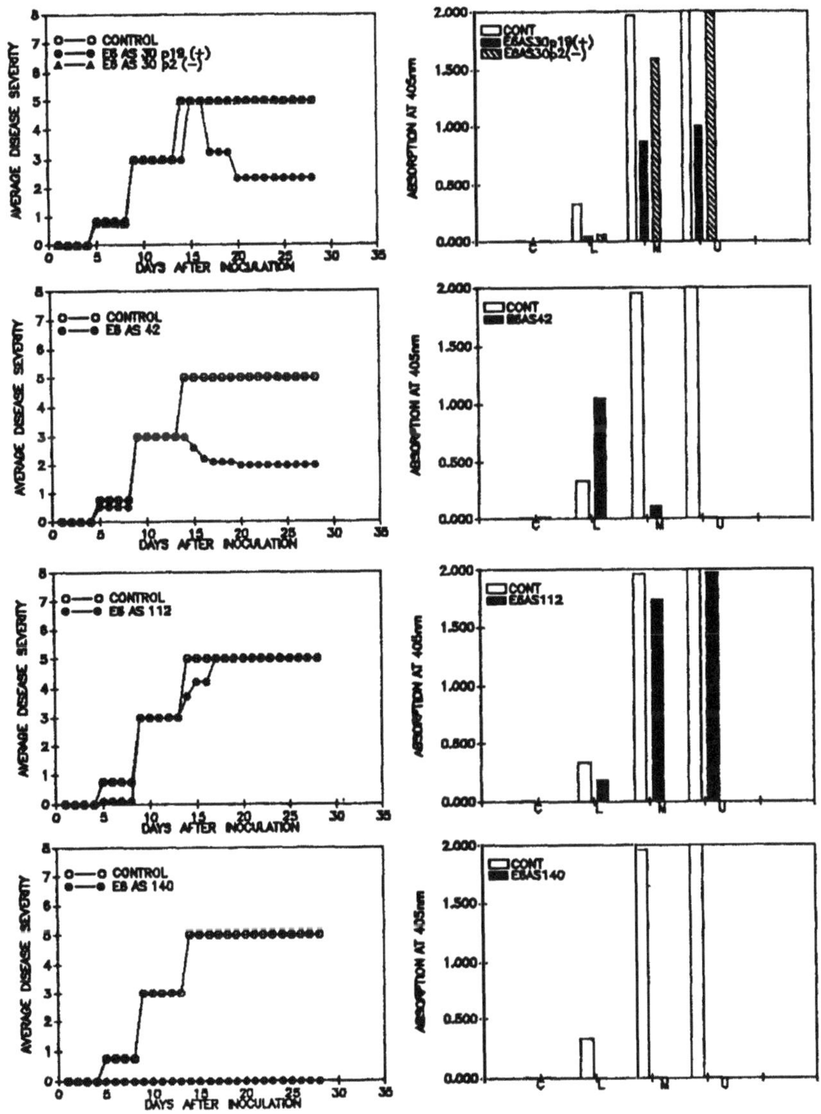

Figure 1. **Symptom Severity (left panels) and Relative Virus Titer (right panels) in AS Plants and Controls Inoculated with 100ug/ml BYMV Ideal A. Open circles (bars) represent non-transgenic controls, and filled circles (bars) the transgenic lines. AS(-) segregant AS 30 R2 population 2 is shown by triangles and hatched bars (top panels). Un-inoculated control (C), and lower (L), middle (M), and upper (L) leaves of representative inoculated plants were tested by ELISA.**

With reduced levels of virus reaching developing tissues, the balance between replication and inhibition of either replication or movement is tipped towards resistance; and in due time total recovery may result, as observed in AS 42.

Inoculation of the AS transformants with purified RNA yielded equivalent results to virus inoculation - that is, absence of any infection in AS 140, infection of AS 112, partial recovery from symptoms in AS 30, and total recovery from infection in later growth of AS 42 (data not shown). When each of the AS transformants was challenged with PeMV or TuMV at 100ug/ml, no delay of symptom expression was observed, nor any apparent symptom moderation. This suggests, as might be expected, that the mechanisms of AS protection for potyviruses are through direct RNA:RNA interaction. The lack of resistance to PeMV or TuMV also would be expected for an RNA:RNA interaction, as there is only limited sequence homology between BYMV and these viruses at the nucleotide level.

The AS construct used contains the complete 3'-non-coding sequence as well as a portion of the coding region from the CP gene. There are two obvious and distinct possibilities for AS mechanisms: 1) by annealing to incoming viral RNA, the portion from the CP coding region has the potential to disrupt translation of the CP region of the genomic polyprotein; and 2) the 3'-non-coding region may block binding of the viral polymerase and, thus, prevent formation of the (-) strand template for replication. Another possible mechanism of AS action that has been postulated is the rapid degradation of RNA:RNA duplexes (Crowley *et al.*, 1985); but the stability of dsRNA and its consequent utility as a diagnostic tool argue against this being the case, although the single-stranded portion of a viral:AS hybrid might be more susceptible to ribonuclease attack. The blocking of replicase activity has greater potential for complete inhibition of infection; if any replication takes place there will be an increase in the number of genomic RNA molecules to be inhibited. This may explain the superior resistance of AS 140 (apparent immunity to BYMV infection) and the complete recovery seen with AS 42, over the partial resistance and reduction in TEV titer with a TEV AS RNA that did not include the 3'-non-coding region (Lindbo and Dougherty, 1992a). To distinguish between translational inhibition and replicase blocking we have now sub-divided the construct into the coding region (able to inhibit translation and potentially to limit polymerase procession) and the non-coding region (able to affect polymerase binding but not translation). Removal of the 3'-terminal 117 nucleotides from an TMV AS construct that conferred a low level of resistance abolished the protection (Powell *et al.*, 1989). We predict that the non-coding region of our current effective AS construct will prove to be more effective than the CP region.

BYMV Challenge Inoculation of CP-1 and CP-2 Plants: The R2 progeny of two separate transformants of CP-1 (CP-1 30 and CP-1 79), and three of CP-2 (CP-2 12, CP-2 37, and CP-2 44) were challenged with BYMV Ideal A at different concentrations. When two populations of CP-1 30 were challenged with 100ug/ml of BYMV, only 3 of 8 plants (pop. G), and 7 of 8 plants (pop. H) became infected, initially showing symptoms indistinguishable from the controls. About 17 days post inoculation, however, all infected plants began to recover from symptoms, developing the same type of interveinal chlorosis in upper leaves as observed on the AS transformants AS 30 and AS 42. The CP-1 30 populations recovered totally from symptoms in later growth; and no virus was detected in the symptomless tissue by ELISA, dot-blot hybridization, or bioassay. At the end of the 4 week assay period transgene CP expression was not significantly detectable in the recovered plants (see also Table 2 for lack of detectable CP in older uninoculated plants) but was detectable in the uninoculated controls and the uninfected plant of population H (Figure 2); the ELISA reaction was due to the transgene CP and not challenge virus, because the reactions of MAb's PTY 1 (reactive with both BYMV-GDD and BYMV Ideal A) and PTY 43 (reactive only with the transgene CP, BYMV-GDD; see Table 1) were very similar, and bioassay was negative. In an experiment where CP-1 30 pop. G was challenged with either 2ug/ml or 20ug/ml BYMV, none of the inoculated plants became infected. Thus, this transformant showed both resistance to infection and recovery from infection once the initial resistance was overcome. The second CP-1 transformant, CP-1 79, became infected by 2ug/ml of BYMV, but plants began to show the interveinal chlorosis instead of mottle/mosaic starting at about 11 days post-inoculation (dpi) and symptomless upper leaves from about 15 dpi. In this instance, resistance to replication and/or movement has been separated from resistance to infection.

Similar separations of the two types of resistance were observed in CP-2 transformants (Figure 2). CP-2 12 plants showed resistance to infection (only four of eight plants infected by 100ug/ml BYMV), but no reduction in symptom severity or virus titer in the infected plants. CP-2 37 plants all became infected, followed by symptom amelioration without full recovery; the virus titer in leaves showing interveinal chlorosis was not significantly reduced. CP-2 44 plants all became infected, developed interveinal chlorosis, and subsequently symptomless leaves; no virus was detectable by ELISA (Figure 2) or recoverable by bioassay from the symptomless leaves. These plants thus showed no resistance to infection, but did show resistance to replication or systemic movement. The additional 14 amino acids from nopaline synthase present in CP-1 did not apparently have any adverse affect on resistance. A PPV CP construct with additional amino acids from the NIb gene did not confer any protection (Regner *et al.*, 1992). The reason for this difference is not obvious.

Heterologous Challenges and Chimeric CP's: CP-1 and CP-2 plants were also challenged with PeMV and TuMV at 100ug/ml. In this experiment, no evidence was obtained for protection against either PeMV or TuMV at this high inoculum concentration, although resistance to, or recovery from, infection with 100ug/ml of BYMV was observed.

In additional experiments, CP-1 (CP-1 30 and CP-1 79), BY/Pe (BY/Pe 9, BY/Pe 25, and BY/Pe 30), and BY/ZY (BY/ZY 7 and BY/ZY 43) plants were challenged with 2ug/ml and 20ug/ml each of BYMV Ideal A, PVY, and TuMV. Some plants in these experiments were lost due to a TMV contamination, so the results are incomplete and must be regarded as preliminary; the TMV contamination was not present in any of the viral inocula, as no TMV local lesions were induced in *Nicotiana glutinosa* plants, nor local or systemic symptoms in *Nicotiana tabacum* cv. Burley 21 inoculated with more concentrated inocula from the same viral preparations.

No plants of CP-1 30 were infected by BYMV, and CP-1 79 plants became infected but recovered from infection in the upper leaves (no symptoms or ELISA-detectable virus; Table 3). Neither CP-1 30 nor CP-1 79 showed any resistance to PVY or TuMV.

BY/ZY 7 plants showed minimal resistance to BYMV; one plant of five inoculated with 2ug/ml developed a milder mosaic than the other plants and the non-transgenic controls, and ELISA from lower, middle, and upper leaves suggested that there was a delay in viral accumulation (Table 3). BY/ZY 43 plants were infected by both BYMV concentrations, but plants recovered through the same progression of decreasing interveinal chlorosis to symptomless and virus-free upper leaves (Table 3) as observed with AS, CP-1, and CP-2 plants. Neither BY/ZY 7 nor BY/ZY 43 plants showed significant resistance to PVY or TuMV (data not shown).

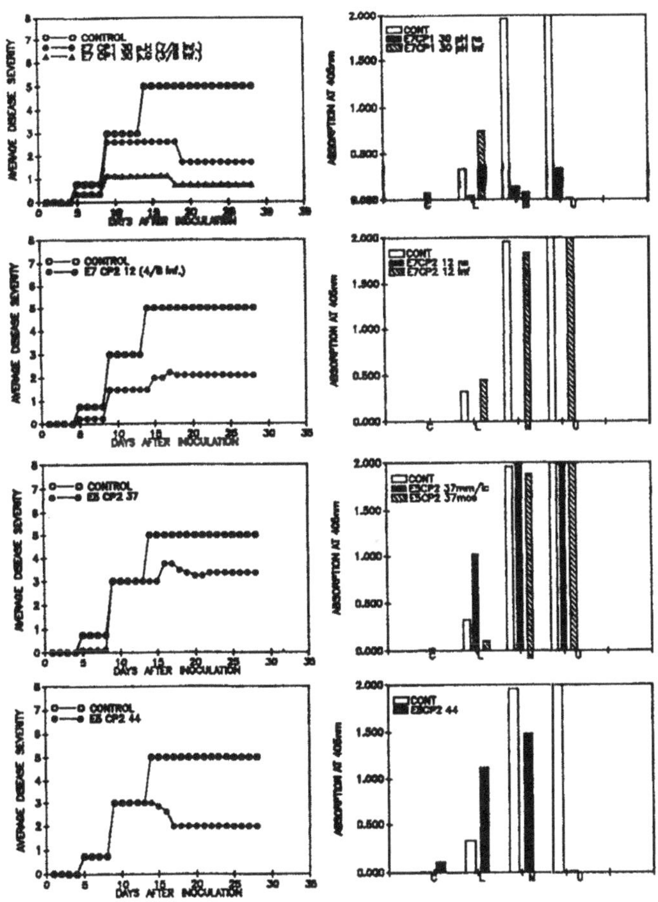

Figure 2. Symptom Severity (left panels) and Relative Virus Titer (right panels) in CP Plants and Controls Inoculated with 100ug/ml of BYMV Ideal A. Open circles (bars) represent non-transgenic controls, and filled symbols (bars) the transgenic lines. Where some plants escaped infection (CP-1 30, CP-2 12), the number of plants infected/number inoculated is indicated. Un-inoculated control (C), and lower (L), middle (M), and upper (U) leaves of representative plants were tested by ELISA; for CP-1 30 and CP-2 12 symptomless plants (ns; solid bars) and infected plants (inf; hatched bars) were tested separately, as were CP-2 37 plants with mild mosaic plus interveinal chlorosis (mm/ic; solid bars) or mosaic (mos; hatched bars). Note that the CP detected in the un-inoculated (CP-1 30, CP-2 44) and un-infected inoculated plants (CP-1 30, ns) is from the transgene (see text).

Only one BY/Pe 9 plant became infected after inoculation with 2ug/ml of BYMV; all of the plants inoculated with 20ug/ml were killed by TMV contamination. The BYMV-infected plant developed a very mild mosaic and had a very low virus titer as estimated by ELISA (Table 3). BY/Pe 30 plants were not infected by 2ug/ml BYMV and were infected by, but recovered from, inoculation with 20ug/ml; recovery was complete in upper leaves produced about 25 dpi, following the same pattern of decreasing interveinal chlorosis noted in other transformants. BY/Pe 25 plants did not show significant resistance to either BYMV or PVY.

BY/Pe 9 was partially resistant to PVY; two of three plants infected by 2ug/ml developed only a mild mosaic, compared to the mosaic, leaf roll, and green vein-banding of other transformants. One of three BY/Pe 9 plants infected by 20ug/ml PVY initially developed mosaic, leaf roll, and green vein-banding but subsequently produced leaves with a milder interveinal chlorosis and eventually symptomless leaves by 28 dpi. BY/Pe 30 showed the greatest resistance to PVY, as only a single plant was infected by 2ug/ml, developing a very mild mosaic from which upper leaves recovered. Four of five plants were infected by 20ug/ml PVY and developed typical symptoms; but all showed reduced interveinal chlorosis by 20-22 dpi, and total absence of symptoms in leaves produced from 23-25 dpi. These leaves were virus-free as assessed by ELISA (Table 3).

Results of TuMV challenge were less clear; TMV contamination killed several plants inoculated with TuMV. However all surviving non-transgenic (3/3), CP-1 30 (5/5), CP-1 79 (3/3), BY/ZY 7 (4/4), and BY/ZY 43 (4/4) plants inoculated with 20ug/ml TuMV became TuMV-infected. In contrast, only three of five BY/Pe 9 plants, three of four BY/Pe 25, and four of five BY/Pe 30 plants became infected with TuMV at 20ug/ml; two of the BY/Pe 30 plants that were infected had at least a 7 day delay in symptom expression compared to the other plants. The BY/Pe chimeric CP thus has some apparent resistance to TuMV.

Representative plants of each transformant with milder symptoms or symptomless upper leaves were further analysed by ELISA of lower, middle, and upper leaves to determine the relative virus levels. In each case where milder symptoms were observed, ELISA values were

Table 3. Virus and CP Levels in Control and Representative Inoculated Plants Relative to Leaf Position and Symptoms, Detected by ELISA with MAb PTY 1. Reactions of non-inoculated controls reflect the expression of CP from the transgene, confirmed by reactivity with MAb PTY 43 (data not shown).

Plant Inoculum	Lower	Middle	Upper	Symptoms
NT				
NI control	0.01	0.01	0.01	NS
2 ug/ml BYMV	0.31	0.15	0.31	mm
CP-1 79				
NI control	0.01	0.02	0.04	NS
2 ug/ml BYMV	0.09	0.02	0.01	mm/IC/RAT
BY/ZY 7				
NI control	0.01	0.01	0.01	NS
2 ug/ml BYMV	0.01	0.19	0.40	vmm
2 ug/ml BYMV	0.33	0.09	0.48	mm
BY/ZY 43				
NI control	0.01	0.02	0.05	NS
2 ug/ml BYMV	0.06	0.30	0.02	mm/IC/RAT
BY/Pe 9				
NI control	0.02	0.02	0.05	NS
2 ug/ml BYMV	0.02	0.02	0.04	vmm
20ug/ml PVY**	0.04	0.18	0.02	mm/GVB/RAT
BY/Pe 30				
NI control	0.06	0.09	0.15	NS
20ug/ml BYMV	0.09	0.37	0.02	mm/IC/RAT
20ug/ml PVY**	0.07	0.33	0.02	mm/GVB/RAT

Leaf samples harvested 30 days after challenge; values are absorbance at 405 nm. NT = non-transgenic; NI = not inoculated; NS = no symptoms; mm = mild mosaic; vmm = very mild mosaic; IC = interveinal chlorosis; RAT = recovery at top (symptomless upper leaves); GVB = green vein banding.

** These transformants were the only ones in which recovery from PVY symptoms were observed.

reduced for at least one leaf position, and where upper leaves were symptomless (recovery at the top of the plant) ELISA values were insignificant (Table 3). Interestingly, the expressed CP in the tranformants was consistently less in the symptomless recovered leaves of the inoculated plants than in the non-inoculated control; the reason is unclear.

It is of interest that only the BY/Pe construct conferred resistance of any kind to PVY; and further that there was apparent resistance to TuMV, which is not closely related to BYMV or PVY, only when the PeMV C-terminus and 3' were present in the construct. Although the gene used for the chimera was from a virus described as PeMV (Dougherty *et al.*, 1985a), sequence comparisons indicated that it was very closely related to PVY (Shukla *et al.*, 1988). More recent work has shown that a California isolate of PeMV ($PeMV_c$) has a sequence quite distinct from PVY, and that $PeMV_c$ and PeMV NC 165 are serologically closely related and distinct from PVY-3 (Jordan and Hammond, 1991; Vance *et al.*, 1992). In light of these results, the BY/Pe construct should perhaps be better described as a chimera between BYMV and PVY. The resistance to PVY may be primarily attributable to the presence of the PeMV (PVY) 3'-terminal domains of the construct; however, whether amino acid sequence or the 3'-non-coding nucleotide sequence is responsible cannot be determined on the basis of the current results. Even if the C-terminal amino acid sequences confer resistance to PVY (and TuMV), there remains the question of why BYMV resistance is retained, as the BYMV C-terminus was replaced. This suggests that more than a single domain may be necessary to confer protection, and that there may be cooperative effects of N-terminal and C-terminal domains that determine some of the specificity of protection.

Separation of Mechanisms: Results gained from both AS and CP transformants show that there are two mechanisms, or at least two differential levels of protection, conferred in each case. These are best described as resistance to initial infection, and resistance to replication and/or systemic movement (leading to partial or complete recovery from symptoms and infection in later growth). In the case of the AS transformants, it is quite possible that these two types of resistance are different levels of effectiveness of the same mechanism, as no transformants have yet been observed in which initial resistance to infection is overcome, but is followed by recovery.

In CP-expressing plants, however, resistance to infection could be overcome (typified by CP-1 30) and yet plants could then recover from initial levels of CP; by the time infected plants recover from symptoms, no CP could be detected in the symptomless leaves, although detectable in equivalent leaves of uninoculated plants and un-infected inoculated plants in the same experiment (Figure 2).

While Lindbo and Dougherty (1992a; 1992b) have shown that an untranslatable TEV CP gene transcript is more effective than AS or CP expression, and that deleted CP's were more effective than a full-length CP, a fundamental difference exists between all of their constructs and ours. Our constructs all contain the complete viral 3'-end and a short poly(A) tail (either from BYMV or the 3'-end of the chimeric constructs), whereas those of Lindbo and Dougherty each contain viral sequence only from the coat protein gene (Lindbo and Dougherty, 1992a; 1992b). They propose that a dysfunctional sense RNA interferes through production of a stable RNA:RNA hybrid; the difference between translated and untranslated constructs (only five nucleotide positions) may be sufficient to prevent stable hybrid formation with the functional construct through ribosome binding and procession along the transcript (Lindbo and Dougherty, 1992b). They do not anticipate any involvement of any viral or host replication factors in the resistance precisely because their constructs *lack* the viral 3'-sequence and because of the extreme difference in protection between translatable and untranslatable constructs differing by so few nucleotides. We also propose that stable RNA:RNA hybrid formation causes resistance; but in the case of our AS plants we believe it is the *presence* of the 3'-region that interferes so effectively with viral replication, presumably by blocking polymerase binding.

There are now reports that CP, mutant CP's, untranslatable RNA's, AS RNA (see **Introduction** and **Discussion** above), and native or mutated replicase genes (Golemboski *et al.*, 1990; Braun and Hemenway, 1992; Zaitlin, 1993; MacFarlane and Davies, 1992) can confer resistance to plant viruses to various levels. It is, thus, apparent that there are many different mechanisms that can interfere with virus replication even within one virus group, as well as differences in the mechanisms between virus groups. We are concentrating upon the interactions responsible for AS and CP resistance to potyviruses; with better understanding of the mechanisms and any contribution of tissue and temporal specificity of expression to resistance, it will be possible to manipulate constructs and to select at an earlier stage those transformants which are likely to confer most effective resistance. This will be of significant value when transformation of commercially important crops becomes more routine.

Acknowledgements

We wish to thank Ramon Jordan for supplying MAb's PTY 1, PTY 10, and PTY 43; Hei-ti Hsu for MAb TBV 27C2H2; Bill Dougherty for the PeMV clone; Rebecca Grumet for the ZYMV clone; and Marc Bauduy and Phuong Bui for excellent technical assistance.

References

Abou Haidar, M. and Erickson, J.W. 1985. Structure and *in vitro* assembly of papaya mosaic virus. In: *Molecular Plant Virology.* J.W. Davies (ed.), pp. 85-121. CRC Press, Boca Raton, Florida.

An, G., Ebert, P.R., Nitra, A., and Ha, S.B. 1988. Binary vectors. In: *Plant Molecular Biology Manual.* S.B. Gelvin and R.A. Schilperoort (eds.). Kluwer Academic, Dordrecht: A3/1-13.

Angenent, G.C., van de Ouweland, J.M.W., and Bol, J.F. 1990. Susceptibility to virus infection of transgenic tobacco plants expressing structural and non-structural genes of tobacco rattle virus. Virology 175:191-198.

Beachy, R.N. 1991. Mechanisms of virus coat protein-induced cross protection. Third International Congress of Plant Molecular Biology, Tucson, Arizona.

Beachy, R.N., Loesch-Fries, S., and Tumer, N.E., 1990. Coat protein-mediated resistance against virus infection. Ann. Rev. Phytopathol. 28:451-474.

Benfey, P.N. and Chua, N.-H. 1989. Regulated genes in transgenic plants. Science 244:174-181.

Braun, C.J. and Hemenway, C.L. 1992. Expression of amino-terminal portions or full-length viral replicase genes in transgenic plants confers resistance to potato virus X infection. The Plant Cell 4:735-744.

Christie, S.R. and Crawford, W.E. 1978. Plant virus range of *Nicotiana benthamiana.* Plant Dis. Reptr. 62:20-22.

Crowley, T.E., Nellen, W., Gomer, R.H., and Firtel, R. 1985. Phenocopy of discoidin I-minus mutants by antisense transformation in *Dictyostelium.* Cell 43:633-641.

Cuozzo, M., O'Connell, K.M., Kaniewski, W., Fang, R.X., Chua, N-H., and Tumer, N.E., 1988. Viral protection in transgenic tobacco plants expressing the cucumber mosaic virus coat protein or its antisense RNA. Bio/Technology 6:549-557.

Dougherty, W.G., Allison, R.F., Parks, T.D., Johnston, R.E., Feild, M.J., and Armstrong, F.B. 1985a. Nucleotide sequence at the 3'-terminus of pepper mottle virus genomic RNA: Evidence for an alternative mode of potyvirus capsid protein gene organization. Virology 146:282-291.

Dougherty, W.G. and Carrington, J.C. 1988. Expression and function of potyviral gene products. Ann. Rev. Phytopathol. 26:123-143.

Golemboski, D.B., Lomonossoff, G.P., and Zaitlin, M. 1990. Plants transformed with a tobacco mosaic virus non-structural gene sequence are resistant to the virus. Proc. Natl. Acad. Sci. U.S.A. 87:6311-6315.

Grumet, R. and Fang, G. 1990. cDNA cloning and sequence analysis of the 3'-terminal region of zucchini yellow mosaic virus RNA. J. Gen. Virol. 71:1619-1622.

Hammond, J. and Chastagner, G.A., 1988. Natural infection of tulips with turnip mosaic virus and another potyvirus isolate distinct from tulip breaking virus in the U.S.A. Acta Hort. 234:235-242.

Hammond, J. and Hammond, R.W. 1989. Molecular cloning, sequencing and expression in *Escherichia coli* of the bean yellow mosaic virus coat protein gene. J. Gen. Virol. 70:1961-1974.

Hammond, J. and Kamo, K.K. 1993. Transgenic coat protein and antisense RNA resistance to bean yellow mosaic potyvirus. Acta Hort. 36:171-178.

Hammond, J. and Kamo, K.K. 1994. Potyvirus resistance derived from antisense RNA and native or chimeric coat protein genes. Acta Hort. (in press).

Hammond, J. and Lawson, R.H. 1988. An improved purification procedure for preparing potyviruses and cytoplasmic inclusions from the same tissue. J. Virol. Methods 20:203-217.

Hammond, J., Jordan, R.L., and Kamo, K.K. 1990. Use of chimeric coat protein constructs and deletion mutants to examine potyvirus structure and coat protein mediated resistance. Phytopathology 80:1018.

Hemenway, C., Fang, R-X., Kaniewski, W., Chua, N-H., and Tumer, N.E. 1988. Analysis of the mechanism of protection in transgenic plants expressing the potato virus X coat protein or its antisense RNA. EMBO J. 7:1273-1280.

Horsch, R.B., Fry, J.E., Hoffman, N.L., Eichholtz, D., Rogers, S.G., and Fraley, R.T. 1985. A simple and general method for transferring genes into plants. Science 227:1229-1231.

Hsu, H-T., Franssen, J.M., van der Hulst, C.T.C., Derks, A.F.L.M., and Lawson, R.H. 1988. Factors affecting epitope specificity of monoclonal antibodies to tulip breaking virus. Phytopathology 78:1337-1340.

Jefferson, R.A., Kavanagh, T.A., and Bevan, M.W. 1987. GUS fusions: B-glucuronidase as a sensitive and versatile gene fusion marker in higher plants. EMBO J. 6:3901-3907.

Jordan, R. 1992. Potyviruses, monoclonal antibodies and antigenic sites. Arch. Virol. (Supplementum 5; Potyvirus Taxonomy) 81:95.

Jordan, R.L. and Hammond, J. 1991. Comparison and differentiaton of potyvirus isolates and identification of strain-, virus-, subgroup-specific and potyvirus group-common epitopes using monoclonal antibodies. J. Gen. Virol. 72:25-36.

Lawson, C., Kaniewski, W., Haley, L., Rozman, R., Newell, C., Sanders, P., and Tumer, N.E. 1990. Engineering resistance to mixed virus infection in a commercial potato cultivar: Resistance to potato virus X and potato virus Y in transgenic Russet Burbank. Bio/Technology 8:127-134.

Lindbo, J.A. and Dougherty, W.G. 1992a. Pathogen-derived resistance to a potyvirus: Immune and resistant phenotypes in transgenic tobacco expressing altered forms of a potyvirus coat protein nucleotide sequence. Mol. Plant-Microbe Interact. 5:144-153.

Lindbo, J.A. and Dougherty, W.G. 1992b. Untranslatable transcripts of the tobacco etch virus coat protein gene sequence can interfere with tobacco etch virus replication in transgenic plants and protoplasts. Virology 189:725-733.

Ling, K., Namba, S., Gonsalves, C., Slightom, J.L., and Gonsalves, D. 1991. Protection against detrimental effects of potyvirus infection in transgenic tobacco plants expressing the papaya ringspot virus coat protein gene. Bio/Technology 9:752-758.

Lutcke, H.A., Chow, K.C., Mickel, F.S., Moss, K.A., Kern, H.F., and Scheele, G.A. 1987. Selection of AUG initiation codons differs in plants and animals. EMBO J. 6:43-48.

MacFarlane, S.A. and Davies, J.W. 1992. Plants transformed with a region of the 201-kilodalton replicase gene from pea early browning virus RNA 1 are resistant to virus infection. Proc. Natl. Acad. Sci. U.S.A. 89:5829-5833.

Mackenzie, D.J. and Tremaine, J.H. 1990. Transgenic *Nicotiana debneyii* expressing viral coat protein are resistant to potato virus S infection. J. Gen. Virol. 71:2167-2170.

McGarvey, P. and Kaper, J.M. 1991. A simple and rapid method for screening transgenic plants using the PCR. BioTechniques 11:428-432.

Namba, S., Ling, K., Gonsalves, C., Slightom, J.L., and Gonsalves, D. 1992. Protection of transgenic plants expressing the coat protein gene of watermelon mosaic virus II or zucchini yellow mosaic virus against six potyviruses. Phytopathology 82:940-946.

Nelson, R.S., Powell-Abel, P., and Beachy, R.N. 1987. Lesions and virus accumulation in inoculated transgenic tobacco plants expressing the coat protein gene of tobacco mosaic virus. Virology 158:126-132.

Oeller, P.W., Lu, M-W., Taylor, L.P., Pike, D.A., and Theologis, A. 1991. Reversible inhibition of tomato fruit senescence by antisense RNA. Science 254:437-439.

Powell, P.A., Sanders, P.R., Tumer, N., Fraley, R.T., and Beachy, R.N. 1990. Protection against tobacco mosaic virus infection in transgenic plants requires accumulation of coat protein rather than coat protein RNA sequences. Virology 175:124-130.

Powell, P.A., Stark, D.M., Sanders, P.R., and Beachy, R.N. 1989. Protection against tobacco mosaic virus in transgenic plants that express tobacco mosaic virus antisense RNA. Proc. Natl. Acad. Sci. U.S.A. 86:6949-6952.

Quacquarelli, A. and Avgelis, A. 1975. *Nicotiana benthamiana* Domin. as host for plant viruses. Phytopathol. Mediterr. 14:36-39.

Register, J.C. and Beachy, R.N. 1988. Resistance to TMV in transgenic plants results from interference with an early event in infection. Virology 166:524-532.

Regner, F., da Camara Machado, A., Laimer da Camara Machado, M., Steinkellner, H., Mattanovich, D., Hanzer, V., Weiss, H., and Kattinger, H., 1992. Coat protein mediated resistance to plum pox virus in *Nicotiana clevelandii* and *benthamiana*. Plant Cell Rep. 11:30-33.

Rezaian, M.A., Skene, K.G.M., and Ellis, J.G. 1988. Anti-sense RNA's of cucumber mosaic virus in transgenic plants assessed for control of the virus. Plant Mol. Biol. 11:463-471.66.

Shukla, D.D., Thomas, J.E., McKern, N.M., Tracy, S.L., and Ward, C.W. 1988. Coat protein of potyviruses. 4. Comparison of biological properties, serological relationships, and coat protein amino acid sequences of four strains of potato virus Y. Arch. Virol. 102:207-219.

Stark, D.M. and Beachy, R.N. 1989. Protection against potyvirus infection in transgenic plants: Evidence for broad spectrum resistance. Bio/Technology 7:1257-1262.

Van Dun, C.M.P., Bol, J.F., and van Vloten-Doting, L. 1987. Expression of alfalfa mosaic virus and tobacco rattle virus coat protein genes in transgenic tobacco plants. Virology 159:299-305.

Vance, V.B., Jordan, R., Edwardson, J.R., Christie, R., Purcifull, D.E., Turpen, T., and Falk, B. 1992. Evidence that pepper mottle virus and potato virus Y are distinct viruses: Analysis of the coat protein and 3'-untranslated region of a California isolate of pepper mottle virus. Arch. Virol. (Supplementum 5; Potyvirus Taxonomy) 337-345.

Wisniewski, L.A., Powell, P.A., Nelson, R.S., and Beachy, R.N. 1990. Local and systemic spread of tobacco mosaic virus in transgenic tobacco. Plant Cell 2:559-567.

Zaitlin, M. 1993. Molecular strategies for the control of plant virus diseases. Acta Hort. 336:63-68.

Zimmern, D. 1987. The nucleotide sequence at the origin for assembly on tobacco mosaic virus RNA. Cell 11:463-482.

The Specificity and Significance of Heterologous Encapsidation of Virus and Virus-Like RNA's

B. W. Falk, B. K. Passmore, M. T. Watson, and L.-S. Chin
Department of Plant Pathology
University of California
Davis, California 95616

Heterologous encapsidation interactions frequently occur between co-infecting viruses or viruses and subviral RNA's and result in the production of chimeric progeny virions which exhibit characteristics of both of the co-infecting parental viruses, or virus and subviral RNA. Heterologous encapsidation interactions occur naturally in virus-infected crop and weed plants and are a natural part of virus-virus and virus-plant interactions. The biological significance of heterologous encapsidation interactions is that the chimeric virus progeny frequently exhibit new or altered biological properties differing from those of the parental viruses. These include altered or enhanced vector transmissibility and plant host range. We discuss some examples of heterologous encapsidation interactions and, specifically, some interactions between beet western yellows virus and a unique RNA, the ST9 associated RNA. We also discuss some possible epidemiological considerations in regards to heterologous encapsidation interactions and virus coat protein-mediated genetically-engineered resistance.

Introduction

A variety of interactions can potentially occur when two or more viruses co-infect the same host. Of these, structural interactions which result in the production of hybrid progeny virions composed of components derived from more than one of the co-infecting parental viruses have been described for plant, animal, and bacterial viruses (for reviews, see Rochow, 1977; Dodds and Hamilton, 1976). Progeny virions resulting from these interactions can exhibit a mixed phenotype where the virion capsid is composed of protein subunits derived from at least two co-infecting parental viruses, or they may exhibit the phenotype of one parental virus and yet contain the genetic material derived from another co-infecting virus. Several terms have been used to describe such interactions; the most general being heterologous encapsidation (Rochow, 1977; Dodds and Hamilton, 1976). However, it is generally accepted that phenotypic mixing is used to describe heterologous encapsidation interactions which result in progeny virions exhibiting a mixed phenotype or a chimeric capsid composed of protein subunits from at least two co-infecting viruses. Genomic masking or transcapsidation are to be used when the progeny exhibit the phenotype of one parental virus but contain the genetic material (genome) of a second co-infecting virus (Rochow, 1977; Dodds and Hamilton, 1976) (Figure 1).

HETEROLOGOUS ENCAPSIDATION INTERACTIONS

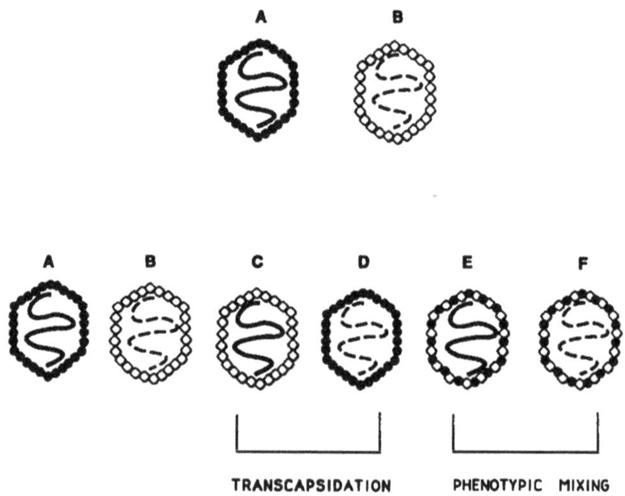

Figure 1. Diagram Depicting Possible Results of Heterologous Encapsidation Interactions When Two Viruses Co-Infect the Same Cell (Based on Rochow, 1977). Outer circles or squares represent capsid protein subunits and inner solid or dashed lines are the viral genome. Upper virions labelled A and B are the parental co-infecting viruses. Lower row (A - F) show possible progeny virions. A and B are parental types. C and D show virions with the genome of one of the parental viruses encapsidated by protein subunits derived from the other (transcapsidation). E and F exhibit phenotypically mixed capsids composed of protein subunits from both of the parental viruses.

In recent years it has become evident that the virion capsid of plant viruses is a biologically important component of the virion. The virion capsid functions more than just to protect the viral genome. It has been shown to have roles in determining, at least in part, the vector transmission specificity of plant viruses and/or facilitating the long distance (systemic) movement of various viruses within infected plants (Rochow, 1970; Gildow, 1987; Atreya *et al.*, 1990; Hull, 1989; Maule, 1991; Dawson *et al.*, 1988). Evidence has shown that heterologous encapsidation interactions can affect both of these properties. Heterologous encapsidation interactions have been reported to result from mixed infections of plant viruses representing several different plant virus groups (see Table 1); and in almost all cases the interactions were discovered because the progeny virions from the doubly-infected plants exhibited altered biological properties.

The fact that mixed infections of plant viruses in nature are extremely frequent (Zink and Duffus, 1972; Davis and Mizuki, 1987; Duffus, 1963) suggests that the opportunity for heterologous encapsidation interactions is commonplace. Yet, we know very little about the natural incidence of heterologous encapsidation interactions in plant virus epidemiology. Similarly, the specificity and significance of these interactions among plant viruses has been examined only to a limited extent, even in the laboratory. Also of significance in this regard is that one of the recent and successful approaches to engineering plants for resistance against viruses has been to engineer plants so that they express functional viral capsid proteins (Powell-Abel *et al.*, 1986; Beachy *et al.*, 1990). It has been questioned whether this additional source of viral capsid protein might pose new risks in regards to affecting the potential for heterologous encapsidation interactions in genetically-engineered plants (De Zoeten, 1991; Hull, 1990). As the coat protein-mediated resistance strategy is proving useful against more and more viruses, it is likely that this strategy will play a role in future efforts to engineer crop plants for resistance against viruses. Thus, the potential of this virus disease control strategy in influencing biologically important heterologous encapsidation interactions should be considered.

In this paper we review information on heterologous encapsidation interactions which occur between plant viruses and between plant viruses and subviral RNA's. We also will discuss some characteristics of a novel interaction between a virus and a subviral RNA; and, finally, we will discuss some factors relating to the biological significance and epidemiological considerations of heterologous encapsidation interactions.

Heterologous Encapsidation Interactions

The first documentation for heterologous encapsidation interactions between co-infecting plant viruses involved two distinct barley yellow

Table 1. Examples of Plant Virus Heterologous Encapsidation Interactions.

Viruses or RNA[A]	Host Plant	Effects[B]	References
RPV + MAV	Cereals	Aphid transmission specificity	Rochow, 1970; Rochow, 1982
RPV + PAV	Cereals	None	Creamer and Falk, 1990; Wen and Lister, 1991
RPV + RMV	Cereals	Aphid transmission specificity	Wen and Lister, 1991; Rochow, 1982
RPV + SGV	Cereals	Aphid transmission specificity	Rochow, 1982
PAV + MAV	Cereals	Aphid transmission specificity	Hu *et al.*, 1988; Rochow, 1982
PAV + RPV	Cereals	Aphid transmission specificity	Creamer and Falk; 1990
PAV + MAV	Cereals	Aphid transmission specificity	Wen and Lister, 1991; Rochow, 1982
PAV + RMV	Cereals	Aphid transmission specificity	Rochow, 1982
RMV + RPV	Cereals	Aphid transmission specificity	Rochow, 1982
RMV + MAV	Cereals	Aphid transmission specificity	Wen and Lister, 1991; Rochow, 1975; ?? 1982
BWYV + LSMV	Dicots	Aphid transmissibility	Falk *et al.*, 1979
CRLV + CMoV	Dicots	Aphid transmissibility	Waterhouse and Murant, 1983
BWYV + ST9aRNA	Dicots	Aphid transmissibility systemic movement	Falk and Duffus, 1984; Passmore *et al.*, 1993
PEMV RNA1 + RNA2	Dicots	Aphid transmissibility	Demmler and De Zoetm, 1991
ZYMV + PRSV	Cucurbits	Aphid transmissibility	Bourdin and Lecoq, 1991
PeMV + TEV	Dicots	None	Nelson and Wheeler, 1979
BSMV + TMV	Barley	Systemic movement?	Dodds and Hamilton, 1974
BMV + BSMV	Barley	None	Peterson and Brakke, 1973
TMV-OM + TMV-T	Tobacco	None	Otsuki and Takebe, 1978
Ω '-CAT-OAS	Transgenic tobacco[c]	None	Sleat *et al.*, 1988

Table 1. continued next page

Table 1. Continued

Viruses or RNA[A]	Host Plant	Effects[B]	References
TMV DT1 and DT1G	Transgenic tobacco[D]	Systemic movement	Osbourn *et al.*, 1990
TMV U3/12CPfs	Transgenic tobacco[E]	Systemic movement	Holt and Beachy, 1991
CP1 CyRSV	Transgenic tobacco[F]	Systemic movement	Dalmay *et al.*, 1992
PVY[C]	Transgenic potato[G]	None	Farinelli *et al.*, 1992

[A] Heterologous encapsidation interactions occurred as the second virus or RNA listed acquiring the complete or partial phenotype of the virus listed first. Viruses and RNA's are: RPV, PAV, RMV, SGV and MAV are corresponding isolates of barley yellow dwarf luteovirus. BWYV and LSMV are beet western yellows luteovirus and lettuce speckles mottle virus, respectively. CRLV and CMoV are carrot redleaf luteovirus and is carrot mottle virus, respectively. ST9aRNA is the BWYV ST9 associated RNA. PEMV and RNA2 are pea enation mosaic virus and the PEMV RNA2, respectively. ZYMV and PRSV are zucchini yellow mosaic potyvirus and papaya ringspot potyvirus, respectively. PeMV and TEV are pepper mottle potyvirus and tobacco etch potyvirus, respectively. BSMV and TMV are barley stripe mosaic hordeivirus and tobacco mosaic tobamovirus, respectively. BMV is brome mosaic bromovirus. CyRSV is cymbidium ringspot tombusvirus.

[B] Aphid transmission specificity refers to altered aphid transmission of the second virus listed. Aphid transmission denotes that the second virus or RNA acquires aphid transmissibility only as a result of being in the mixed infection with the first. Systemic movement indicates that the second virus or RNA does not move systemically very readily, or at all in the given host unless co-infecting with the first virus listed. None indicates that only heterologous encapsidation resulted, and no biological effect(s) was detected.

[C] The transgenic tobacco was engineered so as to produce transcripts containing reporter genes (i.e., CAT) and the TMV origin of assembly (OAS).

[D] The transgenic tobacco was engineered so as to produce functional TMV coat protein. TMV strains DT1 and DT1G are TMV isolates defective in production of functional coat protein (DT1) or do not produce detectable coat protein (DT1G).

[E] The transgenic tobacco was engineered so as to produce functional TMV coat protein. TMV U3/12CPfs was engineered so as to be defective in its coat protein.

[F] The transgenic tobacco was engineered so as to produce functional CyRSV coat protein. CyRSV mutant CP1 produces a defective coat protein.

[G] The transgenic potatoes were engineered so as to produce functional PVY[N] coat protein. Heterologous encapsidation was detected by ELISA and electron microscopy.

dwarf luteoviruses (BYDV's). Rochow (1970) observed that a loss of aphid vector specificity resulted when the MAV and RPV BYDV's co-infected oat (*Avena sativa*) seedlings in greenhouse studies. Although MAV is specifically transmitted by aphids of the species *Sitobion avenae*, and RPV is specifically transmitted by aphids of the species *Rhopalosiphum padi*, both RPV and MAV were efficiently transmitted by *R. padi* from plants infected with both of these viruses (Rochow, 1970). Rochow went on to study this altered aphid vector specificity and showed that it was correlated with transcapsidation of the MAV genomic RNA within capsids composed of RPV protein subunits (for example, see Figure 1). This work demonstrated that heterologous encapsidation interactions can occur when two plant viruses co-infect a plant, but it also showed that there is specificity in the heterologous encapsidation interactions. He obtained no evidence suggesting phenotypic mixing interactions between RPV and MAV, and there was no evidence that the RPV genome could become encapsidated within MAV capsids.

These data were also of great significance in that they provided some understanding of the specificity of plant virus transmission by aphid vectors. The data suggested that the virion capsid most likely specifically interacts with some part aphid vector to, at least in part, determine aphid transmissibility. As such, heterologous encapsidation events between aphid-transmitted plant viruses are potentially of great biological significance.

Other examples of heterologous encapsidation interactions have also been described. Dodds and Hamilton (1974) found that heterologous encapsidation of tobacco mosaic tobamo virus (TMV) RNA occured in barley (*Hordeum vulgare*) plants when they were doubly-infected with barley stripe mosaic hordei virus (BSMV). BSMV causes a systemic infection and replicates to a relatively high titer in barley plants. On the other hand, TMV does not normally systemically infect barley plants but replicates to a limited extent in the inoculated leaves (Dodds and Hamilton, 1972). However, when barley plants are co-infected by TMV and BSMV, both viruses replicate to high titers and move systemically within the plants (Dodds and Hamilton, 1974; Dodds and Hamilton, 1972). Interestingly, in the systemically-infected leaves, up to 8.5 percent of the virus particles were phenotypically BSMV yet contained TMV RNA. In this example, heterologous encapsidation interactions occurred in only one direction, BSMV capsid protein on TMV RNA, and chimeric particles accounted for a significant amount of the progeny virions. However, this example showed some additional important properties.

In the BSMV-TMV interactions, heterologous encapsidation occurred between two unrelated viruses in distinct virus groups. Secondly, the co-infection and possibly the heterologous encapsidation, allowed TMV to systemically invade the tissues of a plant which is not a typical systemic host.

Since these early reports of heterologous interactions between plant viruses, several additional examples resulting from mixed infections of related and/or unrelated plant viruses have been identifed (see Table 1). Most were initially suspected because of the altered vector transmission characteristics of progeny virions from double vs. single infections, and examples of both transcapsidation and phenotypic mixing have now been documented (Wen and Lister, 1991; Hu *et al.*, 1988; Creamer and Falk, 1990; Falk *et al.*, 1979; Waterhouse and Murant, 1983).

Evidence for Heterologous Encapsidation Interactions as a Natural Part of Plant Virus and Subviral RNA Epidemiology

Most examples of heterologous encapsidation interactions are from laboratory or greenhouse studies. However, mixed infections of plant viruses, and of plant viruses and subviral RNA's (i.e., satellite RNA's) are very common in nature (Zink and Duffus, 1972; Davis and Mizuki, 1987; Duffus, 1963; Waterworth *et al.*, 1979; Francki, 1985). In California, we have found individual pepper (*Capsicum annuum*) plants infected by as many as 6 viruses (Falk, unpublished). Thus, the potential for heterologous encapsidation interactions to naturally occur in virus-infected plants is likely great. We now know that they do, and that for some viruses and subviral RNA's (i.e., satellite RNA's), heterologous encapsidation events are a natural part of their epidemiology and survival.

We recently developed a method to directly isolate and analyze doubly-infected plants for virions which had undergone heterologous encapsidation events (Creamer and Falk, 1990). Monoclonal antibodies are used to trap virions onto ELISA plates. The RNA is then extracted from the trapped virions, blotted onto nylon membranes; and this RNA is analyzed using specific cDNA probes (Figure 2). By assaying various field samples of naturally-infected cereals and grassy weeds, heterologous encapsidation interactions were detected in some of these plants. We found evidence for transcapsidation of MAV BYDV RNA by RPV BYDV capsids in doubly-infected oats, suggesting that the original heterologous encapsidation events observed by Rochow do occur naturally in field plants. We also detected transcapsidation of PAV BYDV RNA by RPV BYDV capsids in doubly-infected *Aegolops ovata and Aegolops cylindrica* (Creamer and Falk, 1990). This latter transcapsidation event was not suspected previously because it is not detectable by altered aphid transmissibility as both PAV and RPV are transmissible by the same aphid vector, *R. padi.* This methodology provided a relatively sensitive and accurate means to test doubly-infected plants for the presence of heterologous encapsidation events, and the data were the first evidence for transcapsidation between BYDV's occurring naturally in field plants (Creamer and Falk, 1990). This methodology also has been recently applied to the study of BYDV heterologous encapsidation interactions in a series of greenhouse studies. Results indicated that phenotypic mixing as well as transcapsidation events can readily occur between various BYDV's and confirm previous work showing that both types of events can alter aphid transmissibility of these viruses (Wen and Lister, 1991).

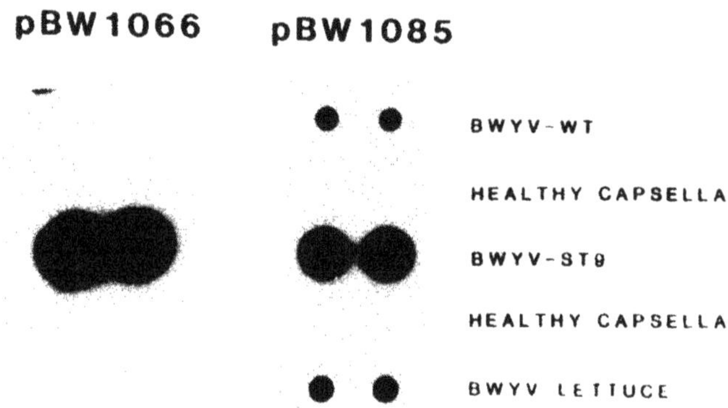

Figure 2. Results of an Immunohybridization Assay on *Capsella bursa-pastoris* Plants Infected by Various Beet Western Yellows Virus (BWYV) Isolates (Creamer, 1989). Samples were: BWYV-WT = common BWYV; healthy Capsella = healthy *C. bursa-pastoris*; BWYV-ST9 = the ST9 isolate of BWYV; BWYV lettuce = a lettuce isolate of BWYV. Plants were triturated, and the extracts were incubated in BWYV monoclonal antibody-coated polystyrene microtiter plates. After washing, the antibody-trapped samples were heated at 70C and the samples were blotted onto nylon membranes. Samples were then analyzed by hybridization analysis using ^{32}-P-labelled probes to the BWYV genomic RNA (pBW1085) or the BWYV ST9 aRNA (pBW1066).

Further evidence supporting the idea that heterologous encapsidation interactions are a natural part of plant virus epidemiology is that some viruses and several subviral RNA's are naturally vector-transmitted only as a result of heterologous encapsidation interactions. These viruses and/or subviral RNA's are thus dependent upon co-infecting plants with a second virus and being co-transmitted with the second virus by its natural vector to new plants.

Many of these mixed infections are specific and cause specific plant diseases which are not caused by infections of only one of the viruses. Carrot motley dwarf, groundnut rosette, tobacco yellow vein, and lettuce speckles are examples of diseases caused by co-infection of two specific viruses (Falk *et al.*, 1979; Waterhouse and Murant, 1983; Hull and Adams, 1968; Adams and Hull, 1972; Falk and Duffus, 1981). In each of these cases, one of the viruses is a luteovirus (e.g., beet western yellows luteovirus (BWYV) for lettuce speckles); and the second virus is a mechanically-transmissible virus (e.g., lettuce speckles mottle virus (LSMV) for lettuce speckles) which is only aphid-transmitted from plants which are co-infected with the corresponding luteovirus. The second virus can systemically infect plants in the absence of the helper luteovirus, and thus is referred to as being helper-dependent for aphid transmission (but not for replication). It has been demonstrated for LSMV and carrot mottle virus (CMoV) that when either of these viruses co-infects plants with the corresponding luteovirus (BWYV for LSMV and carrot redleaf luteovirus [CRLV] for CMoV), a significant portion of their RNA becomes encapsidated within capsids composed of the corresponding luteovirus protein. Only this transcapsidated RNA is aphid-transmissible by the corresponding luteovirus aphid vector. There also is some evidence suggesting that the transcapsidation of at least CMoV RNA is not specific. For example, when plants were doubly-infected with CMoV and BWYV, some CMoV RNA became encapsidated within BWYV capsids and, as such, became transmissible by the BWYV aphid vector, *Myzus persicae*, instead of the CRLV aphid vector, *Cavariella aegopodii* (Waterhouse and Murant, 1983).

Besides gaining aphid transmissibility, viruses such as LSMV appear to gain a greater plant host range when they are aphid-transmitted versus when they are mechanically transmitted (Falk *et al.*, 1979). This expanded host range is a function of transcapsidation, at least indirectly. Two possible explanations are: 1) that the transcapsiated, aphid-transmitted LSMV is more efficiently inoculated to host plants than when the virus is mechanically inoculated; and\or 2) that transcapsidated LSMV RNA generated in the double infections is able to systemically infect plants while unencapsidated LSMV RNA cannot move systemically through these hosts. While there is no direct evidence for the latter possibility, there is for the former. The host range of aphid-transmitted LSMV includes plants species which are not hosts for BWYV, thus supporting the idea that aphid inoculation of the transcapsidated LSMV increases the host range as opposed to mechanical inoculation. However, the recent data supporting the role of the virion capsid in affecting or facilitating systemic virus spread within virus-infected plants suggests that heterologous encapsidation interactions might also play a role in allowing chimeric virions to move in some plant species.

The Beet Western Yellows Virus ST9-Associated RNA: A Novel Type of Heterologous Encapsidation Interaction

Several examples also exist for the heterologous encapsidation of subviral RNA's by the protein capsid of supporting helper viruses. The plant virus satellite RNA's (i.e., the satellite RNA's of cucumber mosaic virus [CMV sat RNA] and tobacco ringspot virus [sat TobRV RNA]) are dependent upon their respective helper viruses not only for replication (Waterworth *et al.*, 1979; Francki, 1985) but also for encapsidation. The latter almost surely serves to facilitate spread of the satellite RNA to new plants via the natural vector of the helper virus. This ensures that these RNA's are transmitted from plant-to-plant along with their helper viruses, and thereby co-infect plants with their helper viruses.

We have recently characterized a novel type of subviral RNA, the ST9-associated RNA (ST9 aRNA), which is dependent upon a helper virus, BWYV, for aphid transmission and very likely for systemic spread within the plant host as well. Evidence suggests that the ST9 aRNA gains both of these properties as a result of transcapsidation of its RNA within a protein capsid derived from BWYV.

The ST9 isolate of BWYV (ST9 BWYV) was originally isolated from broccoli (*Brassica napus* L.) plants showing uncharacteristic reddening symptoms (Falk and Duffus, 1984). Compared to typical BWYV isolates, ST9 BWYV causes severe symptoms in plants of several species, including the BWYV indicator plant, *C. bursa-pastoris* (shepherd's purse) (Figure 3). When virions are purified from ST9 BWYV-infected plants, they contain two RNA's. One is the BWYV genomic RNA of ca. 5,600 nt, and the second is the 2,843 nt ST9 aRNA (Falk and Duffus, 1984; Chin *et al.*, 1992; Falk *et al.*, 1989).

Both are encapsidated within capsids composed of BWYV proteins (Figure 4). The ST9 aRNA does not contain detectable nucleotide sequence homology to the BWYV genomic RNA (Falk *et al.*, 1989) and was believed by us to likely represent a satellite-like RNA (Falk and Duffus, 1984).

We now know that the ST9 aRNA is not a satellite-like RNA. We have determined its complete nucleotide sequence to be 2,843 nt (Chin *et al.*, 1993). The ST9 aRNA sequence contains 3 large and 1 small open reading frames (ORF's) (Figure 5). The ST9 aRNA nucleotide sequence shares significant homology with the carmo-like viruses; but not with BWYV. Similarly, the deduced amino acid sequences of the ST9 aRNA ORF 2 and 3 putative proteins show significant similarity with the RNA-dependent RNA polymerases of these viruses (Chin *et al.*, 1993). The ORF 3 region contains the GDD motif, characteristic of RNA plant virus RNA-dependent RNA polymerases, as well as conserved amino acids surrounding this region.

When transcripts representing the complete ST9 aRNA sequence are translated *in vitro*, proteins corresponding to the individual ORF's as well as specific ORF combinations are detected. Our data suggest that ORF's 1, 2, and 3 can be translated continuously, at least *in vitro*, to yield a protein of ca. 85,000. To accomplish this, the amber termination codon at the end of ORF 1 must be read-through to give continuous translation; and a frameshift event would occur in the ORF 2 and 3 overlapping region (Figure 5). The ORF 1 amber codon shares consensus amino acids characteristic of read-through codons found in many carmo-like viruses, as well as TMV and beet necrotic yellow vein virus and is likely read-through (Chin *et al.*, 1993). It is likely that the ORF 2-ORF 3 frameshift occurs at least *in vitro*; however, no typical frameshift sequences have so far been identified in the 91 nucleotide overlap between ORF's 2 and 3.

Recently, we have been investigating the relationships and interactions which occur between the ST9 aRNA and the BWYV genomic RNA when they simultaneously infect the same host. When protoplasts of *Nicotiana tabacum* are electroporated with purified ST9 BWYV virions or virion RNA's, both RNA's replicate and both are packaged into capsids composed of BWYV capsid protein (Passmore *et al.*, 1993; Sanger *et al.*, 1994). Virions can be recovered from these protoplasts, and when fed to aphids (*M. persicae*) both RNA's can be transmitted to plants. When *C. bursa-pastoris* plants are inoculated in this way, they show the characteristic phenotype of a ST9 BWYV infection; severe stunting and yellowing (Figure 6).

In order to more clearly assess the interactions between these RNA's, we constructed full-length cDNA clones for both the ST9 aRNA and the ST9 BWYV genomic RNA (BWYV gRNA) (Passmore *et al.*, 1993). When transcripts were generated from these clones *in vitro*, capped transcripts of the BWYV gRNA and uncapped or capped transcripts of the ST9 aRNA were infectious in protoplasts. As would be expected, the BWYV gRNA transcripts replicated in protoplasts whether or not the ST9 aRNA transcripts were present. Virions were produced in these infections, and they were recovered and transmitted to plants using *M. persicae* (Passmore *et al.*, 1993; Sanger *et al.*, 1994).

Interestingly, the ST9 aRNA transcripts were also competent for replication when electroporated into protoplasts, whether in the presence or absence of the BWYV gRNA transcripts (Passmore *et al.*, 1993). Because the ST9 aRNA can replicate when it infects plant cells in the absence of BWYV, the ST9 aRNA cannot be called a satellite RNA. However, as would be expected when only the ST9 aRNA transcripts are inoculated to protoplasts, no virions are formed; and no aphid transmission to plants results from these protoplast preparations. Replication can be detected because the ST9 aRNA increases over time in the electroporated protoplasts, and both (+) and (-) sense forms can be detected by northern hybridization analysis (Passmore *et al.*, 1993).

By using this system we also have assessed the specificity of the ST9 aRNA for transcapsidation interactions with other BWYV isolates and for its role in the severe symptom development observed in whole plant infections. When ST9 aRNA transcripts were co-electroporated into *N. tabacum* protoplasts along with virion RNA from a mild BWYV isolate (mild isolates lack the ST9 aRNA), both RNA's replicated. We also were able to transmit both RNA's from protoplast extracts by using *M. persicae*, suggesting that the ST9 aRNA can become transcapsidated when it co-infects cells with BWYV isolates other than that with which it occurs naturally. *C. bursa-pastoris* plants inoculated in this manner showed a severe phenotype, more characteristic of the infection by the ST9 BWYV isolate (Sanger *et al.*, 1994) (Figure 6). Infection by the mild BWYV alone caused only typical symptoms, as did the ST9 gRNA alone, supporting that co-infections of *C. bursa-pastoris* plants by BWYV and the ST9 aRNA are necessary for the severe symptom phenotype.

If co-replication of the ST9 aRNA and BWYV is what determines the severe symptom phenotype seen in infected *C. bursa-pastoris* plants, what possible interactions between these RNA's give rise to

Figure 3. ***C. bursa-pastoris*** **Plants Showing Symptoms Resulting from Infection by Common (Mild) Beet Western Yellows Virus (BWYV) and Infection by the ST9 Isolate of BWYV. Healthy** ***C. bursa-pastoris*** **plant is labelled as check. Genomic-T indicated plant infected with only the genomic RNA of BWYV.**

ST9 Isolate of Beet Western Yellows Virus :

ST9 BWYV genomic RNA

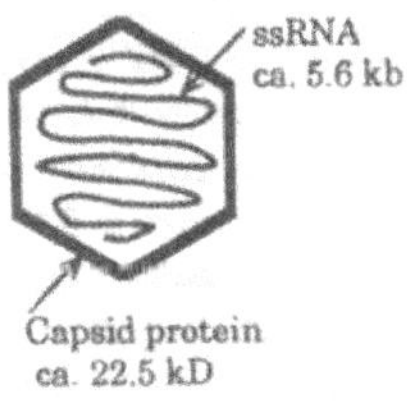

Virion : 25 nm in dia.
one species of capsid protein (22.5 kD)
possible readthrough protein of capsid protein (70 kD)

Genome : positive-sense ssRNA (ca. 5.6 kb)
5' : genome-linked protein (VPg)
3' : non-polyadenylated

Transmission : circulative non-propagative by aphids

Tissue tropism : Phloem limited

ST9-associated RNA

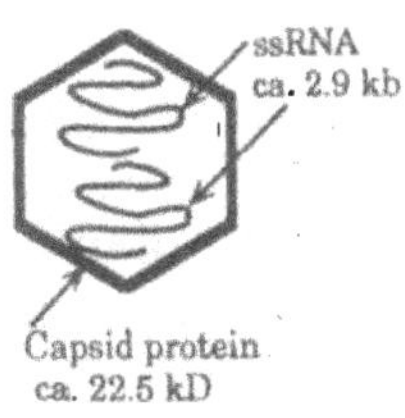

Virion : 25 nm in dia.
one species of capsid protein (22.5 kD)
possible readthrough protein of capsid protein (70 kD)

Genome : positive-sense ssRNA (ca. 2.9 kb)
5' : genome-linked protein (VPg) (?)
3' : non-polyadenylated

Transmission : circulative non-propagative by aphids

Tissue tropism : Phloem limited

Figure 4. Model Depicting the Composition of Virus Particles Associated with the ST9 Isolate of Beet Western Yellows Virus (BWYV).

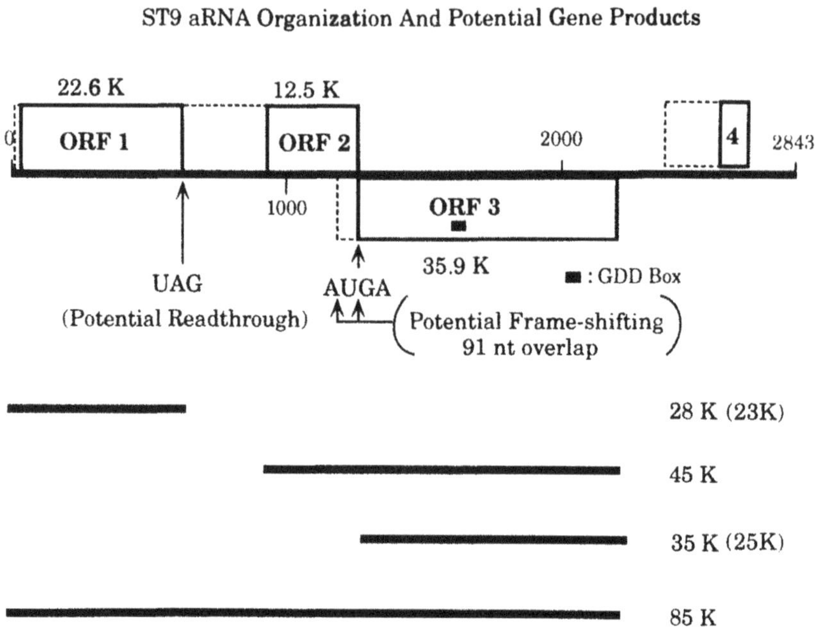

Figure 5. Schematic Diagram Showing the Potential Organization of the 2,843 Nucleotide Beet Western Yellows Virus ST9 aRNA and its Potential Protein Products. Open boxes depict open reading frames within the nucleotide sequence, and numbers above or below the boxes indicate sizes of putative proteins encoded by given ORF's. A potential read-through occurs at the ORF 1 terminus (UAG) and no additional in-frame stop codons occur between ORF 1 and the ORF 2 terminus. ORF 3 is "open" for 91 nucleotides upstream from its first AUG codon, and overlaps with the 3'-region of ORF 2 for 91 nucleotides. ORF 3 contains the GDD motif characteristic of RNA virus RNA-dependent RNA polymerases. Solid lines at the bottom indicate relative sizes of proteins which can be detected *in vitro* when full-length transcripts and transcripts representing the respective regions of the ST9 aRNA are translated *in vitro*. The 85K protein would likely result from read-through at the ORF 1 terminus and a frameshift event occurring somewhere in the ORF 2-ORF 3 overlapping region.

these symptoms? Our previous studies showed that approximately 10 times the amount of BWYV antigen (and BWYV virions) occurs in ST9 BWYV-infected *C. bursa-pastoris* plants versus plants infected by typical BWYV isolates lacking the ST9 aRNA (Falk and Duffus, 1984; Falk *et al.*, 1989). Therefore, it seems likely that BWYV must replicate

Figure 6. *C. bursa-pastoris* Plants Infection by Mild Beet Western Yellows Virus Plus the ST9 aRNA Transcript (mild+ST9-T), Infection by Only the Genomic RNA of BWYV (genomic-T), Beet Western Yellows Mild Isolate (BWYV Mild), and the ST9 Isolate of BWYV (BWYV ST9).

in more cells, or replicate to a higher titer per cell in ST9 BWYV-infected *C. bursa-pastoris* plants than in *C. bursa-pastoris* plants infected by typical BWYV isolates.

One aspect of replication in more cells would be if the ST9 aRNA somehow facilitates movement of BWYV from phloem tissues of infected plants into the mesophyll tissues. Movement of phloem-limited luteoviruses and geminiviruses into mesophyll tissues has been reported in several specific mixed infections (Barker 1987; Carr and Kim, 1983). However, we found that BWYV virions only occurred within phloem tissues of *C. bursa-pastoris* plants infected by ST9 BWYV or by typical BWYV isolates (Sanger *et al.*, 1994). We have found that co-infection of protoplasts by BWYV and the ST9 aRNA causes an increase in level of BWYV replication (Passmore *et al.*, 1993). Thus, it seems possible that the increased BWYV titer in ST9 BWYV-infected plants could be due to increased replication per cell.

We also have been able to show that the ST9 aRNA replicates when mechanically inoculated to *C. bursa-pastoris* plants (Passmore *et al.*, 1993). However, our data suggest that the ST9 aRNA fails to move systemically from the point of inoculation. This suggests that to systemically move through plants, the ST9 aRNA must be encapsidated, or that the ST9 aRNA lacks the ability to move from cell-to-cell in mesophyll tissues (those inoculated via mechanical inoculation), but can move systemically through phloem tissues when transcapsidated particles are inoculated into vascular tissues by aphids.

Possible Considerations in Regards to Heterologous Encapsidation Interactions and Transgenic Plants Expressing Viral Capsid Proteins

The above examples support the contention that heterologous encapsidation events occur in nature and play a natural role in plant virus epidemiology. As mixed infections of plant viruses of all types are common in nature (Zink and Duffus, 1972; Davis and Mizuki, 1987; Duffus, 1963; Falk, unpublished), it is quite likely that there are many as yet unrecognized examples of heterologous encapsidation interactions which naturally occur between plant viruses. However, the research that has been done so far also clearly shows that heterologous encapsidation interactions are specific and do not occur in most mixed infections. Now, with the likelihood being great that genetically-engineered plants expressing viral capsid proteins will play a role in future plant virus disease control strategies, the question is whether the use of viral capsid protein-expressing transgenic plants might increase the possibility for heterologous encapsidation interactions to occur; and if the possibility is increased, whether it poses a significant risk. At the present time we don't know the answers, but it is important to consider both parts of this question.

It has already been demonstrated that heterologous encapsidation events can occur in transgenic plants which express viral capsid proteins, with the transgenic plant being the source of viral capsid protein (Osbourn *et al.*, 1990; Sleat *et al.*, 1988; Dalmay *et al.*, 1992; Holt and Beachy, 1991; Farinelli *et al.*, 1992). There also are examples which clearly show that within transgenic plants the expressed viral capsid proteins can assemble into empty capsids and/or pseudovirions (Bertoli *et al.*, 1991). Therefore, it is reasonable to assume that heterologous encapsidation interactions will likely occur in some situations. However, the consequences of such interactions are as yet unknown.

It seems that if heterologous encapsidaton events occur in transgenic plants as a result of virus infection in the field, there are at least two biologically significant outcomes which we can consider.

These two are:

1) That heterologous encapsidation events in transgenic plants could serve to alter or facilitate vector transmissibility of the new progeny virions (those generated as a result of heterologous encapsidation).

2) That heterologous encapsidation events in transgenic plants could serve to facilitate systemic movement of the resulting progeny virions within the transgenic plant, a plant species in which the normal virus (that not resulting from heterologous encapsidation interactions) does not readily move systemically.

If the first scenario were to occur, and a virus were to gain vector transmissibility via heterologous encapsidation within the transgenic plant, would the potential for new disease development be great? It is impossible to know the answer to this until it has been tested; and factors such as the cropping situation, geographic location, type of vector and its abundance, local crops and other factors all would play a role in determining the possible outcomes, as they do in normal plant virus epidemiology. However, several hypothetical scenarios can still be envisioned.

Altered Vector Transmission and Disease Development within the Transgenic Crop: If vector transmission of a plant virus is altered or facilitated as a result of heterologous encapsidation interactions resulting from infection of a capsid protein-expressing transgenic plant, it is presently unknown as to whether this scenario would contribute to significantly greater virus spread and disease development within the genetically engineered crop. If vector spread is facilitated via heterologous encapsidation interactions, then the resulting chimeric vector-transmissible virions would contain capsid proteins derived from the capsid protein-expressing transgenic plant. If these are subsequently vector transmitted to another capsid protein-expressing plant within the same field (secondary spread, see Figure 7), they may or may not be able to infect such a plant. The capsid protein expressing plants are resistant to homologous viruses which have the same phenotype as do the capsid protein molecules expressed by this plant (Osbourn *et al.*, 1990). Results with TMV capsid protein mutants have shown that these mutants can become encapsidated by capsid protein molecules expressed in transgenic plants (Osbourn *et al.*, 1990). However, when the resulting transcapsidated virions were extracted from these plants, they were unable to initiate systemic infections in healthy capsid protein-expressing transgenic plants. This supports the idea that secondary spread of chimeric virions might not occur within the transgenic crop, as the plants would be resistant to the chimeric virions.

However, there also is evidence suggesting that capsid protein mediated transgenic resistance might not be as effective against virions resulting from heterologous encapsidation interactions if the encapsidated nucleic acid is not closely related to the virus from which the capsid protein gene was obtained. Evidence in this regard comes from experiments done using Sunn-hemp mosaic tobamovirus (Cc TMV) and tobacco plants transgenic for U1 TMV capsid protein (Osbourn *et al.*, 1989). When CcTMV RNA was transcapsidated within U1 TMV capsids, the chimeric virions were able to infect and replicate within protoplasts from the transgenic plants and replication levels were comparable to when CcTMV RNA was used as inoculum (Osbourn *et al.*, 1989). However, similar experiments showed that gene expression of reporter genes (as mRNA's) encapsidated in U1 TMV capsids were only 1 - 3 percent as active as in protoplasts from non-transformed plants (Osbourn *et al.*, 1989). The authors suggested that transgenic coat protein-mediated resistance may not be as effective when the challenging inoculum is of the same phenotype as is the transgenically expressed protein but that the RNA is not closely related to the original virus. However, this has not been tested at the whole plant level; and in the above example there was no evidence for systemic movement of CcTMV in the transgenic tobaccos. It seems much more likely that if secondary spread occurs, it will occur via the vector of the non-transcapsidated virus; and this vector will spread wild-type virions which will undoubtedly make up the majority of progeny virions in the infected plant.

Altered Vector Transmission and Disease Development in Another Crop: It also is possible that if heterologous encapsidation interactions were to occur in capsid protein-expressing transgenic plants, the resulting chimeric virions might be transmitted by a "new" vector to another crop composed of host plants which are not transgenic. In this scenario, the transgenic crop would serve as a new virus resevoir (for the chimeric virus), contributing virus spread to a new crop. However, spread of chimeric virions to the second crop would be only primary spread; that is, spread from the genetically-engineered plants where transcapsidation took place (Figure 7). Once the chimeric or transcapsidated virions infect a new, non-transgenic host they will again resort to the phenotype determined by the viral nucleic acid, as the only capsid protein source in these plants is now their own genome. For these viruses to spread secondarily through the crop, they would now have to be spread by their original vectors which may or may not be present.

Potential Epidemiological Considerations In Regards to Heterologous Encapsidation Interactions In Transgenic Plants

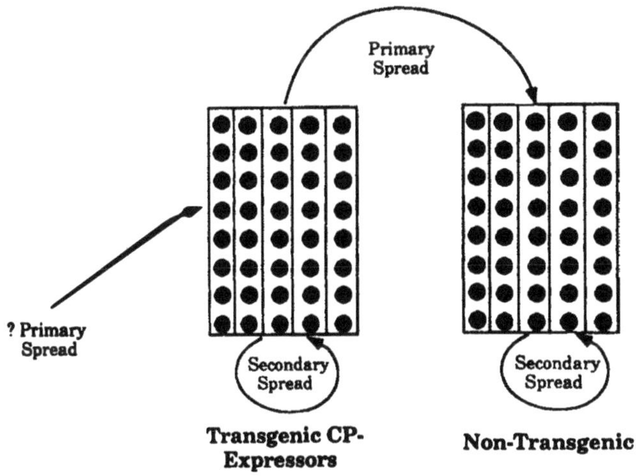

Figure 7. Diagram of Potential Vector-Mediated Virus Spread. Primary spread is initial spread into the given crop. Secondary spread is the subsequent spread of virus throughout the given crop. Transgenic CP-Expressors indicates a hypothetical field composed of coat protein-expressing transgenic plants. Non-transgenic indicates a hypothetical field of non-transgenic plants.

For many plant viruses, especially those of annual crops, the most common and economically important form of virus spread is secondary spread (Simons, 1959; Adlerz, 1978). Primary spread generally involves few or a limited number of plants and, in most cases, does not result in economically important losses. Secondary spread can be rapid and involves spread from the initial, primary infected plants to the large number of remaining healthy plants. Thus, in the above example if the transgenic plants serve as sources for primary spread of chimeric virions to another crop, disease losses and virus incidence from the primary spread would likely be limited in scope. Secondary spread now could only occur if the natural vector(s) of the wild-type virus are already present. However, if the natural vector is already present, then it also is possible that the natural vector could provide for primary as well as secondary spread; and both forms of spread could be wild-type virus.

Disease Development Resulting From New Systemic Spread Within Transgenic Plants: The virion capsid protein has been shown to play a role in long distance systemic movement of viruses within infected plants (Hull, 1989; Maule, 1991; Dawson *et al.*, 1988). Thus, it seems likely that heterologous encapsidation interactions could possibly facilitate long distance, systemic movement of chimeric virions within capsid protein expressing transgenic plants. Experiments already done with TMV and cymbidium ringspot tombusvirus (CyRSV) capsid protein transgenic plants have shown that TMV and CyRSV mutants defective in functional capsid proteins can replicate, become transcapsidated, and systemically move within the transgenic but not non-transgenic plants (Osbourn *et al.*, 1990; Dalmay *et al.*, 1992). However, the experiments discussed above which involved Cc TMV and U1 TMV transgenic tobaccos (Osbourn *et al.*, 1989) seem to suggest that more than just transcapsidation will be required for a virus to be able to move in a plant which is not normally a systemic host. The wild-type virus will most likely need to have the ability to move from cell-to-cell in the given host, and this form of virus movement generally does not involve the virion capsid (Hull, 1989; Maule, 1991; Dawson *et al.*, 1988).

Conclusions

Future plant virus disease control strategies will include the use of genetically engineered plants, and transgenic plants expressing viral capsid proteins will be one of the approaches used. Heterologous encapsidation interactions in transgenic plants may occur in some situations; however, the important question is whether the results from such events will differ from those which occur naturally. Mixed infections of plant viruses are already very common in crop and weed plants; and, thus, the potential for heterologous encapsidation interactions has always been and will continue to be a natural component of plant virus epidemiology and plant virus disease incidence. However, it seems to us that the potential risks from such interactions occurring specifically in transgenic plants and increasing plant virus disease severity will likely not be nearly as great as that which occurs naturally or so much as to preclude their use. We must test this virus disease control strategy at the field level and monitor disease development within the transgenic crop and surrounding crops. This will be the only way to scientifically assess potential risks.

It also is very likely that the more we learn about vector transmission of plant viruses, local and systemic aspects of virus spread within plants, virus replication and virion assembly, the better our chances will be to disarm components of genetically engineered resistance approaches which might have potential risks associated with them. For example, in recent years much has been learned about the molecular basis of potyvirus transmission by aphids. Recent work has identified an amino acid triplet which appears to be conserved among potyviruses and plays a direct role in determining potyvirus aphid transmissibility (Atreya *et al.*, 1990; Atreya *et al.*, 1991; Gal-on *et al.*, 1992). Experiments can now be done testing whether this region of the potyvirus capsid protein can be mutated so as to prevent aphid transmission and yet still provide genetically-engineered resistance if produced in transgenic plants. Recent work with tobacco etch potyvirus (TEV) has shown that when less-than-full-length regions of the TEV capsid protein are expressed in transgenic plants, some resistance, as measured by decreased time to symptom development, can be obtained (Lindbo and Dougherty, 1992 (see mutant XN/C #61)), suggesting that this approach is feasible.

Also, in many cases, titers of capsid proteins produced in transgenic plants are much lower than in non-transgenic plants infected by the same virus (Barker *et al.*, 1992). This also will contribute to a decreased chance for heterologous encapsidation interactions between the low titer proteins and the RNA of an infecting virus. If there is limited transcapsidation within the transgenic plant (low titer of the transcapsidated RNA within the virus population), this also could lessen the chances that the transcapsidated RNA will be vector transmitted. It is well documented that low virus titer in virus-infected plants decreases the probability for aphid transmission of viruses transmitted in a nonpersistent manner (Simons, 1958).

Thus, it is our contention that the potential risks due to possible heterologous encapsidation interactions are not so great as to warrant not using viral capsid protein mediated resistance as a strategy to control virus diseases. However, we should monitor disease carefully when transgenic plants are used in the field and be aware for possible abnormalities in plant virus epidemiology. In addition, by continuing efforts directed towards gaining a better understanding of vector transmission of plant viruses, we will be able to direct research efforts towards obtaining coat protein-mediated transgenic virus resistance in ways which will further decrease the potential for biologically significant heterologous encapsidation interactions to occur.

References

Adams, A.N. and Hull, R. 1972. Tobacco yellow vein, a virus dependent on assistor viruses for its transmission by aphids. Ann. Appl. Biol. 71:135-140.

Adlerz, W.C. 1978. Secondary spread of watermelon mosaic virus 2 by *Anuraphis middletonii*. J. Economic Entomology 71:531-533.

Atreya, P.L., Atreya, C.D., and Pirone, T.P. 1991. Amino acid substitutions in the coat protein result in loss of insect transmissibility of a plant virus. Proc. Nat. Acad. Sciences, U.S.A. 88:7887-7891.

Atreya, C.D., Raccah, B., and Pirone, T.P. 1990. A point mutation in the coat protein abolishes aphid transmissibility of a potyvirus. Virology 178:161-165.

Barker, H. 1987. Invasion of non-phloem tissue in *Nicotiana clevelandii* by potato leafroll luteovirus is enhanced in plants also infected with potato Y-potyvirus. J. Gen. Virol. 68:1223-1227.

Barker, H., Revy, B., Kumar, A., Webster, K.D., and Mayo, M.A. 1992. Restricted virus multiplication in potatoes transformed with the coat protein gene of potato leafroll luteovirus: Similarities with a type of host gene-mediated resistance. Ann. Appl. Biol. 120:55-64.

Beachy, R.N., Loesch-Fries, S., and Tumer, N.E. 1990. Coat protein-mediated resistance against virus infection. Annu. Rev. Phytopathol. 28:451-474.

Bertoli, D.J., Harris, R.D., Edwards, M.L., Cooper, J.I., and Hawes, W.S. 1991. Transgenic plants and insect cells expressing the coat protein of arabis mosaic virus produce empty virus-like particles. J. Gen. Virol. 72:1201-1209.

Bourdin, D. and Lecoq, H. 1991. Evidence that Heteroencapsidation between two potyviruses is involved in aphid transmission of a non-aphid transmissible isolate from mixed infections. Phytopathology 81:1459-1464.

Carr, R.J. and Kim, K.S. 1983. Evidence that bean golden mosaic virus invades non-phloem tissue in double infections with tobacco mosaic virus. J. Gen. Virol. 64:2489-2492.

Chin, L-S., Foster, J.L., and Falk, B.W. 1993. The beet western yellows virus ST9-associated RNA shares structural and nucleotide sequence with the with carmo-like viruses. Virology 192:473:482.

Creamer, J.R. 1989. Aphid vector specificity and transcapsidation in barley yellow dwarf virusew. PhD Thesis. University of California, Davis, CA.

Creamer, R. and Falk, B.W. 1990. Direct detection of transcapsidated barley yellow dwarf luteoviruses in doubly infected plants. J. Gen. Virol. 71:211-217.

Dalmay, T., Rubino, L., Burgyan, J., and Russo, M. 1992. Replication and movement of a coat protein mutant of cymbidium ringspot tombusvirus. Molec. Plant Microbe Interactions 5:379-383.

Davis, R.F. and Mizuki, M.K. 1987. Detection of cucurbit viruses in New Jersey. Plant Disease 71:40-44.

Dawson, W.O., Bubrick, P., and Grantham, G.L. 1988. Modifications of tobacco mosaic virus coat protein gene affecting replication, movement and symptomatology. Phytopathology 78:783-789.

De Zoeten, G.A. 1991. Risk assessment: Do we let history repeat itself. Phytopathology 81:585-586.

Demmler, S.A. and De Zoeten, G.A. 1991. The nucleotide sequence and luteovirus-like nature of RNA 1 of an aphid non-transmissible strain of pea enation mosaic virus. Journal of General Virology 72:1819-1834.

Dodds, J.A. and Hamilton, R.I. 1972. The influence of barley stripe mosaic virus on the replication of tobacco mosaic virus in *Hordeum vulgare* L. Virology 50:404-411.

Dodds, J.A. and Hamilton, R.I. 1974. Masking of the RNA genome of tobacco mosaic virus by the protein of barley stripe mosaic virus in doubly infected barley. Virology 59:418-427.

Dodds, J.A. and Hamilton, R.I. 1976. Structural interactions between viruses as a consequence of mixed infections. Adv. Virus Research 20:33-86.

Duffus, J.E. 1963. Incidence of beet viruses in relation to overwintering beet fields. Plant Dis Reporter 47:428-431.

Falk, B. W., Chin, L.S., and Duffus, J.E. 1989. Complementary DNA cloning and hybridization analysis of beet western yellows luteovirus RNA's. J. Gen. Virol. 70:1301-1309.

Falk, B.W. and Duffus, J.E. 1981. Epidemiology of persistent helper-dependent aphid transmitted virus complexes. In: *Plant Diseases and Vectors: Ecology and Epidemiology*, Chapter 5. K. Maramorosch and K. F. Harris (eds.), pp. 162-179. Academic Press, New York .

Falk, B.W. and Duffus, J.E. 1984. Identification of small single- and double-stranded RNA's associated with severe symptoms in beet western yellows virus-infected *Capsella bursa-pastoris*. Phytopathology 74:1224-1229.

Falk, B.W., Duffus, J.E., and Morris, T.J. 1979. Transmission, host range and serological properties of the viruses causing lettuce speckles disease. Phytopathology 69: 612-617.

Farinelli, L., Malnoe, P., and Collet, G.F. 1992. Heterologous encapsidation of potato virus Y strain O (PVY^{O}) with the transgenic coat protein of PVY strain N (PVY^{N}) in *Solanum tuberosum* cv. Bintje. Bio/Technology 10:1020-1025.

Francki, R.I.B. 1985. Plant virus satellites. Ann. Rev. Microbiol. 39:151-174.

Gal-on, A., Antignus, Y., Rosner, A., and Raccah, B. 1992. A zucchini yellow mosaic virus coat protein gene mutation restores aphid transmissibility but has no effect on multiplication. J. Gen. Virol. 73:2183-2187.

Gildow, F.E. 1987. Virus-membrane interactions involved in circulative transmission of luteoviruses by aphids. Current Topics in Vector Research 4:93-120.

www.ingramcontent.com/pod-product-compliance
Lightning Source LLC
LaVergne TN
LVHW051925120926
847902LV00009B/73